Water and Wastewater Management

Volume-1

D. R. Khanna
R. Bhutiani
Gagan Matta
Vikas Singh

2024

Daya Publishing House®
A Division of
Astral International Pvt. Ltd.
New Delhi – 110 002

First Published, 2010

Reprinted, 2024

ISBN: 978-81-7035-748-3 (HB)

Disclaimer:

Every possible effort has been made to ensure that the information contained in this book is accurate at the time of going to press, and the publisher and author cannot accept responsibility for any errors or omissions, however caused. No responsibility for loss or damage occasioned to any person acting, or refraining from action, as a result of the material in this publication can be accepted by the editor, the publisher or the author. The Publisher is not associated with any product or vendor mentioned in the book. The contents of this work are intended to further general scientific research, understanding and discussion only. Readers should consult with a specialist where appropriate.

Every effort has been made to trace the owners of copyright material used in this book, if any. The author and the publisher will be grateful for any omission brought to their notice for acknowledgment in the future editions of the book.

Published by : **Daya Publishing House®**
A Division of
Astral International Pvt. Ltd.
– ISO 9001:2015 Certified Company –
4736/23, Ansari Road, Darya Ganj,
New Delhi-110 002
Ph. 011-43549197, 23278134
E-mail: info@astralint.com
Website: www.astralint.com

Printed at : **Replika Press Pvt. Ltd.**

Preface

Indian rivers are known to be sacred for us because a large number of populations depend directly or indirectly on it for their daily use. Water resources are of critical importance to both natural ecosystems and human development. But industrialization, urbanization and rapid increase in population has resulted in the degradation of surface as well as ground water quality. Various industrial processes coupled with domestic activities often produce harmful discharges which ultimately found their way to the water resources. Water pollution not only affects inhabited plants and animals, individuals, species and populations but also to the natural biotic communities as well.

This book presents an overview of wastewater pollution and its abatement techniques which are being used in the present world. This book will also result in the preservation of the precious data of various significant water resources which could be helpful for present and future both. Through this book we have tried to determine the present status of water resources, their conservation strategies *etc.* We are grateful to the eminent scientists, researchers, scholars *etc.*, for contributing their research work for the benefits of humankind. We acknowledge for their great support and hope best for this book in fulfilling the purpose of making awareness among social community.

Editors

Contents

Water and Wastewater Management in 2 Vols (2011) *Pages 1–9*
Editors: **D.R. Khanna, R. Bhutiani & Gagan Matta**
Published by: **DAYA PUBLISHING HOUSE, NEW DELHI**

Chapter 1

Plankton Ecology in Two Groundwater Irrigated Paddy Fields, Indira Nagar, Miran Sahib, Jammu

S.P.S. Dutta, M. Khajuria, P. Dogra and A. Singh
Department of Environmental Sciences,
University of Jammu, Jammu

ABSTRACT

Plankton ecology, in relation to abiotic characteristics of water in two groundwater irrigated paddy fields, Indira Nagar, Miran Sahib, Jammu, was undertaken during June, 2002 to September, 2002 and has been described. Phytoplankton, qualitatively, has shown the dominance of Bacillariophyceae (7 genera) followed by Cyanophyceae (6 genera) and Chlorophyceae (5 genera). Among zooplankton, there is dominance of Protozoa (19 species) followed by Arthropoda (6 species and 3 larvae), Rotifera (6 species) and Nematoda.Planktonic diversity and density remained low during first observation (preparation of fields). Coefficient of correlation (r) of phytoplankton and zooplankton, with the various water characteristics is insignificant.

Keywords: *Phytoplanktonic diversity, Paddy field, Correlation coefficient.*

Introduction

The area under paddy culture is 1,04,000 ha in Jammu and 1,63,000 ha in Kashmir and quality water (surface and subsurface) is available for irrigation. These seasonal wetlands support a diversified aquatic biota and have not received much scientific attention. Among the earlier workers Goyal *et al.* (1984), Rathor and Mir (1987) and Dutta *et al.* (2002 a,b and 2004) have worked on the plankton ecology of paddy fields. The present work was undertaken to add to the existing knowledge of planktonic ecology from the area and groundwater irrigated paddy fields were selected in Indira Nagar.

Topography of the Area

Indira Nagar, Miran Sahib, Jammu, the area of present study, is situated at a distance of 10 km from the University campus. The area is plain and is suitable for agriculture. The texture of the soil is soft and loamy. Annually two crops *viz.* wheat during winter and paddy during summer-monsoon are cultivated. Three varieties of paddy *viz.* China, Ratna and local basmati are cultivated in the area. In the field of present study (each of 0.1 hectare size), local basmati is cultivated. Irrigation of these fields is done through undergroundwater.

Materials and Method

Sampling of Water

Water samples, were collected fortnightly during paddy season *viz.* June-September 2002 in 5 litres plastic containers and analyzed for abiotic characteristics by standard methods (APHA,1998).

Sampling and Analysis of Plankton

Plankton were collected by filtering 2 litres of water through a planktonic net (No.25) and analysed in the laboratory (Nair *et al.*, 1971; Dutta, 1983; Kudo, 1986 and Ward and Whipple, 1992) and counted by drop count method. The results are expressed as number per litre (n/l).

Coefficient of correlation (r) of phytoplankton and zooplankton, with various abiotic parameters, is calculated by the formula:

$$\text{Coefficient of correlation (r)} = \frac{\sum xy - \bar{x}\sum y}{\sqrt{\left(\sum x^2 - x\sum x\right)\left(\sum y^2 - y\sum y\right)}}$$

where,

x and y are the values of either species or physico-chemical parameters and *x* and *y* are their means.

Results and Discussion

Water Characteristics

The results of various physico-chemical characteristics of water are summarized in Table 1.1. Due to shallowness, water temperature closely followed the air temperature and varied between 25.6°C-33.7°C and 27.0°C-34.0°C, respectively. Highest record of air and water temperature is seen during 3rd observation. Due to mixing of hot soil and water during preparation of field, water temperature was higher than air temperature, during first observation. Depth in these paddy fields varied between 2.5-10.0 cm and recorded its maximum value during 1st observation, when large amount of water is required and is available, before seeding transplantation. Turbidity varied between 21.25-262.60 NTU and remained high during 1st, 2nd and 3rd observation, with maximum record on 2nd observation. Agitation of sediments during seedling transplantation may explain maximum record of turbidity on 2nd observation. pH varied between 7.54-8.66. Its wide variation has correlation with the presence or absence of CO_2 and CO_3^{2-}. Direct relationship of pH with CO_3^{2-} may explain the highest record of pH (8.66) during 3rd observation, when CO_3^{2-} was present. Similarly, an inverse relationship of pH with free CO_2 may explain the low record of pH (7.76) during 7th observation, when free CO_2 was highest. An inverse relationship of pH with CO_2 and direct with CO_3^{2-} is already on record (Welch, 1952; Hutchinson, 2004; Reid and Wood, 1976; Goldman and Horne, 1983; Jhingran, 1991 and Wetzel 2000). Dissolved oxygen ranged between 1.96-6.53 mg/l. Its highest record on 4th observation coincided with lowest record of free CO_2 and has correlation with high rate of photosynthesis, due to phytoplanktonic increase, indicated by floating algal mats and low shade caused by newly transplanted seedlings and penetration of light upto the bottom. Free CO_2 varied between 5.50-39.92 mg/l and remained high due to irrigation of the field by the groundwater. Its highest and lowest record is observed during 7th and 4th observation, respectively. Presence of water in pools and decaying dead organic matter may explain the highest record of CO_2 on 7th observation. Carbonate is seen only once during 3rd observation (13.8 mg/l). Bicarbonate recorded its minimum (101.24 mg/l) record on 4th observation. Maximum record of bicarbonate (416.00 mg/l) during 1st observation has correlation with mixing of sediments, decayed decomposed vegetation, marls (dead molluscan shells) with water during preparation of field for seedling transplantation. Chloride recorded its highest (36.41 mg/l) and lowest (8.80 mg/l) value during 7th and 1st observation, respectively. Presence of water in pools may explain the highest record of chloride during 7th observation. Calcium, magnesium and total

Table 1.1: Mean Results of Physico-Chemical Parameters of Water from Two Paddy Fields, Indira Nagar (Miran Sahib), Jammu (June 2002 to September 2002)

Water Parameters	*Observation*							
	1	*2*	*3*	*4*	*5*	*6*	*7*	*8*
	29/6/02	*01/7/02*	*16/7/02*	*27/7/02*	*14/8/02*	*2/9/02*	*16/9/02*	*28/9/02*
Air temperature (°C)	30	31	34	31.5	27	31	31	30.2
Water temperature (°C)	31	30	33.7	31.65	25.6	29.6	30.3	29.8
Depth (cm)	10	5.5	4.3	3.7	4.75	2.95	2.5	2.8
Turbidity (N.T.U.)	157.7	262.6	240	21.25	22.3	25.6	26.5	45
T.D.S. (ppm)	73	173.65	203.9	91.6	122.6	164.1	176.75	199.1
pH	7.67	7.6	8.66	8.09	7.54	7.67	7.765	7.73
Electrical Conductivity (mMho cm)	0.474	0.347	0.414	0.4	0.380	0.452	0.357	0.385
Salinity (ppt)	0.4	0.3	0.4	0.3	0.25	0.3	0.4	0.4
D.O. (mg/l)	3.17	5.95	4.82	6.53	4.79	4.2	1.96	4.44
Free CO_2 (mg/l)	15.8	19.9	–	5.5	13.77	15.29	39.92	18.01
CO_3^{2-}	–	–	13.8	–	–	–	–	–
HCO_3 (mg/l)	416	224.3	119.11	101.24	193.77	217.35	402.46	297.65
Cl^- (mg/l)	8.8	11.3	17.2	14.78	9.75	10.74	36.41	17.87
Ca^{++} (mg/l)	DNA*	43.47	40.36	47.94	43.9	56.61	65.29	64.68
Mg^{++} (mg/l)	DNA*	17.5	17.80	18.26	13.87	23.33	25.4	12.37
Total Hardness (mg/l)	DNA*	180.67	145.4	194.77	117.97	150.89	167.81	212.24
Na^+ (mg/l)	56	26.75	15.75	24.5	23.25	18.50	31.25	24
K^+ (mg/l)	6.5	1	1.25	1.75	1.25	2.25	8.75	16.5

* DNA: Does not analyzed.

hardness varied between 40.36-65.29 mg/l, 12.37-25.4.00 mg/l 117.97-167.81 mg/l, respectively. Highest record of calcium, magnesium and total hardness during 7th observation may be explained by presence of water in pools. Sodium varied between 15.75-56.00 mg/l with its highest and lowest record on 1st and 3rd observation, respectively. Mixing of sediments, cow dung and decayed crop residue during preparation of fields may explain the highest record of sodium during 1st observation. Potassium recorded its maximum (16.5 mg/l) and minimum (1.00 mg/l) value during 8th and 2nd observation, respectively. Highest record of potassium during 8th observation may be explained by low consumption of potassium by paddy, at the time of ripening and presence of water in pools

Planktonic Analysis

Phytoplankton, comprising of 18 genera, has shown the dominance of Bacillariophyceae (7 genera) followed by Cyanophyceae (6 genera) and Chlorophyceae (5 genera). Qualitatively, various phytoplanktonic genera remained irregular. Among the various genera of Bacillariophyceae, *Navicula* and *Fragillaria* are seen seven times, *Cymbella* six times, *Amphora* thrice, *Pinnularia* and *Gomphonema* twice and *Nitzchia* only once (Table 1.2). Among the various Cyanophycean genera, *Oscillatoria* is seen seven times, *Phormidium* six times, *Anabaena* five times, *Nostoc* and *Spirulina* four times and *Microcystis* thrice (Table 1.2). Various genera of Chlorophyceae also made their irregular presence. *Cosmarium* is observed six times, *Spirogyra* five times, *Chlorella* four times, *Closterium* thrice and *Scenedesmus* only once. An overall analysis has shown minimum and maximum phytoplanktonic diversity during third and second observation, respectively (Table 1.2). Quantitatively, total phytoplankton varied between nil-2868 n/l. The order of qualitative dominance of different phytoplanktonic group is seen as Bacillariophyceae (nil-2544 n/l)>Cyanophyceae (nil-615 n/l) and Chlorophyceae (nil-140 n/l). Analysis of coefficient of correlation (r) of Bacillariophyceae, Cyanophyceae and Chlorophyceae with various physico-chemical characteristics of water is insignificant (Table 1.4). This indicates that no single factor is a strong determinant for phytoplanktonic abundance in these paddy fields. A total of 33 zooplanktonic species and 3 larvae, seen in these paddy fields (Table 1.3), have shown the qualitative dominance of Protozoa (19 spp.) followed by Arthropoda (6 spp. and 3 larvae), Rotifera (7 spp.) and Nematoda. Protozoa, the most dominant group, is represented by *Sarcodina* (17 spp.) and *Ciliata* (2 spp.) only. Qualitatively, *Sarcodina* is seen during all the eight observations. Among its various genera *Diffugia* and *Centropyxis* are observed during all the eight observations, *Arcella* seven times, *Lespuereusia* twice and *Pyxidicula* only once.

Among ciliates *Vorticella* is seen thrice and *Paramecium* twice.Total protozoans, quantitatively, varied between 105-988 n/l. Among protozoans, quantitatively, there is dominance of *Sarcodina* (105-988 n/l) and sub-dominance of Ciliata (nil-45 n/l).Quantitatively total protozoans recorded trimodal increase. These observed first peak during 2nd observation, second during 4th observation and third during 8th observation (Table 1.3). Rotifers also made their irregular presence in the planktonic samples collected from these paddy fields. All the rotifers, *viz. Lecane, Brachionus, Trichocerca, Notholca* and *Philodina* are observed only once (Table 1.3). Quantitatively, total rotifers varied between nil-65 n/l. Analysis of coefficient of correlation of rotifers, with various physico-chemical characteristics, reveals insignificant relationship (Table 1.5) Arthropods in these paddy fields are represented by two classes, *viz.* Crustacea and Insecta. Crustacea has shown the presence of three orders *viz.* Copepoda, Ostracoda and Cladocera. Among copepods, *Mesocyclops, Leuckartii* and *M. hyalinus* are observed thrice. *Nauplius* and *Metanauplius* larvae of copepods made their record four times and twice, respectively. Among ostracods, *Cypris* is observed twice and *Potamocypris* only once. *Moina* and *Alona*, only cladocerans seen in these paddy fields, were observed only once (Table 1.3). Chironomous larva, a dipteran, is seen only twice.

Table 1.2: Mean Results of Phytoplanktonic Variations (n/l) in Two Paddy Fields, Indira Nagar (Miran Sahib) Jammu

Phytoplanktonic Organisms	*Observation*							
	1	*2*	*3*	*4*	*5*	*6*	*7*	*8*
	29/6/02	*01/7/02*	*16/7/02*	*27/7/02*	*14/8/02*	*2/9/02*	*16/9/02*	*28/9/02*
I. Cyanophyceae								
Nostoc spp.	–	–	23	110	13	–	42	–
Anabaena spp.	–	60	–	180	13	10	30	–
Phormidium spp.	–	–	13	10	68	20	23	48
Oscillatoria spp.	–	40	23	205	42	125	113	168
Spirulina spp.	–	–	–	35	43	23	–	36
Microcystis spp.	–	120	–	75	–	–	30	–
Total	–	*220*	*58*	*615*	*179*	*238*	*238*	*252*
II. Bacillariophyceae								
Fragillaria spp.	–	155	43	70	25	108	65	24
Navicula spp.	–	1384	39	150	100	135	110	96
Gomphonema spp.	–	32	–	–	–	–	15	–
Cymbella spp.	–	80	8	25	–	20	81	96
Pinnularia spp.	–	896	–	–	–	–	81	–
Nitzchia spp.	–	–	–	–	–	20	–	–
Amphora spp.	–	–	–	20	17	8	–	–
Total	–	*2544*	*90*	*265*	*142*	*291*	*319*	*216*
III. Chlorophyceae								
Chlorella spp.	–	64	8	20	–	15	–	–
Closterium spp.	–	–	7	–	25	–	–	24
Cosmarium spp.	–	40	35	45	17	55	23	–
Spirogyra spp.	–	–	13	75	–	68	9	36
Scendesmus spp.	–	–	–	–	–	–	23	–
Total	–	*104*	*63*	*140*	*37*	*138*	*55*	*60*
Total Phytoplankton	–	*2868*	*211*	*1020*	*358*	*667*	*612*	*528*

–: Absent.

Quantitatively, arthropods varied between 7-174 n/l and recorded increase during 2nd, 4th, 6th and 8th observation. Water nematodes and unidentified eggs made their appearance twice in these paddy fields (Table 1.3). Quantitatively, total zooplankton varied between 147-1088 n/l and recorded a trimodal increase. These recorded first peak during 2nd, second during 4th and third during 8th observation. Qualitatively, phytoplankton and zooplankton recorded slight increase on 2nd observation and can be attributed to their transfer alongwith seedlings. In paddy culture seedlings are grown about a month before their transplantation. There is daily irrigation of these seedlings and is optimum for planktonic growth. Presence of shelled forms of zooplankton during 1st observation has its correlation with resistance of shells to biodegradation and chemical corrosion. This may explain their dominance in these paddy fields. An increase in zooplanktonic diversity after irrigation of fields has its correlation with hatching of cysts/spores produced by various planktonic forms during drying of fields.

Table 1.3: Mean Result of Phytoplanktonic Variations (n/l) in Two Paddy Fields, Indira Nagar (Miran Sahib, Jammu) (June 2002 to September 2002)

Zooplanktonic Organisms	*Observation*							
	1	*2*	*3*	*4*	*5*	*6*	*7*	*8*
	29/6/02	*01/7/02*	*16/7/02*	*27/7/02*	*14/8/02*	*2/9/02*	*16/9/02*	*28/9/02*
Phyllum: Protozoa								
Class: Sarcodina								
Order: Testacida								
Difflugia tuberculata (Wallich)	110	196	13	100	25	43	23	–
D. pyriformes Perty	33	124	13	5	25	–	8	36
D. corona Wallich	–	48	–	55	13	20	–	–
D. accuminata Ehren	75	112	36	60	–	8	39	72
D. lebes Penard	–	60	–	5	–	–	23	–
D. urceolata Carter	–	20	–	25	–	–	8	–
D. lobostomo Leidy	–	–	8	–	–	–	–	
Total Difflugia	*218*	*560*	*67*	*250*	*63*	*71*	*101*	*108*
Centropyxis ecornis (Ehren)	33	40	35	125	–	18	45	60
C. spinosa (Cash and Hoppinson)	–	–	–	25	–	–	–	–
C. *aculeate* (Ehren)	33	208	7	150	17	18	9	84
C. aerophila deflandre	–	–	–	40	–	–	–	–
C. constricta (Ehren)	18	112	8	20	–	–	8	–
Total Centropyxic	*84*	*360*	*50*	*360*	*17*	*36*	*62*	*144*
Arcella discoides Ehren	–	–	–	45	–	8	23	36
A. vulgaris Ehren	–	20	30	–	25	30	–	–
A. arcelloides Penard	–	32	–	10	–	–	–	24
Total Arcella	–	52	30	55	25	38	23	60
Lesquereusia spiralis Ehren	–	16	–	–	5	–	–	–
Pxydicula spp.	–	–	–	5	–			
Total Sarcodina	*302*	*988*	*147*	*675*	*105*	*145*	*186*	*312*
Class: Ciliata								
Paramecium Hill	–	–	–	–	–	23	27	–
Vorticella Linnaeus	–	–	–	–	–	15	18	12
Total Ciliata	–	–	–	–	–	38	45	12
Total Protozoa	302	988	147	675	105	183	231	324

Contd...

Table 1.3–Contd...

Zooplanktonic Organisms	*Observation*							
	1	*2*	*3*	*4*	*5*	*6*	*7*	*8*
	29/6/02	*01/7/02*	*16/7/02*	*27/7/02*	*14/8/02*	*2/9/02*	*16/9/02*	*28/9/02*
Phyllum: Rotifera								
Class: Mongononta								
Order: Ploima								
Lecane spp.	–	–	–	25	–	–	–	–
Trichocerca spp.	–	–	–	–	9	–	–	–
Notholca Gosse	–	–	–	–	–	–	8	–
Platyias spp.	–	–	–	–	9	–	–	–
Brachionus quadridentata Hermann	–	–	–	–	–	23	–	–
Order: Bdelloida								
Philodina Ehren	–	–	–	40	–	–	–	–
Total Rotifers	–	–	–	*65*	*18*	*23*	*8*	–
Phyllum: Arthropoda								
Class: Crustacea								
Order: Copepoda								
Mesocyclops leuckartii (claus)	–	–	–	5	–	–	–	24
M. hyalinus (Rehberg)	–	–	–	–	9	–	–	48
Nauplius larva	–	40	7	20	–	–	–	24
Metanauplius larva	–	–	–	–	–	8	–	12
Order: Ostrocoda								
Cypris Muller	–	20	–	–	15	–	–	–
Potamocypris Brady	–	40	–	–	–	–	–	–
Order: Cladocera								
Alona spp.	–	–	–	–	–	–	–	48
Moina Baird	–	–	–	–	–	23	–	–
Total Crustacea	–	*100*	*7*	*25*	*24*	*31*	–	*156*
Class: Insecta								
Order: Diptera								
Chironomous larva	–	–	–	–	–	–	24	18
Total Arthropods	–	*100*	*7*	*25*	*24*	*31*	*24*	*174*
Water Nematodes	15	–	–	–	–	–	–	12
Eggs	–	–	–	5	–	–	–	108
Total Zooplankton	*317*	*1088*	*154*	*765*	*147*	*237*	*263*	*618*

Table 1.4: Co-efficient of Correlation (r) Between Phytoplankton and Physico-chemical Parameters of Water in Two Paddy Fields (Miran Sahib), Jammu

Phytoplanktonic Groups	*Water Parameters*																
	Water Temp.	*Depth*	*Turbidity*	*pH*	*Elec. Cond.*	*Salinity*	*T.D.S.*	*DO*	*Free CO_2*	*CO_3^{2-}*	*HCO_3^-*	*Cl^-*	*Ca^{++}*	*Mg^{++}*	*Total Hardness*	*Na^+*	*K^+*
Bacillariophyceae	–005	0.04	0.55	–0.28	–0.54	–0.31	0.23	0.40	0.19	–0.22	–0.08	–0.13	–0.28	–0.02	0.22	–0.08	–0.26
Chlorophyceae	0.15	–0.56	–0.16	0.12	–0.14	–0.47	0.15	0.58	–0.20	0.33	–0.61	–0.07	–0.08	0.34	0.32	–0.61	–0.35
Cyanophyceae	–0.07	–0.52	–0.50	0.01	–0.32	–0.41	–0.18	0.52	–0.06	0.37	–0.45	0.13	0.05	0.53	–0.35	–0.06	
Total phytoplankton	*–0.08*	*–0.04*	*–0.09*	*0.18*	*–0.58*	*–0.40*	*0.18*	*0.51*	*0.15*	*–0.11*	*–0.20*	*0.26*	*0.30*	*0.30*	*0.56*	*–0.48*	*–0.04*

Table 1.5: Co-efficient of Correlation (r) Between Zooplankton and Physico-chemical Parameters of Water in Two Paddy Fields, Indira Nagar, (Miran Sahib), Jammu

Zooplanktonic Groups	*Water Parameters*																
	Water Temp.	*Depth*	*Turbidity*	*pH*	*Elec. Cond.*	*Salinity*	*T.D.S.*	*DO*	*Free CO_2*	*CO_3^{2-}*	*HCO_3^-*	*Cl^-*	*Ca^{++}*	*Mg^{++}*	*Total Hardness*	*Na^+*	*K^+*
Protozoa	0.16	0.12	0.39	–0.14	0.40	–0.25	–0.16	0.6	–0.19	0.08	–0.18	–0.18	–0.24	–0.09	0.56	–0.06	–0.19
Rotifera	–0.06	–0.27	–0.54	0.10	0.05	–0.52	–0.49	0.49	–0.55	0.50	–0.53	–0.09	–0.12	–0.13	0.12	0.23	–0.35
Arthropoda	–0.13	0.08	0.29	–0.19	–0.44	–0.19	0.49	0.22	0.03	–0.33	0.08	0.02	0.41	–0.54	0.69	–0.22	0.63
Nematoda	0.06	0.56	0.04	–0.06	–0.23	0.46	0.55	–0.31	0.03	–0.21	0.60	–0.20	–0.48	–0.82	–0.48	0.70	0.67
Total Zooplankton	*0.10*	*0.09*	*–0.18*	*0.46*	*0.46*	*–0.19*	*–0.04*	*0.60*	*–0.22*	*0.02*	*–0.15*	*–0.15*	*–0.09*	*–0.24*	*0.71*	*0.06*	*0.02*

Insignificant results of coefficient of correlation of phytoplankton and zooplankton (Tables 1.4 and 1.5), with various physico-chemical characteristics of water, indicate that no single factor is a strong determinant for planktonic abundance in these paddy fields of Indira Nagar, Jammu.

References

APHA, 1998. *Standard Methods for the Examination of Water and Wastewater*, 20th edition.

Dutta, S.P.S., 1983. On some freshwater Rhizopoda from Jammu. Part 1. *Jammu Univ. Review*, 1: 91–94.

Dutta, S.P.S., Dogra, K. and Tak, A.H., 2002a. Qualitative and quantitative study of plankton in some paddy fields, Maralia Morh Jammu (J&K). In: *Nat. Seminar on Environmental Pollution and Fishery Management*, 8–10 Aug., School of Studies in Zoology, Vikram Univ., Ujjain.

Dutta, S.P.S., Kiran, K., Sharma, A., Khatoon, N. and Mir, A.M., 2002b. Ecology of plankton in some paddy fields, Miran Sahib, Jammu (J&K). In: *Nat. Seminar on Recent Research Trends in Life Sciences.* August, 26–28, Centre of Research for Development, The University of Kashmir, p. 32 (2–7 Abstract).

Dutta, S.P.S., Mir, A.M., Kiran, K., Sharma, A., Khatoon, N. and Slathia, D., 2004. Ecology of plankton in some paddy fields of Miran Sahib, Jammu (J&K). *Aquacult.*, 5(1):1–10.

Goldman and Horne, A.J., 1983. *Limnology*. McGraw Hill, International Book Company, New Delhi.

Goyal, S.K., Sharma, B.M and Gupta, R.S., 1984. Algal flora of rice field soils of J&K state. *Phykos*, 23(1 and 2): 57–64.

Hutchinson, G.E., 2004. *A Treatise on Limnology, Vol. 1 Part 2: Chemistry of Lakes*. Wiley, New York.

Jhingran, V.G., 1991. *Fish and Fisheries of India*. Hindustan Publishing Corporation, India.

Kudo, R.R., 1986. *Protozoology*. Books and Periodicals Corporation, India.

Nair, K.N., Das, A.K. and Mukherjee, B.N., 1971. On some freshwater Rhizopoda and Helizoa (Protozoa) from Kolkata and its environs. Part–I. *Res. Zool. Surv., India*, 65: 1–6.

Rather, M.D. and Mir, A.M., 1987. Non symbiotic nitrogen fixation by some BGA from paddy field soil of Kashmir. *Phykos*, 26: 63–66.

Reid, G.K. and Wood, R.D., 1976. *Ecology of Inland Waters and Estuaries.* D. Van Norstrand Company, New York.

Ward, H.B. and Whipple, G.C., 1992. *Freshwater Biology*, 2nd edn. John Wiley and Sons, Publ. by International Books and Periodicals Supply Service, New Delhi.

Welch, P.S., 1952. *Limnology*. McGraw Hill Book Co. Inc., New York, 539 pp.

Wetzel, R.G., 2000. *Limnology: Lakes and River Ecosystem*, 3rd edn. Academic Press, p. 1–841.

Water and Wastewater Management in 2 Vols (2011) *Pages* **10–14**
Editors: **D.R. Khanna, R. Bhutiani & Gagan Matta**
Published by: **DAYA PUBLISHING HOUSE, NEW DELHI**

Chapter 2

A Preliminary Study of Zooplankton Diversity of Ramala Lake District Chandrapur, Maharashtra

Rajlaxmi R. Kulkarni[1], Suresh Zade[2] and S.B. Patharde[3]
[1]*Department of Zoology, S.P. College, Chandrapur*
[2]*P.G. Department of Zoology,*
[3]*P.G. Department of Sericulture, Manoharbhai Patel College, Sakoli Dist. Bhandara, M.S.*

ABSTRACT

The present study reports the zooplankton community of Ramala lake in Chandrapur city, Maharashtra for period of 12 months from June 2006 to May 2007. The study showed much diversity. The lake is one of ancient, historical, man made water body, constructed by Gond Raja of Chandrapur. Monthly variation in the abundance of zooplankton were studied during the study period of 12 months. Total 18 zooplankton species were recorded, out of which 10 rotifera species,02 copepoda species, 01 ostracoda and 05 cladocera species were found. The rotifer Brachionus was dominated throughout the study period.

Keywords: Zooplankton, Copepoda, Rotifera, Ramala lake.

Introduction

Zooplankton are most fascinating group of microorganism found in aquatic body. Diversity refers to the range of variation of difference among some set of entities, zooplankton diversity thus refers to variety within their community. They include a varied assemblage of taxonomically unrelated microscopic organism. Their common ecological characteristic being their habitat, they are found freely drifting in the epilimnion. They are abundant in shallow water areas of reservoir. They play a vital role as primary consumer which feed upon phytoplankton. The occurrence and abundance of zooplankton in a pond depends on its productivity which in turn is influenced by physico-chemical parameters and nutrients. Zooplankton has been used as indicators of water quality, trophic status and pollution level. They provide food for fishes in the freshwater ponds and play major role in the fish production. Zooplankton has been an interesting subject of study in India and several workers worked on such as Ganapati (1943), Sharma and Patnaik (1985), Chandrashekhar and Kodarkar (1996), Pawar and Madlapure (2002), Patil *et al*. (2005).

This chapter deals with diversity of zooplankton in Ramala lake which was studied for a period of twelve months from June 2006 to May 2007. The lake is one of chief lakes, within the municipal limits of Chandrapur city, running along the north, east section of the historical city wall of Gondraja's fort. It is situated 761 feet above the sea level in 19° 57′ North latitude and 79° 22′ East longitude.

Materials and Method

The present study was conducted for the period of twelve months from June 2006 to May 2007. Monthy samples of zooplankton were collected from three different sampling sites, S_1, S_2 and S_3, far apart from each other. The zooplankton were collected with the help of a plankton net of standard bolting silk cloth No. 25 (mesh size 0.03 – 0.04 mm) by filtering 50 litres of water by a plastic container of 05 liters capacity. The collected samples then centrifuged and volume was adjusted to 30 ml. The samples then preserved by using 4 per cent formaline solution. Detailed microscopic examination of zooplankton under compound binocular microscope has been carried out. As per as possible the zooplankton were identified up to species level. Preliminary identification of zooplankton was made by using standard key and other literature like Ward and Whipple (1992), Tonapi (1980) Battish (1992) Chandrashekhar and Kodarkar (1995) and taxonomic notes on rotifers, copepods *etc.* (IAAB Publication). For the quantitative analysis of plankton a Sedgwick Rafter counting cell of 1 ml capacity was used.

Results and Discussion

The zooplankton of Ramala lake composed of four major groups (classes) rotifer, ostracoda, copepoda and cladocera. The number of species that were identified during the study period were rotifer 10 species, copepods 02 species, ostractoda 01 species and cladocera 05 species. The rotifers were found to be dominant. It was noted that the total number of zooplankton varied from 24 to 65 number per litre at sampling site I, 05 to 45 number per lit. at sampling site III. The zooplankton population showed depletion in the month of Dec.–March and maximum number of zooplankton was found to be in the month of July –August.

The seasonal variation of zooplankton in the order of abundance at three sites were as follows:-

S_1: Cladocera > Rotifera > Copepoda > Ostracoda

S_2: Rotifera > Cladocera > Copepoda > Ostracoda

S_3: Rotifera > Cladocera > Copepoda > Ostracoda

Rotifera

The monthly average and total number of rotifers varied from 02 to 19 per liter at S_1, 01 to 15 at S_2 and 03 to 20 at S_3. Rotifer populations are very useful in indicating the water quality particularly in pollution studies. Rotifers play an important role as grazers, suspension feedar and predators within

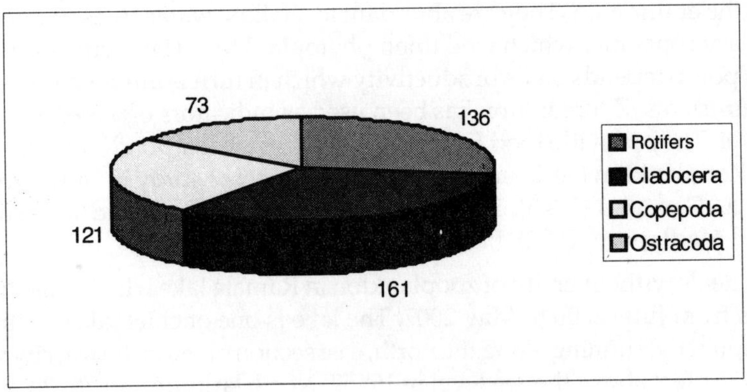

Figure 2.1: Sampling Site S_1

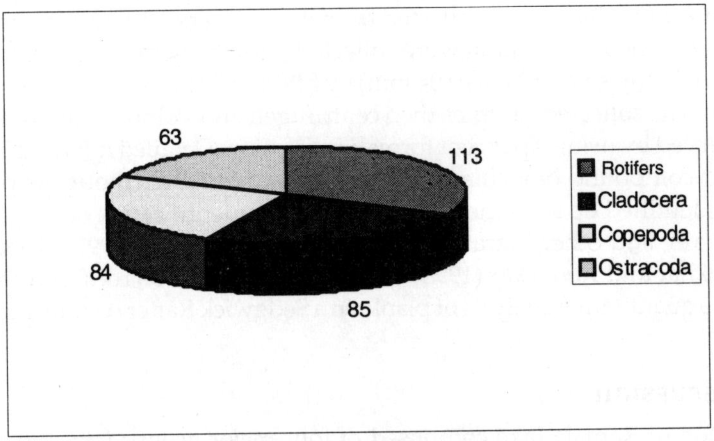

Figure 2.2: Sampling Site S_2

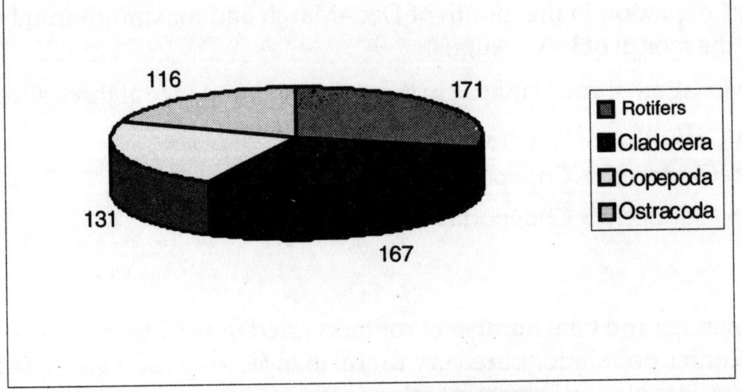

Figure 2.3: Sampling Site S_3

zooplankton community. Rotifer species exhibit marked differences in their tolerance and adaptability to changes in physico-chemical and biological parameters. Rotifers are commonly termed as 'Wheel-Animalcules'. The significance of rotifera is based on their abundance and their role in aquatic food web. According to Peijler (1957) there is no direct effect of pH on rotiferan population. Edmondson (1959) and Baker (1979) observed that the high rotifer population in winter could be attributed with favorable temperature and availability of abundant food. Chandrashekhar (1996) observed that in summer and monsoon, the factors like water temperature, and availability of abundant food. Chandrashekhar (1996) observed that in summer and mansoon, the factors like water temperature, turbidity, dissolved oxygen play an important role in controlling the rotifer population. In present study, the rotifer population was found to be maximum in the month of July 2006 at station S_3.

Copepoda

The monthly average and total number of copepods varied from 02 to 20 number per lit. at S_1, 02 to 14 at S_2 and 04 to 22 per lit. at S_3. Water temperature and availability of food organism affect the copepods. Allan (1976) observed the inverse, relationship between high population of rotifers and cladocera and low population of copepods. Sharma and Hussain (2001) have reported low copepode population in summer season. The quantitative dominance of copepods is also reported by Sharma and Hussain (2001). The copepod population was found maximum in August (19/lit) at S_1 sampling station whereas in May the number was comparatively less. The water with copepod abundance were regarded to be at a lower trophic stage than those with rotifer abundance (Yousuf, 1988). Lakes rich in organic matter support higher number of cychopoids (Subbamma, 1992), suggesting their prepoderence in higher trophic state of water.

Cladocera

Most of the cladocera species are primary consumers and feed on microscopic algae and fine particulate matter in the detritus. The monthly average and total number of cladocera varied from 02 to 19 per liter at S_1, 01 to 12 at S_2 and 04 to 20 at S_3. Quadari and Yusuf (1980) investigated the influence of some physico- chemical parameters like temperature, dissolved oxygen, turbidity on the population density and diversity of cladocera. The monsoon population was observed maximum in the present study, this may be due to favourable condition of temperature and availability of abundant food in the form of bacteria, nanoplankton and suspended detritus or due to thick deposit of organic matter in aquatic ecosystem. Usha Choubey (1997) found the similar result *i.e.* high density of cladocera in the month of July.

Ostracoda

The monthly average and total number of ostracoda varies from 01 to 18 per liter at S_1, 07 to 12 per lit. at S_2 and 02 to 20 per liter at S_3. The abundance of ostracodes provides a good food for aquatic organism. In the present study only one species of ostracoda was found *i.e.* cypris, having the maximum number in the month of July.

List of Zooplankton Species

Rotifer

Asplanchana spp., *Asplachana intermedia, Keratella tropia (with one spine), Keratella* spp. *Brachionus calyciforous, B. caudatus, B.durgae, B. bidentata, Beauchampiella cudactylotum, Filinia* spp., *Tricocera* spp.

Cladocera

Moina micrura, Moinodaphnia macleayi, Chydorus spheericus, C. faviformis, Ceriodaphnia spp.

Copepoda

Cyclop, Mesocyclops spp.

Ostracoda

Cypris.

References

APHA, 1998. *Standard Method for the Examination of Water and Wastewater*. American Public Health Association, Washington D.C., p. 874.

Annapurna, C. and Chatterjee, Tapas, 1999. The freshwater ostracodas (crustacea, ostracoda) from Dhanbad, Bihar. *J. Aqua. Biol.*, 14(1 and 2): 11–15.

Battish, S.K., 1992. *Freshwater Zooplankton of India*. Oxford and IBH Publishing Co., New Delhi.

Chandrashekhar, S.V.A. and Kodarkar, M.S., 1994. Biodiversity of zooplankton in Saroornagar Lake, Hyderabad. *J. Aqua. Biol.*, 9(1 and 2): 30–33.

Dhanapathi, M.V.S.S.S., 2000. *Taxonomic Notes on the Rotifers from India (from 1889– 2000)*. Indian Association of Aquatic Biologists (IAAB), Hyderabad, A.P., p. 178.

Edmondson, W.T., 1959. Rotifera. In: *Freshwater Biology*, (Ed.) W.T. Edmondson. John Wiley and Sons, Inc., New York, London, p. 420.

Nayar, C.K.G., 1970. Studies on the Rotifer Population of the ponds at Pilani, Rajasthan. *J. 2001 Soc. India*, 2: 21–34.

Pawar, S.K., Madlapure, V.R. and Pulle, J.S., 2003. Study of zooplanktonic community of Sirur dam water, near Mukhed in Nanded district. (M.S.). *Indian J. Aqua. Biol.*, 18(2):37– 40.

Salaskar Promod, 1996. Environmental studies of Powai lake. *Ph.D. Thesis*, Uninversity of Bombay.

Ward, W.B. and Whipple, G.C., 1992. *Freshwater Biology*, (Ed.) W.T. Edmondson.

Water and Wastewater Management in 2 Vols (2011) *Pages* **15–24**
Editors: **D.R. Khanna, R. Bhutiani & Gagan Matta**
Published by: **DAYA PUBLISHING HOUSE, NEW DELHI**

Chapter 3

An Appraisal of a Wastewater Treatment Plant for Pulp and Paper Mill

V.P. Deshpande
National Environmental Engineering Research Institute, Nagpur

ABSTRACT

Due to stringent requirements of regional regulatory agencies as also about environmental protection, there is a growing concern that is being shown in implementing wastewater management schemes. However, the approach towards solving wastewater management problem is still not that serious as a result, wastewater treatment plants are inadequately designed and constructed. Moreover, after commissioning of treatment plant, there are problems of poor operation and maintenance. All these finally adversely affect the treated wastewater quality, thus, defeating the basic objective of environmental protection. A field study conducted on one of the pulp and paper manufacturing industry has been presented in the paper alongwith shortcomings and possible remedial measures for achieving improvements.

Keywords: Wastewater management, Pulp and paper mill, Appraisal of treatment plant, Appropriate strategies, Discernible improvement.

Introduction

One of the water polluting industry is pulp and paper manufacturing industry. To meet the ever increasing demand of paper for various uses, there has been phenomenal growth in pulp and paper manufacturing industry. As a result, large number of pulp and paper making plants have been established. Depending on the raw materials and process technology adopted as also production capacity the paper mills have been grouped in 3 categories namely small,medium and large. While large and medium paper mills use conventional raw materials such as bamboo and wood, the small pulp and paper mills are based on cheaper raw materials like bagasse, waste paper, waste rags *etc.* These raw materials are the waste products and are relatively economical as compared to bamboo and wood. Moreover use of bagasse and waste paper helps in conservation of forest resources, thereby minimizing the adverse impacts on ecology and biodiversity of the region.

Generally it is seen that there is a tendency in industrial owners to concentrate on production with the sole aim of achieving maximum profits from the investments made while little or no attention is paid to the safe disposal of wastewaters into environment. It is only in recent years that due to stringent regulatory requirements, lot of awareness has been generated amongst the industries and public at large towards environmental protection.

One of the paper mill located in Central India was using bagasse and waste paper as a raw material for pulp and paper making and had installed a wastewater treatment plant for the treatment of combined wastewaters. Due to inadequate design and improper operation and maintenance, the treatment plant was not functioning well and as a result the desired standards for treated wastewater could not be met. In the present chapter, an attempt has been made to pin-point the discrepancies in design and operational problems and inadequacies in the satisfactory performance of the wastewater treatment plant. Possible solutions to remedy these problems have also been delineated.

Manufacturing Process

The paper mill had an installed capacity to manufacture 30 MT of Kraft paper per day. Unconventional raw materials such as bagasse, wheat straw, waste paper, and hessein bags were being used as main raw material. Bagasse from sugar mills or wheat, straw was used upto 70-80 per cent while the remaining 30-20 per cent was from waste paper, gunny bags or hessein bags.

Bagasse cooking is done in 10 per cent caustic solution in separate digesters rotating at 0.5 RPM. Steam is injected at 170°C and at 100 psi pressure. Digestion is continued for 4-5 hours. Pulping of gunny bags or hessein bags is carried out in separate digesters with 6 per cent caustic solution for a period of 6 hours at 170°C temperature and at 100 psi pressure. The pulp is then passed through sand trap and screens and subjected to washing at 2-3 stages. At this stage, black liquor is drained out. Pulp from waste paper is prepared in hydro-pulper unit and cleaned by passing through sand trap and screen.

In stock preparation section, washed pulps namely rag pulp, waste paper pulp and bagasse pulp are mixed together in a specified proportion and passed on to blending chest. Chemicals are added to it to achieve strength and durability. The pulp is then sent for blending for providing desired pulp to water ratio before sending it to paper machine. The paper machine consists of moving wire mesh and rotary driers. Steam is injected into the driers to remove moisture from the sheet of paper. Finished kraft paper is finally cut to the desired size and it is made ready for marketing. The moisture content of finished paper varies from 2-8 per cent by weight with an average value of 4-5 per cent. Schematic block diagrams for waste paper pulp, bagasse pulp and paper making are depicted in Figure 3.1.

Water Consumption

The paper mill used to draw raw water from the nearby perennial river to the tune of 4500 m^3/d. The water received adequate treatment in conventional water treatment plant alongwith ion exchange for removal of dissolved solids before sending it to steam generation plant.For washing of pulp and floor washing and other miscellaneous uses, raw water was directly being used. The break-up of water consumption is as follows:

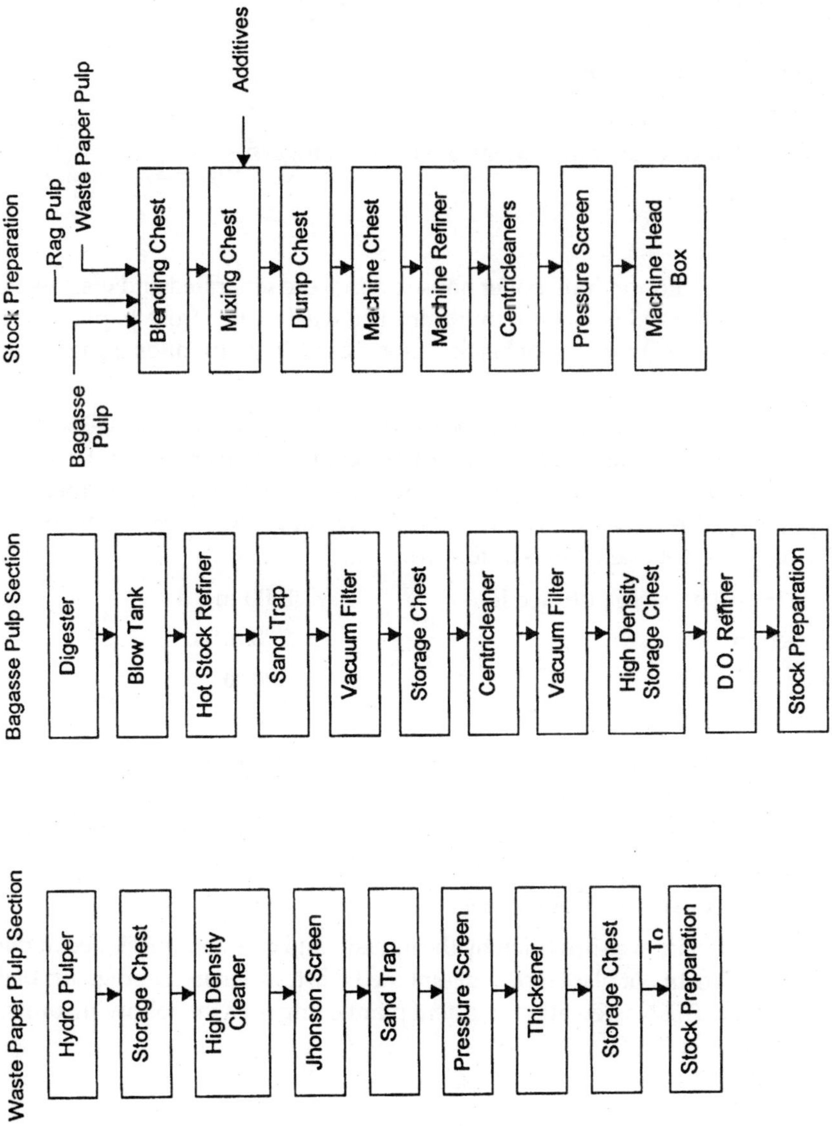

Figure 3.1: Schematic Block Diagram for Pulp and Paper Manufacture

- Process 4000-4500 m^3/d
- Water Consumption per MT 130-150^3/MT/d

Wastewater Sources and Quantity

During pulp and paper production, wastewater is emanated from different sources as indicated below:

- Black liquor from cooking section
- Pulp washing
- Boiler blowdown and regenerant water from ion exchangers
- Pulp thickener
- Paper machine

During the field visit, black liquor drain boiler house drain were separated and wastewaters from other sources was being discharged in a common drain. The wastewater from paper machine was being recycled back for it's reuse. A portion of black liquor was utilized for quenching the burnt hot ash from the boiler.

The wastewater originating from different process sections was between 3000-3600 m^3/d with an average quantity of 3250 m^3/d. Wide variations in wastewater quantity were encountered depending on number of process units working at a time. Due to absence of facilities for flow measurements, actual quantities flowing in different drains could not be ascertained. However, the break-up of total wastewater quantity could be obtained and is as follows:

Black liquor from 2 bagasse pulp section	1000 m^3/d
Rag plant section	700 m^3/d
Boiler blowdown	200 m^3/d
Waste paper pulp section	650 m^3/d
Excess wash water from paper machine	500 m^3/d
Floor washings and misc. wastewater	200 m^3/d
Total Quantity	*3250 m^3/d*

Existing Wastewater Treatment Facilities

A wastewater treatment plant was provided by the industy and was commissioned at the time of initiation of production. The treatment plant receives combined wastewater and segregation of different wastewater was nonexistent. The wastewater treatment plant comprises the following units:

- Screen
- Grit Chamber
- Grease Trap

Primary Settling Tank

- Oxidation Hopper Ditch–6 Nos
- Secondary Settling Tank–2 Nos

- ☆ Aerated Lagoon
- ☆ Single Drying Beds–2 Nos

Schematic treatment flowsheet of the WWTP is shown in Figure 3.2.

The wastewater treatment plant was designed based on assumed characteristics and kinetic constants for biological treatment units. Design basis of the WWTP was as follows:

- ☆ Flow: 3000 m^3/d
- ☆ pH: 9.5 to 9.8
- ☆ BOD_5, (20°C): 700 to 800 mg/l
- ☆ S.S.: 1000 to 1200 mg/l

The combined wastewater after treatment in screen, grit chamber and grease trap is passed on to primary settling tank with 3 hours detention time where bulk of the suspended solids settle down and the clear supernatent is led to oxidation ditch and aerated lagoon in equal proportion for biological treatment. The wastewater after oxygenation in both the units is passed on to hopper bottom setting tank of 3.5 hours detention time. The wastewater from secondary settling tank after separation of biological solids is stored in a sump wherefrom it is pumped for utilization on land as a source of irrigation water for sugarcane fields.There is no direct discharge of treated wastewater into near by water body.

Field Observations and Critical Appraisal of WWTP

With a view to plan the strategies for achieving significant improvements in the quality of treated wastewater, a field visit was made for onsite observations on the WWTP wherein each unit of the treatment plant was seen for technical appraisal. The observations made and shortcomings noticed in the WWTP are discussed below.

Screen

- ☆ Spacing of screen bars was not uniform as a result of which some of the floating matter was getting passed on to grit chamber.
- ☆ Removal of screenings from the screen was not being done at regular intervals.Due to this, heaps of screenings were observed in nearby area.

Grit Chamber

- ☆ Grit in huge quanity was seen deposited at the bottom of grit chamber. This had resulted in carry over of gritty material to subsequent treatment units. This had also caused reduction in designed detention time.

Grease Trap

- ☆ Foam was observed to be floating over the liquid surface in the grease trap.

Primary Settling Tank

- ☆ Due to operation of different pulp sections for meeting the requirements, some of wastewater discharges were intermittent as a result, there were wide fluctuations in the wastewater

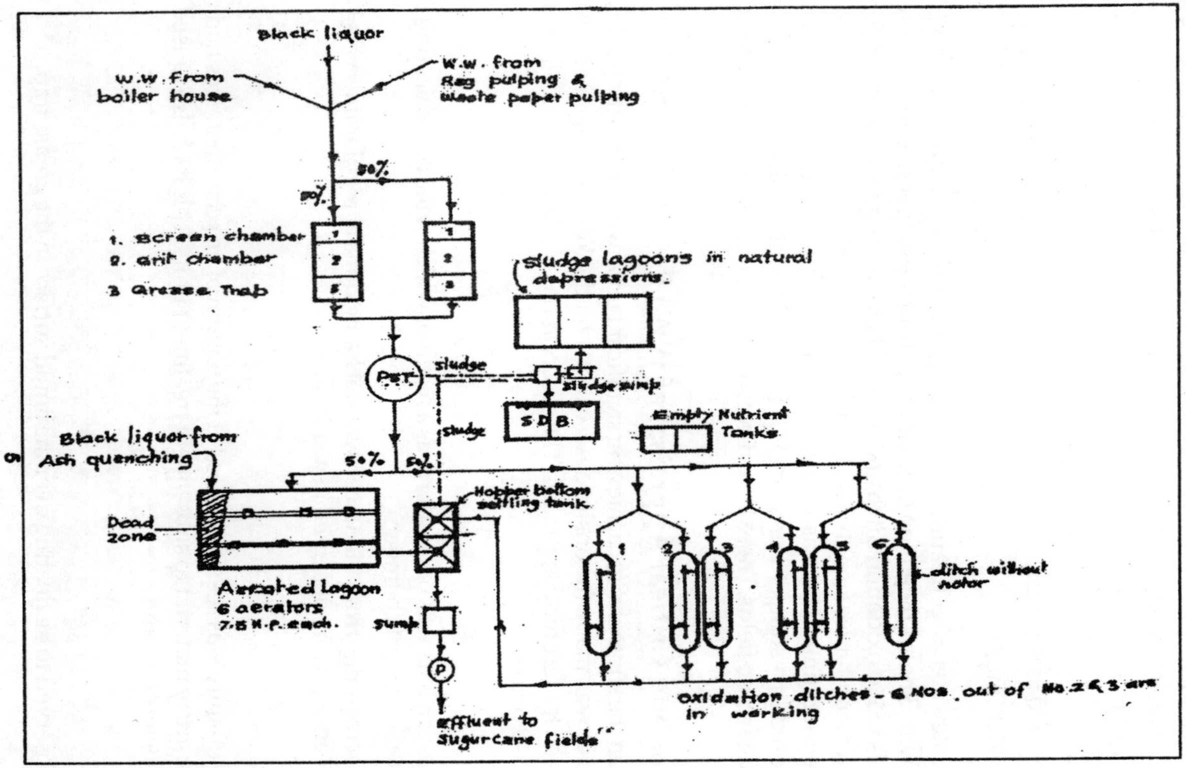

Figure 3.2: Schematic Flowsheet for the Existing Wastewater Treatment Plant for Pulp and Paper Mill

quantity. In absence of equalization tank, it used to impose shock loads on the treatment plant. In particular the primary settling tank was seen adversely affected as observed from the persistence of suspended solids in the supernatant.

Oxidation Ditch

- ☆ There were six units of the oxidation ditch working in parallel. The ditches were constructed in brick masonry. Due to poor workmanship and poor quality construction, masonry walls of 3 ditches got collapsed and bottom of all ditches got severely damaged.
- ☆ Out of 6 ditches, only 2 were in working condition at the time of field visit. This has resulted in inadequate treatment to the primary treated wastewater.
- ☆ Each ditch was provided with 2 horizontal rotors of 3.7 KW (5HP) capacity and out of 10 rotors, only 4 were working.
- ☆ The depth of immersion of rotor blades was not uniform in case of rotors in working condition and it was on higher side as compared to desirable value of 16 cm for proper oxygenation to take place.
- ☆ 2 RCC tanks were seen provided for nutrient (N and P) storage, mixing and addition but these were completely empty with the result that the biological treatment units did not receive nutrient dozing even it is required.
- ☆ Arrangement for recirculation of sludge was non-existent. As a result, desired MLSS concentration could not be maintained in the mixed liquor. This had also disturbed the organic loading to the ditch resulting in poor treatment efficiency.

Appraisal of the design of oxidation ditch indicated F/M=0.1, MLSS=5000 mg/l, oxygen requirement= 2 kg of O_2 per kg BOD applied and HRT=26 Hrs. Out of these parameters, MLSS concentration of 5000 mg/l appeared to be on much higher side, particularly in view of the absence of sludge recirculation facility. Spot testing of mixed liquor from one of the ditch indicate MLSS concentration of 2700 mg/l which not only disturbed the organic loading but also resulted into poor treatment efficiency and poor quality of treated effluent.

Aerated Lagoon

- ☆ Aerated lagoon of size 60mx50mx3.5m+1m freeboard was constructed in earthwork with sloping sides.
- ☆ The lagoon was designed on the basis of arbitrarily chosen value of the system rate constant as 0.234 per day and a detention time of 4 days was provided.
- ☆ For oxygenation of the contents of lagoon, 6 aerators of 7.5 HP each were provided. The alignment and spacing of the aerators was not proper since the adjacent aerators were moving in the same direction with depth of immersion varying between 10 to 40 cm. Due to this, even after operating all the aerators, desired oxygenation was not taking place as could be seen from D.O. value between 1.2-1.5 mg/l. At the corners of the lagoon, dead pockets had formed due to insufficient oxygenation. Moreover, depth of lagoon itself being 3.5m was causing low D.O. in bottom layers.
- ☆ The black liquor used for quenching the hot ash from boiler was seen getting percolated from the sides and bottom of lagoon due to non-provision of impermeable lining on the sides and bottom of the lagoon which was adversely affecting the functioning of the unit.

Secondary Settling Tank

- 2 units of the SST in the form of hopper bottom settling tank with size of each as 12mx 6mx 3m were provided. The oxygenated wastewater from aerated lagoon and oxidation ditch was getting treated in the tank.
- Lot of scum was seen floating over the liquid suface of SST and disludging was not done at regular intervals resulting into sludge deposition within the tank itself and consequently there was reduction in designed detention time.
- The oxidation ditch was designed for 5000 mg/l MLSS concentration while in the aerated lagoon, MLSS concentration in the range of 300-500 mg/l was observed. With these concentrations, the effective MLSS of the combined oxygenated wastewater works out to 2700 mg/l. With such MLSS, for effective removal of S.S. to take place, the SST should have been designed based on solids loading rate with an average value of 4.5 kg MLSS/m^2/d. Instead, the SST was designed based on the assumed value of surface overflow rate as 20 m^3/m^2/d. This was certainly affecting the treatment efficiency of SST.

Sludge Drying Beds

- At the WWTP, 2 sludge drying beds of size 25mx8m were provided. With the sludge from PST with 5 per cent solids content and biological sludge from SST with 1 per cent solids content, the area of sludge drying beds(400 m^2) was inadequate for effective drying and the removal of sludge in semi-solid state once a week was being practiced.
- The design of drying beds based on 5 days drying period and 60 cm depth of sludge. Both these value were inadequate since for effective drying, a dry cycle period of 10 days and a depth of sludge of 30 cm is considered. Due to this shortcoming at the design stage itself, the drying of sludge was not proper.

Strategies for Improvements in the Functioning of Wastewater Treatment Plant

In view of the problems in the design, construction and operation of the WWTP as also shortcomings noticed during field visit, there was a need to formulate most appropriate strategies for achieving discernible improvement in the performance of WWTP with the sole objective of meeting the regulatory requirements. Unitwise suggestions/modifications were suggested as mentioned here under.

Screen

- Spacing of screen bars need to be made uniform by rearranging the bars so as to have a spacing of 100 mm centre to centre.
- The screening needs to be stored separately for drying then dispose of by burning or burying in low lying areas.

Grit Chamber

- Removal of settled grit from the grit chamber needs to be done at regular intervals.The grit removed need be disposed off in low lying area.
- For improving functioning of grit chamber, horizontal velocity of flow need to be controlled by installing velocity control device such as parshall flume or proportional flow weir at the outlet of chamber.

Grease Trap

- ☆ The floating foam over the liquid surface need to be removed with the help of nylon scoops or by spraying anti-foaming agent on the surface.

Primary Settling Tank (PST)

- ☆ In view of variations in the wastewater flow, there is a need to even out variations by providing equalization basing of adequate capacity. This needs to be provided with mixing arrangement (0.02-0.04HP/1000 gallons volume) for preventing solids from settling within the equalization tank itself.
- ☆ A pump of suitable capacity (15 HP) needs to be provided for pumping the equalized flow at a constant rate. This will help in improving the performance of subsequent treatment units.
- ☆ Desludging from the PST needs to be done at regular intervals. This will aid in the improvement of existing efficiency.

Oxidation Ditch

- ☆ The walls and bottom of oxidation ditch (6 Nos) need to be repaired and reconstructed with sturdy construction. To prevent percolation of wastewater, plastering with waterproof cement be provided to all surfaces coming in contact with the liquid.
- ☆ It needs to be ensured that all the 6 ditches alongwith their rotors are kept in operation simultaneously for effective treatment. The depth of submergence of rotor blades needs to be kept at uniform depth of 16 cm for achieving proper aeration of the contents.
- ☆ The success of the biological treatment of wastewater depends on maintenance of BOD:N:P ratio as 100:5:1 and if the wastewaters are found deficient in nitrogen or phosphorus, these are required to be supplemented externally in the term of urea for nitrogen and superphosphate for phosphorus.

With a view to maintain the F/M ratio at the designed level, it is necessary to keep MLSS concentration at a particular level with variations in a narrow range. For this, biological sludge from secondary settling tank needs to be recirculated back to oxiation ditch at suitable recirculation ratio. This ratio needs to be adjusted based on experience with the operation of ditch.

Aerated Lagoon

- ☆ The aerated lagoon should be provided with impermeable lining on the sides and bottom. This will help in preventing seepage from the lagoon and consequently protect the undergroundwater resources from pollution.
- ☆ The alignment of the aerators should be so selected that the adjacent aerators rotate in opposite direction and the circle of influence of one aerator touches with the other aerator. For proper splashing action and exposure of wastewater droplets to atmospheric oxygen, it is necessary that the depth of submergence of the aerator blades be kept at 100-150 mm.
- ☆ It is normally seen that aeration of wastewater from pulp and paper mills generates foam in the aeration tank which may become air borne and with the favourable wind velocity, it can get spread in nearby locality, thus disturbing the local residents. Foaming during aeration of pulp and paper mill wastewater takes place due to presence of lignin. To overcome this

problem, antifoaming agents such as ethyl hexanol need to be sprayed over the liquid surface. The dose of this chemical can be fixed after gaining experience on its application.

- In order to see as to whether oxygenation is taking place effectively or not, D.O. levels needs to be monitored at various depths along the length and breadth of the aeration tank and it needs to be seen that the D.O. in the mixer liquor of the aeration tank is maintained in the range of 1.5-2.0 mg/l.

Sludge drying Beds

- The area of existing sludge drying beds was worked out considering only primary sludge quantity, sludge layer thickness of 50 cm and dry cycle period of 5 days. To take care of secondary biological sludge (with 1 per cent solids concentration), additional 10 beds of size 65 m x 50 m x 1.05 m need be provided. This has been worked out based on sludge layer thickness of 25 cm and dry cycle period of 10 days for effective drying of sludge.
- The sludge cake after drying needs to be removed and disposed of at a identified site free from human habitation and other activities.

Conclusion

For meeting the requirements of regional regulatory agencies, it becomes mandatory for all manufacturing industries to provide a well designed, constructed, operated and maintained wastewater treatment plant. However, provision of treatment plant receives last priority since the industry is more interested in maximizing the economic benefits and minimizing the cost of production which also includes cost to be incurred on environmental protection. The field survey conducted on one of the paper mill revealed that there were discrepancies in the design, construction and operation of the wastewater treatment plant. With a view to solve these shortcomings, strategies have been formulated and delineated in the paper.

Water and Wastewater Management in 2 Vols (2011) *Pages 25–29*
Editors: **D.R. Khanna, R. Bhutiani & Gagan Matta**
Published by: **DAYA PUBLISHING HOUSE, NEW DELHI**

Chapter 4

The Impact of Industrial Effluent on Growth of *Cicer arietinum*

Rageni Agarwal
115-Vijay Nagar Colony, Badaun, U.P.

ABSTRACT

The present chapter deals with physico-chemical characteristics of a industrial effluents and effluent treated soil.The physico-chemical characteristics suggested that effluent was under the permissible limit as per ISI- guidelines. Therefore, there is no harm to use this water for irrigation purpose. The pot experiments reflected that there was increase in total biomass and chlorophyll content in effluent irrigated plants. The most encouraging results were observed in 60 per cent of effluent than 40 per cent and 100 per cent. The higher concentration of effluent restricts growth of plants like *Cicer arietinum*.

Keywords: Cicer arietinum, Metalaid, Industrial effluent, Physico-chemical.

Introduction

Metal toxicity has become a problem in marginal cropland because of the increased metal solubility from acid rain. Industrial activities generate large quantities of water containing chemicals, trace elements, effluent containing alkaline, acid calcium carbonate, sodium, potassium, nitrate, aluminium are discharged. For these elements known to be essential in animal human and in plant nutrition. "Micro-nutrients are more useful and explicit term". The micro-nutrients important for plant growth, are Cu, Zn, Fe, Mn, Mo, Co, Na, B and U possibly.

Various types of chemicals are used for forming the product in metal-aid industry. The main chemicals which are used as raw materials are aluminium ash, aluminium powder, sodium,

Silicofloride, nitrate, pf reagent, wooden dust, fly ash, Saw-dust and ash. The treatment used for industrial wastes are almost the same as that of sewage. All types of treatment such as preliminary treatment, primary treatment, secondary treatment, tertiary treatment are also necessary in case of industrial waste treatment. In industrial waste treatment it is impossible to select a well suited treatment for a particular effluent (organic, inorganic and chemical effluent) because of the fact that the quality of wastewater is uniform as well as predictable and the pollutant present is known.

In the present investigation an attempt has been made to analyse the physico-chemical aspect of metal-aid industrial effluent and their effect on the growth parameter of *Cicer arietinum*.

Materials and Method

The materials and method include the following aspects.

- ☆ Location of site
- ☆ Sampling
- ☆ Water analysis
- ☆ Soil analysis
- ☆ Plant growth

Sample of water for various examination were collected from metal aid industry located in Haridwar District. Soil sample was collected from agriculture land.

Water samples were collected in plastic cans for analysis of physico-chemical characteristics of various parameters *i.e.* pH, electrical conductivity, total dissolved solid, total hardness, dissolved oxygen, alkalinity, sodium, potassium and were estimated as per methods of APHA (1998) and Trivedy and Goel (1986).

Soil samples were collected from the agricultural field and were analysed for various parameters *i.e.* pH, electrical conductivity, alkalinity, organic matter, carbon present, sodium, potassium and bicarbonate. The soil samples were air dried crusted and passed through 2µm sieve and stored in polythene bags for analysis.

Results and Discussion

The results of the above study is shown in Tables 4.1–4.6. The quality and composition of effluents depends upon the type of industry. According to Bhatnagar (1993) colour of water is usually estimated by visual method. Colour of metallic industrial effluent depends upon the metal used.

Temperature is basically an important factor for its effect on chemical and biological reaction in water. The mean value of temperature recorded was 27°C ± 0.82°C. Electrical conductivity is an indication of dissolve solid and suspended solids. The mean values of electrical conductivity was 0.255 µmhos/cm ± 0.008 µmhos/cm. The ability to transmit an electrical current depends on the concentration of charged ionic species in the water (Trivedy and Goel, 1986). The mean value of pH of effluent collected from metal aid industry was recorded 8.69 ± 0.660 and this is with in acceptable range for industrial effluent (ISI). Alkalinity of water is its capacity to neutralize a strong acid and is characterised by the presence of hydroxyl (OH^{+}) ions capable for combining with hydrogen ($H^{+)}$ ions (Saxena, 1994). The mean value of alkalinity was 126.8 mg/l ± 3.02 mg/l.

Table 4.1: Values of Some Selected Physico-chemical Parameters of Effluent of Metal Aid Industry (Values are mean±SE for Eight observation each)

Sl.No.	Parameters	Mean Values
1.	Color	Grey
2.	Temperature (°C)	27.0±0.82
3.	Conductivity (m mhos cm)	0.255±0.008
4.	pH	8.69±0.660
5.	Alkalinity (mg/l)	126.80±3.02
6.	Hardness (mg/l)	144.0±0.915
7.	DO (mg/l)	4.738±0.607
8.	BOD (mg/l)	1.538±0.132
9.	Sodium (mg/l)	3.85±0.512
10.	Potassium (mg/l)	20.0±0.895

±: Standard Error.

Table 4.2: Values of Some Selected Physico-chemical Parameters of Soil in Agricultural Field (Values are mean±SE for Eight observation each)

Sl.No.	Parameters	Mean Values
1.	pH	7.56±0.521
2.	Conductivity (m mhos cm)	0.24±0.004
3.	Alkalinity (mg/l)	50.00±0.982
4.	Organic matter (per cent)	1.42±0.019
5.	Organic Carbon (per cent)	0.825±0.019
6.	Sodium (mg/l)	6.35±0.434
7.	Potassium (mg/l)	3.04±0.129
8.	Nitrate (per cent)	0.0065±0.02

±: Standard Error.

Table 4.3: Values of Germination Percentage of Plant of *Cicer arietinum*

Concentration	Control	40 per cent	60 per cent	100 per cent
Germination per cent	8.86±2.65	7.86±2.23	7.86±1.26	6.46±2.04

Table 4.4: Values of Root Length and Shoot Length of Plant *Cicer arietinum* (After 20 days of germination)

Sl.No.	Concentration	Shoot Length (cm)	Root Length (cm)
1.	Control	20.0±0.041	14.60±0.083
2.	40 per cent	15.97±0.028	11.0±0.062
3.	60 per cent	18.70±0.032	16.50±0.070
4.	100 per cent	14.80±0.021	13.1±0.036

Table 4.5: Average Biomass (gm/m²)

Concentration	Control	40 per cent	60 per cent	100 per cent
Fresh X_1	5874.02	4529.37	4543.52	4465.67
Dry X_2	612.17	527.76	537.86	428.16

In the present study, mean value of total hardness has no known adverse effect. Total hardness of water is the sum of alkaline earth metals cations present in it. Hardness is mainly due to presence of carbonate, bicarbonate, sulphate chloride *etc.* (Trivedy and Goel, 1986). Mean value of dissolved oxygen in the sample observed was 4.738 mg/l ± 0.607mg/l and biochemical oxygen demand (BOD) was 1.538 mg/l ± 0.1321 mg/l. BOD can be defined as the rate of removal of oxygen by microorganisms in aerobic degradation of the dissolved or even the particulate organic matter in water and it is used as an index of organic pollution in water (Saxena, 1994).

Table 4.6: Chlorophyll Content of *Cicer arietinum*

Sl.No.	Treatment	Total Chlorophyll (mg/gm)	a Fraction of Chlorophyll (mg/gm)	b Fraction of Chlorophyll (mg/gm)
1.	Control	0.4684	0.3884	0.0800
2.	40 per cent	0.4983	0.4095	0.0888
3.	60 per cent	0.5260	0.4420	0.0840
4.	100 per cent	0.3682	0.3150	0.0532

In the present study, the mean value of sodium was 3.85 mg/l ± 0.512 mg/l, potassium is also a naturally occuring element in the present study the mean value of potassium was 20.00 mg/l ± 0.895 mg/l.

Soil pH strongly influence plant nutrient availability and soil productivity. From the evidence available, neither a high pH above 8.4 nor a low pH below 5 is favourable for maximum field of crop (Puruthi, 1970). The mean value of pH was found 7.56 ±0.521. According to Verma (1994) the favourable pH for growth of plants is 7.8. The mean value of electrical conductivity of soil was found to be 0.24 μmhos/cm ± 0.004 μmhos/cm.

During present study, the mean value of alkalinity was 50.00mg/l ± 0.982 mg/l. In the present study mean value of carbon present is 0.825 per cent ± 0.013 per cent and the mean value of organic matter was 1.42 per cent ± 0.019 per cent.

The value of exchangable Sodium was found 6.35 mg/l ± 0.434 mg/l. Soil temperature also affects the potassium uptake by plants. Reduced temperature showed down plant growth and rate of potassium uptake (Tisdale *et al.*, 1995). Tisdale *et al.* (1995) reported that when the content of sodium ion increased the potassium ions may become less as sodium ion can partially substitute for potassium. The value of exchangeable potassium was 3.04 mg/l ± 0.1289 mg/l in agricultural soil at the time of study. In the present study the mean value of nitrate was 0.0065 per cent ± 0.02 per cent.

The results shows that the germination rate in control was high, where as in 40 per cent and 60 per cent the germination present was slightly less than control and 100 per cent concentration also exhibit moderate per cent for germination of seeds.The shoot length in control was high than in 60 per cent, 40 per cent and 100 per cent concentration. The 60 per cent concentration was more suitable for root length of *Cicer arietinum* in comparison to control, 100 per cent and 40 per cent effluent. The reduction in root and shoot length in seeds treatment with effluent might be due to the stress caused by high salinity Kumar *et al.* (1993).

The control irrigated plants biomass was found high during the course of study, therefore the plants of control sites were found healthy whereas higher concentration of effluent retards the biomass. Effect of tailing water irrigation on biomass of *Vigna radiata, Tritienum aestivum, Brassica compestris etc.* was studied by Thukral (1986). The study revealed that with the increase of effluent concentration, biomass of different parts of the plant decreased gradually. The 60 per cent content shows the highest amount of chlorophyll than 40 per cent. The control shows relatively less values of chlorophyll content. The least values were observed for 100 per cent concentration.

References

APHA, 1998. *Standard Methods for the Examination of Water and Wastewater*, 19th edn. APHA, Washington, D.C.

Bhatnagar, M.K., 1993. A study of physico-chemical properties of industrial effluents and city sewage effected of biogas production, *M.Sc. Dissertation*, Department of Chemistry, Gurukul Kangri University, Haridwar.

Kumar, N.V., Bhagyalakshmi, K.N. and Dhananjaya, M., 1993. Tannery and chromate industrial effluent and pesticide contamination of reservoir and other water bodies: Physico-chemical and ecotoxicological studies. In: *Ecology and Pollution of Indian Lakes and Reservoirs*. Ashish Publishing House, New Delhi.

Puruthi, D., 1970. Determination of pH colorimetric method. *Nature*, 174: 1100.

Saxena, M.M., 1994. *Environmental Analysis Water, Soil and Air*. Agro Botanical Publishers, New Delhi.

Tisdale, S.L., Werner, L.N. and John, L.H., 1995. *Soil Fertility and Fertilizer*. Prentice Hall of India Pvt. Ltd., New Delhi.

Thukral, A.K., 1986. Effect of tailings water irrigation on the biomass of some crop plants. *J. Env. Biol.*, 10(4): 337.

Trivedy, R.K. and Goel, P.K., 1986. Physico-chemical analysis of water. In: *Chemical and Biological Methods for Water Pollution Studies*. Environmental Publication, Karad.

Verma, N., 1994. Comparative studies of different sub standard soil with cultivated land in relation to some physico-chemical component. *M.Sc. Dissertation*, G.K.U., Haridwar.

Water and Wastewater Management in 2 Vols (2011) *Pages 30–34*
Editors: D.R. Khanna, R. Bhutiani & Gagan Matta
Published by: DAYA PUBLISHING HOUSE, NEW DELHI

Chapter 5

Studies on the Sex Ratio of *Chela untrahi* of Bhadra Reservoir, Karnataka, India

B.R. Kiran[1], E.T. Puttaiah[1] and C.S. Anantha[2]
[1]*Department of Environment Science, Kuvempu University, Shankaraghatta*
[2]*Bhadra Fish Seed Farm, Bundh Breeding, B.R. Project*

ABSTRACT

In the present study 1432 specimens of *Chela untrahi* belonging to various size ranges were examined for sexuality at backwaters of Bhadra reservoir. Dominance of male was noticed in most of the size groups. Significant chi-square values were obtained in fishes of size between 81 and 100 mm during 1998-99 and 81 and 115 mm for 1999-2000. The overall male:female ratio was found to be 1:0.5057 and predominance of male was noticed all through the year.

Keywords: Chela untrahi, Sex ratio, Chi-square, Bhadra reservoir.

Introduction

A good deal of work has been done on the reproductive biology of various fishes. Notable among them are: Hickling, 1930; Clark, 1934; Keasteven, 1942; Alikunhi, 1956; Qasim and Qayyum, 1961; Raitt, 1968 and Bagenal, 1978. A knowledge of the sex composition of catches is helpful in understanding whether any differential fishing exists and if solubility what possible bearing it has on the fishable stocks (Kesteven, 1942). It may indicate differences in the growth rate of the two sexes (Qasim, 1966). *Chela untrahi* is included in the sub order-Cyprinoidei under the order-Physostomi and belong to the family-Cyprinidae (Day, 1958). Most of the *Chela* species are inhabitants of tropical and subtropical waters. Bhadra reservoir from where specimens were collected for study serves as a source of fishing, irrigation and for producing electricity. Fishermen and villagers also use it for drinking purpose and pisciculture of locally available fishes and other major carps. The Bhadra reservoir is located at a latitude 13°45′-00″N and longitude 75°-30′-14″ E. No scientific study has been conducted on the sex-ratio of *Chela untrahi* occurring in Bhadra reservoir so far. Hence, the present study was aimed at enhancing the knowledge regarding the sex ratio of *Chela untrahi* of Bhadra reservoir and to enable the formulation of suitable management measures towards rational exploitation and management.

Materials and Method

The samples of *Chela untrahi* were obtained from the backwaters of Bhadra reservoir at Narasimharajapura near Kalgudda. *Chela untrahi* fishes were collected with the help of fishermen by using monofilaments gillnets of size 16 mm. Random samples were taken for sexuality of which 951 were males and 481 were females. The size (total length) range of species varied between 71 mm and 135 mm. The period of study was from July 1998 to June 2000. Fish samples were collected once in a month.

Results and Discussion

The results showed that male was dominant in most of the size groups. Table 5.1 represents the sex ratio of *Chela untrahi* in relation to size groups of fish. Significant chi-square values were obtained in size between 81 and 100 mm during 1998-99 and 81 and 115 for 1999-2000. Tables 5.2 and 5.3 represents data on sex ratio of *Chela untrahi* with respect to different months of year. Predominance of male was noticed all through the year except in August and September 1998 and again August-99 when the ratio was most equal to the hypothetical 1:1 Chi-square (x^2) values indicated that dominance of males in the population was significant during most part of the year and the male : female ratio was found to be 1:0.5057.

Table 5.1: Sex Ratio of *Chela untrahi* in Different Size Groups from July, 1998 to June, 2000

Size Groups (mm)	Year 1998–99					Year 1999–2000				
	Males	Females	Total	X^2 Values	Sex Ratio	Males	Females	Total	X^2 Values	Sex Ratio
71-75	1	0	1	1.0000	1:0.000	0	0	0	0	0
76-80	4	1	5	1.8000	1:0.2500	1	0	1	1.0000	0
81-85	37	17	54	7.40274*	1:0.4594	13	2	15	8.0666*	1:0.1538
86-90	135	49	184	40.1956*	1:0.3629	73	8	81	52.1605*	1:0.1069
91-95	136	78	214	15.7196*	1:0.5735	140	37	177	59.9378*	1:0.2642
96-100	127	51	178	32.4494*	1:0.4015	157	71	228	32.4386*	1:0.4522
101-105	35	27	62	1.0322	1:0.7714	71	59	130	1.1077	1:0.8309
106-110	4	13	17	4.7647*	1:3.2500	15	38	53	9.9811*	1:2.5333
111-115	0	3	3	3.0000	0	0	12	12	12.0000*	0
116-120	0	1	1	1.0000	0	2	8	10	3.6000	1:4.0000
121-125	0	0	0	0	0	0	4	4	4.0000	0
126-130	0	0	0	0	0	0	1	1	1.0000	0
131-135	0	0	0	0	0	0	1	1	1.0000	0
Total	479	240	719	79.4450*	1:0.5010	472	241	713	74.8401*	1:0.5106

* Significant at 5 per cent level.

Table 5.2: Sex Ratio of *Chela untrahi* in Monthly Samples from July, 1998 to June, 1999

Months	*Males*	*Females*	*Total*	*X^2 Values*	*Sex Ratio*
July 1998	18	19	35	0.0270	1:1.0555
August	31	29	60	0.0666	1:0.9355
September	30	30	60	0.0000	1:1.00
October	37	23	60	3.2667	1:0.6216
November	59	14	73	27.7397*	1:0.2372
December	42	27	69	3.2608	1:0.6428
January 1999	45	15	60	15.0000*	1:0.3333
February	45	15	60	15.0000*	1:0.3333
March	47	13	60	19.2666*	1:0.2766
April	48	12	60	21.60000*	1:0.2500
May	34	16	50	6.4800*	1:0.4705
June	43	27	70	3.6571	1:0.6279
Total	479	240	719	79.4450*	1:0.5010

* Significant at 5 per cent level.

It is believed that the following factors might be responsible for sex composition.

1. Segregation of the sexes through various periods of the year including segregation resulting from sex differences in age and size at maturity.
2. Gear selectivity in relation to sex differences in morphology and in physiological activity.
3. Differences in natural and fishing mortality between sexes. Bannet (1962) states "Most studies of sex ratio of the individuals comprising isolated populations of freshwater fishes show more males than females among smaller fishes, but among the older fishes the dominance of females is solubility great as to leave little doubt that the males died off much faster than females".

Data on sex ratio in different sizes showed that males were dominant in size groups other than 106-110, 111-115 and 116-120 mm during 1998-99. But an observation during 1999-2000 revealed that the males were dominant in size groups other than 106-110, 111-115, 116-120, 121-125, 126-130 and 131-135 mm. Austin (1971) studied the sex ratio in fishes of sizes up to 110 mm. He found that above 110 mm size range females were dominant over males. Similar condition was observed during the study period. Significant chi-square values were obtained with sex ratio in size groups of 81-85, 86-90, 91-95, 96-100, 106-110 and 111-115 mm during 1998-2000. Chi-square test was also applied to monthly samples and it was found that at 0.05 probability levels, significant difference was noticed in all the months except July, October and December 1998, June and August 1999 and June 2000. Joseph K. Manissery *et al.*, 1979 studied the sex ratio in *Puntius ticto* in relation to months. According to them, male:female ratio of *Puntius ticto* was 1:0.901 and concluded that there is no significant difference between the observed and the expected values of 1:1 (P=0.05), in the sex ratio of *P.ticto*. However, in the present study there is a significant difference exists in the sex ratio of *Chela untrahi* (1:0.5057).

Table 5.3: Sex Ratio of *Chela untrahi* in Monthly Samples from July, 1999 to June, 2000

Months	*Males*	*Females*	*Total*	*X^2 Values*	*Sex Ratio*
July 1999	16	44	60	13.0666*	1:2.7500
August	30	30	60	0.0000	1:1.0000
September	48	12	60	21.6000*	1:0.2500
October	14	46	60	17.0667*	1:0.2857
November	29	11	40	8.1000*	1:0.3793
December	53	7	60	35.2666*	1:0.1320
January 2000	43	5	48	30.0833*	1:0.1162
February	39	15	54	10.6667*	1:0.3846
March	45	8	53	25.8301*	1:0.1777
April	54	11	65	28.4461*	1:0.2037
May	45	11	56	20.6428*	1:0.2444
June	56	41	97	2.3195	1:0.7321
Total	472	241	713	74.8401*	1:0.5105

* Significant at 5 per cent level.

Sex ratio in *Chela untrahi* at different months and years has been showed that males outnumber females. Similar observation was made by Neelakantan *et al.*, 1980-81 in *Lactarius lactarius.* While Vinci 1984 and Azad 1990 studying on *Silonia childreni* and *Anabas testudineus* reported that females outnumbered males. Nevertheless, Bhimasena Rao and Karamchandani 1986 noticed equal population of male and females in *Ompok bimaculatus* from Kulgarhi reservoir (M.P.). However among the smaller fishes, males constituted the majority (Mc Fadden and Copper 1962, Bailey 1963 and Bhatnagar 1972). Therefore, the observations made in the present study are conformity with the above researchers.

Kesteven 1942 observed the predominance of males in Australian mullet (*Mugil dobula*) which he attributed to differential fishing. This may not be solubility in the case of *Chela untrahi* since no differential fishing has been observed. The apparent variation in the monthly sex ratios may be due to sampling variations. The insignificant differences in the number of individuals of both the sexes during spawning month (June, July, August, September, October, and December) indicated that males and females congregate during the spawning season. Therefore, fish catch should be avoided during aforesaid months.

Acknowledgements

Authors are thankful to Bhadra Fish Seed Farm authorities and Kuvempu University for providing facilities.

References

Alikunhi, K.H., 1956. Observation on the fecundity, larval development and early growth a *Labeo bata* (Hamilton). *Indian J. Fish.*, 3(1): 216–229.

Austin, H.M., 1971. Some aspects of the biology of the rhomboid majarra *Diapterus rhombeus* in Purto Rico. *Bull. Mar. Sci.*, 21: 886–903.

Azad, I.S., 1990. *Proceedings of the Second Indian Fisheries Forum*, May 27–31, pp. 131–134.

Bagenal, T.B., 1978. Aspects of fish fecundity. In: *Ecology of Freshwater Fish Production*, (Ed.) S.D. Gerking. Blackwell Scientific, Oxford, pp. 75–101.

Bailey, M.K., 1963. Age, growth and maturity of round Whitefish of Apostle islands and Ile Royale Regions, Lake Superior. *U.S. Fish and Wildlife, Fishery Bull.*, 63: 63–75.

Bannet, G.W., 1962. *Management of Artificial Lakes and Ponds*. Reinhold Publishing Corporation, New York.

Bhatnagar, G.K., 1972. Maturity, fecundity, spawning season and certain related aspects of *Labeo fimbriatus* (Bloch) of river Narmada near Hashanabad, (M.P). *J. Inland Fish. Soc., India*, 4: 26–37.

Bhimasena Rao, J. and Karamchandni, S.J., 1986. On the spawning biology of *Ompok bimaculatus* (Bloch) from Kulgarhi reservoir of Madhya Pradesh. *J. Inland Fish. Soc. India*, 18(20): 40–47.

Clark, F.N., 1934. Maturity of the California sardine (*Sardinella caerulea*), determined by ova diameter measurement. *Fish. Bull., California*, pp. 1–49.

Day, F., 1958. *The Fishes of India: Being a Natural History of the Fishes Known to Inhabit the Seas and Freshwaters of India, Burma and Ceylon*, Vol. 1. William Dawson and Sons Lie., London, pp. 438–603.

Hickling, C.F., 1930. The natural history of the lake–Part III: Seasonal changes in the condition of the Lake. *Fish. Invest. Lond. Ser.*, 2(12): 1–78.

Joseph K. Manissery, Rao, Dwaraka S. and Vijayalakshmi, S., 1979. Studies on the maturation, spawning and fecundity of *Puntius ticto* from Bellandur tank, Bangalore. *Mysore J. Agric. Sci.*, 13: 207–213.

Kesteven, G.L., 1942. Studies on the biology of Australian mullets. Part I. Account of the fishery and preliminary statement of biology of *Mugil doubla Gunther*. *Coun. Sci. Industr. Res. Bull.*, 157, Melbourne.

McFadden, J.T. and Copper, E.L., 1962. An ecological comparison of sex populations of brown trout (*Salmo trutta*). *Trans. Am. Fish. Soc.*, 91: 53–62.

Neelakantan, B., Kusuma, M.S. and Pai, M.V., 1980–81. Sex ratio of the false travelly *Lactarius lactarius* (Bloch and Schneider). *J. Indian Fish. Assn.*, 10 and 11: 17–20.

Qasim, S.Z., 1966. Sex ratio in fish populations as a function sexual difference in growth rate. *Curr. Sci.*, 35: 140–142.

Qasim, S.Z. and Qayyum, A., 1961. Spawning frequencies, and breeding seasons of some freshwater fishes with special reference to those occurring in the plains of Northern India. *Indian J. Fish.*, 8(1): 25–43.

Raitt, D.F.S., 1968. The population dynamics of the Norway pent in the North sea. *Mar. Res.*, 5: 1–24.

Vinci, G.K., 1984. Some aspects of the biology of *Silonia childreni* (Sykes) from Nagarjunasagar reservoir, Andhra Pradesh, India. *J. Inland Fish Soc. India*, 16(1 and 2): 25–31.

Water and Wastewater Management in 2 Vols (2011) *Pages 35–38*
Editors: **D.R. Khanna, R. Bhutiani & Gagan Matta**
Published by: **DAYA PUBLISHING HOUSE, NEW DELHI**

Chapter 6

Accumulation of Heavy Metals in Crop Plants through Irrigation of Contaminated Groundwater in Panipat Region

D.S. Malik, Rasmi Yadav and Pawan K. Bharti
Department of Zoology and Environmental Sciences, Gurukul Kangri University, Haridwar – 249 404, Uttaranchal

ABSTRACT

In the modern industrialization period, most of water resources had affected enormously by seepage, leaching and intermixing of industrial effluents. The textile effluents had consisting high concentrations of trace heavy metals and through its accumulations in different trophic levels of ecosystem ultimately cause the health hazards among livestock and human beings. The different crop plant samples collected from agriculture fields adjoining of textile industrial effluent flowing channels contained *in situ* these heavy metals *e.g.* Lead (pb), Copper (Cu), Manganese (Mn), Nickel (Ni), Iron (Fe), Cadmium (Cd) and Zinc (Zn). The bioremediation processes are urgently required to combat groundwater pollution.

Keywords: Heavy metals, Textile effluents, Bio-accumulation, Industrial pollution, Crop plants.

Introduction

In India, the industrial effluents have contributed as major source of pollution. Different industries of the country contribute about 16 per cent of the total wastewater generated. The treatment facilities have been installed in less than one third of the polluting industries (Gopal, 1994). Most of effluent treatment plants (ETP) have not performed physical and chemical treatment processes satisfactory due to economic reason on commercial scales.

Textile industrial effluents have accounted to about 15-20 per cent of total wastewater in the country. The textile effluents containing different colours, inorganic and organic chemicals and heavy metals are highly polluted in nature and vary in its composition. The practice of disposing textile wastewater without any treatment affects the aquatic and soil system. In the adjoining agricultural area of textile industries sector, there is immense potential of groundwater being contaminated from effluents.

The present study reveals the accumulation of different heavy metals in crop plants and exhibits the cycles of transfer of heavy metals from industrial effluent to consumer levels and lastly affected the health of livestock and human being.

Materials and Method

Panipat is situated on the national highway of Delhi-Chandigarh in Haryana, which is well known for textile and dye industries. The effluent samples were collected from the outlet of dyeing units and physico-chemical parameters were analyzed (APHA, 1995). Shoot and root samples of different crop plants grown in agriculture fields were also collected for analysis. These samples were washed with tap water, dilute hydrochloric acid, and distilled water and then dried in oven at 67°C and grind in a stainless steel grinder to pass through 16-mesh sieve. Plant samples were digested in HNO_{3-} $HClO_4$ diacid mixture and analyzed by atomic absorption spectrophotometer AAS-4129.

Table 6.1: Physico-chemical Parameters and Heavy Metals of Textile Effluents

Parameters	*Effluent-I*	*Effluent-II*	*Effluent-III*	*Effluent-IV*
EC (ms/m)	5.87	1.50	7.70	4.45
TDS (gm/l)	2.93	7.38	4.09	7.28
pH	7.30	8.00	7.50	8.25
DO (mg/l)	7.70	5.20	1.80	1.30
BOD (mg/l)	31	51	27	51
COD (mg/l)	734	158	137	151
Na (mg/l)	9.90	1.84	3.04	5.27
K (mg/l)	0.36	0.43	0.22	0.53
Ca+Mg (mg/l)	11.20	7.60	5.60	9.00
Cl (mg/l)	2.20	3.20	3.00	6.10
CO_3 (mg/l)	1.20	0.00	0.00	0.00
HCO_3 (mg/l)	11.60	6.50	3.95	4.60
PO_4 (ppm)	4.0	0.0	0.0	1.0
NH_3–N (ppm)	60.90	2.45	2.15	3.50
NO_3–N (ppm)	3.50	8.40	8.20	2.45
Mn (ppm)	0.06	0.01	0.16	0.069
Ni (ppm)	0.02	0.00	0.05	0.00
Fe (ppm)	0.30	0.03	0.12	0.12
Cu (ppm)	0.13	0.02	0.10	0.40
Cd (ppm)	0.00	0.01	0.02	0.01
Pb (ppm)	0.23	0.18	0.59	0.33
Zn (ppm)	0.20	0.03	0.03	0.11

Results and Discussion

The textile effluent had alkaline pH varied 7.3-8.25. Dissolved oxygen of the dye house wastewater was higher in comparison to the sewage water. The impact of dissolved oxygen was also quite obvious on BOD (Biochemical oxygen demand). The COD (Chemical oxygen demand) of the dye wastewater was very high being (137–734 mg/I). Senthinathan and Azeez (1999) and Groff (1993) reported that textile wastewaters may cause problem used as untreated effluent consists higher phosphorous, COD, pH, temperature and absorbable organic halogen. Since NaCI is mostly used for fixing dyes the contents of both CI and Na increased with electrical conductivity.

Concentration of heavy metals in textile effluents as Zn, Ni, Mn and Cd were observed quite low while the Fe, Cu and Pb were analyzed to be slightly higher (Table 6.1). Rahman (1996), Usha (1989) and Khan *et al.* (1995) also found Pb and Cd in excessive amount than those of the permissible limits.

Table 6.2: Heavy Metals in Crop Plants Samples

Crop Plants	*Pb (μg)*	*Cd (μg)*	*Fe (μg)*	*Zn (μg)*	*Mn (μg)*	*Ni (μg)*	*Cu (μg)*
Shoots							
Wheat shoot	10	0.20	144	271	16	15.6	4.0
Barseem shoot	18	0.6	172	264	10	23.2	7.2
Radish shoot	0.0	0.0	126	249	21.4	18.8	0.4
Cauliflower shoot	24	0.0	215	284	23.0	0.0	10.3
Coriender shoot	0.0	0.0	107	266	36.0	15.6	2.8
Roots							
Barseem root	27	0.0	165	220	57.6	30.2	8.3
Radish root	16	0.0	146	136	42.2	0.0	10.4
Cauliflower root	28	0.0	196	156	45.2	0.0	16.2
Coriander root	22	0.0	179	152	36.2	0.0	6.2
Palak root	21	0.0	172	122	35	0.0	12.0

The chemical analysis of the crop plants from the adjoining agricultural fields, which are mostly irrigated by groundwater through bore tube well in Panipat region were found to accumulate the heavy metals like Pb, Cd, Fe, Zn, Mn, Ni, and Cu Contents were with in the prescribed limits excepts lead (Pb). The concentration of Pb in Barseem and Cauliflower shoots was recorded 18 ppm and 24 ppm respectiviely. Moreover Pb contents in roots of crop plants were still higher varied as 16–28 ppm (Table 6.2). It has been earlier reported, by Khan *et al.* (1997) and Navarro (1993) that use of textile industrial wastewater in irrigation, result in increased heavy metals contents in the crop plants. The contents of lead are beyond safe limits and hence its sources need to identified and remediated otherwise it would create health hazards. The effluents also cause various health problems such as skin allergies, lung infection, carcino-mutagenic disorders among the human being through contaminated groundwater supply by municipal tube well water scheme (Kumar 1998 and Sharma *et al.*, 1999).

References

APHA, 1995. *Standard Methods for Examination of Water and Wastewater,* 19th edn. American Public Health Association Inc., New York, pp. 1970.

Gopal, B., 1994. Conservation of inland water in India: An overview. Verh. Internet. Verein. *Hydrobiologia*, 384: 267.

Groff, K.A., 1993. Textile waste. *J. Water Environment Research*, 65: 421.

Khan, T.I. and Jain, V., 1995. Effects of textile industries wastewater on growth and some biochemical parameters of *Triticum aestivum* var. Raj 3077. *J. Environ. Poll.*, 2(2): 47–52.

Khan, T.I. Kaur, N. and Agrawal, M., 1997. Heavy metal analysis of crop plants from agriculture fields of Sanagar town. *J. Environ. Poll.*, 4(1): 35–36.

Navarro, A. and Font, X., 1993. Discriminating different sources of groundwater con tamination caused by industrial waste in the Besos River basin, Barcelona. *J. Applied Geochemistry APPGEY*, 8: 277.

Rahman, A., 1996. Groundwater as a sources of contamination for water supply in rapidly growing mega cities of Asia: Case of Karachi, Pakistan. In: *Proceeding of the 18th Biennial Conference of the International Association of Water Quality*, 34(7–8): 285.

Senthilnathan, S. and Azeez, P.A., 1999. Influence of dyeing and bleaching industries on groundwater of Tripur, Tamil Nadu, India. *Bull. Environ. Contam. Toxicol.*, 62: 330–334.

Sharma, K.P., Sharma, K., Bhardwaj, S.M. and Chaturvedi, R.K., 1999. Environmental impact assessment of textile printing industries in Sanagar, Jaipur: A case study. *J. Ind. Bot. Soc.*, 78: 71–74.

Usha, M., 1989. *Impact Analysis of Industries in Sanagar*. P.G. diploma field study report Indira Gandhi Centre for HEEPS, University of Rajasthan, Jaipur, pp. 65.

Water and Wastewater Management in 2 Vols (2011) Pages 39–45
Editors: D.R. Khanna, R. Bhutiani & Gagan Matta
Published by: DAYA PUBLISHING HOUSE, NEW DELHI

Chapter 7

Influence of Star Paper Mill Effluent on Biomass Production and Yield at Harvesting Time of Different Agricultural Crops

Sangeeta Arora[1], A.K. Chopra[2], G. Prasad[3] and N. Joshi[1]
[1]*Department of Environmental Sciences, Kanya Gurukul Mahavidyalaya, Haridwar*
[2]*Department of Zoology and Environmental Sciences,*
[3]*Department of Botany and Microbiology,*
Gurukul Kangri University, Haridwar

ABSTRACT

Influence of Star paper mill effluent of different concentration on biomass and yield production of four agricultural crop *i.e. Solanum melongena* (Brinjal), *Cicer arietinum* (Black gram), *Glycine max* (Soyabean) and *Lycopersicon esculentum* (Tomato) was investigated in the field condition. Variable behaviour on pattern of seed germination, root length, shoot length, root and shoot ratio was recorded. Similarly all four crops showed variability with concentration of dose and yield production at harvesting stage. Biomass production was better in control (only tap water) than plot irrigated up to 10 per cent with effluent but was found suitable at 25 per cent and after that it was deleterious for brinjal crop. However effluent was suitable at 5 per cent dose of black gram and after that it showed adverse effect. Adverse effect of effluent on biomass production was recorded at all different concentration used in soyabean crop, but effluent above 50 per cent concentration showed a positive response for *L. esculentum* crop. Similarly crop yield in terms of number of fruits/plant and weight of fruits per plant showed variability for their treated dose. Number of fruits and their weight in Brinjal was found better in enhanced trend till 25 per cent and after that deleterious effect was noted while black gram respond positively only at 5 per cent concentration. Similarly maximum yield was found in soyabean at 5 per cent concentration and above that deleterious effect was recorded. Progressive and positive trend at all concentration in weight of tomato fruit per plant was found.

Keywords: Paper mill effluent, Biomass, Agriculture, Vigour index.

Introduction

Due to explosion of population, change in attitude of life and development of high degree of scientific researches have yielded rapid growth of industrialization to fulfill the human need in the last two decades of twentieth century. This phenomenon has generated laterally most serious problem concerned with human health and other welfare. Albeit water consumption in the industries has third rank, but many industries are generating effluent not only of high amount but they contain high level of organic components which act as suitable media for growth and multiplication of both virulent and avirulent micro organism. Both are dangerous, because both are concerned in reducing oxygen level resulting adverse effect on aquatic biota. Pathogenic organisms present in the effluent may cause epidemic if surface water is contaminated with these effluent. Besides this several industries are concerned with such type of product having high toxicity level in their effluent due to existence of chemical or heavy metal salts.

Even after proper treatment, effluent should not be discharged in any aquatic system or near by recreation point in order to control outbreak of any kind of epidemic. But as in our condition where maximum number of cities and industries do not possess adequate treatment plants and it is customy to discharge untreated raw sewage either in aquatic reservoir or on land surface. Since this raw sewage contains high amount of organic component and even treated effluent also possess certain amount of organic components. Both may serve the need of nutritional requirement to the plant as fertilizer. Therefore, in order to find out alternative suitable means of safer sewage disposal as well as minimizing the investment in agricultural cost, the investigation was carried out.

In the light of above and in welfare of the society efforts have been made to utilize the effluent for irrigation in several agricultural crops all over the world, based on the reports made by several workers (Dutta and Boissya 1999; Kumar *et al.*, 1997). But certain industrial effluent may contain toxic components, which may inhibit the germination of seed or retard the growth of crop plant (Rajannan and Oblisami, 1979; Bishop 1983; Sahai *et al.*1986 and Swaminathan *et al.*, 1989). In the present study an effort has been made to find out influence of star paper mill effluent on biomass and yield of certain agricultural crop *i.e. Solanum melongena* (Brinjal), *Cicer arietinum* (Black gram), *Glycine max*(Soyabean) and *Lycopersicon esculentum* (Tomato) at harvesting time.

Materials and Methods

Composite samples were collected from star paper mill, Saharanpur and were brought to laboratory for analysis. The effluent sample was analyzed for various physico-chemical and microbiological parameters as suggested by APHA (1985) and was further used for the treatment of seeds of test crops. Various concentrations of effluent (5 per cent, 10 per cent, 25 per cent, 50 per cent, 75 per cent and 100 per cent) were prepared by using tap water separately. The seeds of uniform size of each crop were selected and surfaces sterilized with 0.1 per cent mercuric chloride for one minute and were washed thoroughly with distilled water to avoid any traces of $HgCl_2$. These surface sterilized seeds were used for the biomass production and yield with different treated/irrigated concentrations of the effluent.

Biomass Production and Yield Under Field Conditions

For determining the biomass of the crops (*Solanum melongena, Cicer arietinum, Glycine max* and *Lycopersicon esculentum*) seven plots (1-7) each of the size 18 X 18 inch^2 were taken. Ten seeds of each crop treated with different concentrations were sown in each plot. The plots (1-7) were irrigated with 0 per cent, 10 per cent, 25 per cent, 50 per cent, 75 per cent and 100 per cent effluent respectively on alternate days. The percentage of seed germination, root length, shoot length, vigour index and biomass were observed after 21 days from the date of sowing. The experiment was conducted four times.

Seven plot of size 18 X 18 inch^2 were used for the study of yield production. In each plot ten seeds were sown and all these plots were irrigated with different concentrations *i.e.* 0 per cent, 5 per cent, 10 per cent, 25 per cent, 50 per cent, 75 per cent and 100 per cent of effluent on alternate days. Biomass production at harvesting stage of different agricultural crops was measured by considering various parameters *i.e.* root length, shoot length, root/shoot ratio *etc.*.Production of yield was recorded in terms of number of fruits on seed production on per plant and weight of the fruit or seeds on per plant. Due to variation in growth rate and germination the yield was recorded at different age in different crops *i.e.* 160 days for *C. arietinum* and for *G.. max,* 180 days for *L. esculentum* and 240 days for *S. melongena.*

Statistical Analysis

The data was statistically analyzed (Levin and Rubin, 1995) using standard deviation and correlation factor.

Results

Results of influence of star paper mill effluent on biomass production and yield have been presented in the Tables 7.1 and 7.2 respectively. Under field conditions, at 21 days data showed variable results on percentage of seed germination and on various parameters related with biomass production (Table 7.1). Maximum percentage of seed germination *i.e.* 35 at 25 per cent concentration in brinjal while 90 at 10 per cent concentration in blackgram and 60 at 100 per cent (undiluted effluent) in tomato than control (Table 7.1). Lower percentage of seed germination was recorded at all the concentrations than compared to control in soyabean.

Maximum root length (0.91 cm), shoot length (1.52 cm), vigour index (54.11) and biomass (47.79 mg/plant) in brinjal while root length (8.15 cm), shoot length (14.68 cm), vigour index (1230.99) and biomass (826 mg/plant) were recorded at 25 per cent concentration. In case of soyabean maximum value of root length (6.82 cm), shoot length (13.13 cm), vigour index (860.27) at 50 per cent concentration but seed germination (85) and biomass (1334 mg/plant) at 5 per cent concentration. In tomato maximum root length (1.35 cm), shoot length (2.42 cm), vigour index (175.95) and biomass (82.91) at 100 per cent concentration of effluent (Table 7.1).

Findings of biomass production and yield of above different crops are presented in (Table 7.2). Maximum value for root length, shoot length, vigour index and biomass were found in the same pattern *i.e.* maximum value at 25 per cent in brinjal at 240 days. But maximum value for root length, shoot length, vigour index and biomass were found in the plot treated with 10 per cent concentration in black gram and soyabean at age of 160 days each (Table 7.2). Value for all above parameters were found in same pattern of biomass production at 21 days in tomato crop at harvesting time of 180 days old plant in 100 per cent concentration.

Table 7.1: Seed Germination and Biomass Production of Four Agricultural Crops at 21 Days Under Field Conditions

Crops	*Parameters*	*Treatments*						
		0	*5*	*10*	*25*	*50*	*75*	*100*
Brinjal	SG (%)	22.5	27.5	30	35	20	20	17.5
	RL (cm)	0.45	0.44	0.65	0.91	0.58	0.36	0.31
	SL (cm)	0.54	0.53	0.71	1.52	1.10	0.76	0.48
	R/S	0.83	0.83	0.91	0.59	0.52	0.47	0.64
	VI	12.6	15.01	21.95	54.11	22.58	15.56	8.71
	BM (mg/plant)	42.93	26.52	28.5	47.79	20.75	19.62	2.85
Black Gram	SG (per cent)	80	86.6	90	83.3	83.3	73.3	50
	RL (cm)	7.85	8.53	8.63	8.15	7.74	4.71	4.45
	SL (cm)	15.8	12.49	15.5	14.68	14.53	12.07	9.89
	R/S	0.49	0.68	0.56	0.55	0.53	0.39	0.44
	VI	1271.85	1090.16	1403.68	1230.99	1218.08	889.44	498.95
	BM (mg/plant)	849.25	1128.75	1008.0	826	723.75	498.95	65.5
Soyabean	SG (per cent)	90	85	80	75	65	55	50
	RL (cm)	4.35	4.68	5.30	6.82	6.82	5.90	2.84
	SL (cm)	12.73	6.07	9.89	10.48	13.13	11.12	8.55
	R/S	0.34	0.77	0.53	0.64	1.92	0.53	0.33
	VI	1150.05	520.63	796.5	792.81	860.27	617.50	430.34
	BM (mg/plant)	1436.5	1334.0	1231.5	1169.0	1039.75	1002.50	462.75
Tomato	SG (per cent)	20	20	30	47.5	50	52.5	60
	RL (cm)	0.46	0.44	0.64	0.86	0.82	1.25	135
	SL (cm)	1.54	2.91	1.41	2.66	2.42	2.09	2.42
	R/S	0.29	0.45	0.21	0.32	0.33	0.51	0.46
	VI	31.26	28.64	63.34	127.21	121.82	128.30	175.95
	BM (mg/plant)	48	26.67	33.10	38.15	79.22	80.32	82.91

SG: Seed Germination; SL: Shoot Length; VI: Vigour Index; RI: Root Length; R/S: Root/Shoot Length; BM: Biomass.

Discussion

Data recorded on biomass production at the age of 21 days and biomass and yield at harvesting stage under field condition treated with different concentration of effluent are presented in (Tables 7.1 and 7.2). Results revealed that no definite pattern for seed germination and biomass production at 21 days of treated effluent concentration could be established among the different used crops *i.e.* brinjal, blackgram, soyabean and tomato. Conclusion based on findings could be established that star paper mill contains varying amount of nutrient in the form of organic components which serve the purpose of measured parameters up to certain extent in general but specially in brinjal and blackgram (Table 7.1). Their nutrient support is found limited upto certain limit in term of their concentration. These findings are in accordance to (Rajanan and Oblisami, 1979; Mishra and Sahoo,1989). However Snehlata

(1991) found deleterious effect of induced cadmium on enzyme activity which ultimately effect seed germination. Nutritional support of effluent is found effective upto definite concentration, beyond that with increased concentration of effluent showed deleterious effect in brinjal and black gram.These may be perhaps either due to higher concentration of nutrients along with other inorganic chemicals which showed their significant injurious effect either influencing the physiological mechanism of the crop plants. Declined trend of biomass production at higher concentration dose may also be associated due to minimization of porosity of the soil and decomposition of suspended and dissolved solids.

Table 7.2: Biomass and Yield Production of Four Agricultural Crops at Harvesting Time Under Field Conditions

Crops	*Parameters*	*Treatments*						
		0	*5*	*10*	*25*	*50*	*75*	*100*
Brinjal	RL (cm)	2.30	2.50	2.55	5.35	2.35	2.05	1.85
(240 days)	SI (cm)	4.85	2.55	4.40	10.25	4.05	3.30	4.95
	R/S	0.47	0.98	0.57	0.52	0.58	0.62	0.37
	No. of B/Plant	1.8	1.8	4.6	4.8	1.6	1.4	1.2
	Yield of B/Plant	200	160	170	200	102	0.233	0.110
	Dry wt. (mg/Plant)	123.66×10^3	172.5×10^3	176.8×10^3	251.33×10^3	144.2×10^3	79.2×10^3	54.5×10^3
Black Gram	RL (cm)	4.55	8.40	7.35	7.30	6.65	5.60	4.95
(160 days)	SI (cm)	15.1	21.50	24.25	15.40	17.00	13.30	15.65
	R/S	0.30	0.39	0.30	0.47	0.39	0.42	0.31
	No. of BG/Plant	33.0	50.6	42.6	41.14	40.3	378	21.2
	Yield of BG/Plant	7.0	14.0	13.0	11.0	10.7	10.3	3.0
	Dry wt. (mg/Plant)	$49^8\times10^3$	$6^7\times10^3$	$54.8^5\times\times10^3$	$45.7^5\times10^3$	$45^4\times10^3$	$3^7\times10^3$	$3^2\times10^3$
Soyabean	RL (cm)	7.35	11.75	10.25	10.20	9.50	9.50	8.85
(160 days)	SI (cm)	20.25	25.55	18.15	17.70	21.85	12.10	18.60
	R/S	0.36	0.45	0.56	0.57	0.43	0.78	0.47
	No. of Sb/Plant	5.88	27.87	20.85	16.62	14.12	11.5	8.33
	Yield of Sb/Plant	2.50	90.0	75.0	75.0	70.0	40.0	15.5
	Dry wt. (mg/Plant)	3×10^3	10.6×10^3	9.3×10^3	7.5×10^3	6.4×10^3	4.5×10^3	2.2×10^3
Tomato	RL (cm)	8.65	5.35	5.50	6.15	6.40	7.80	1170
(180 days)	SI (cm)	29.0	22.10	25.95	21.50	19.05	21.20	26.90
	R/S	0.29	0.24	0.21	0.28	0.33	0.36	0.43
	No. of T/Plant	10.7	8.4	9.83	9.85	10.57	12.0	19.6
	Yield of T/Plant	0.195	0.245	0.494	87.5	93.0	117.85	142.85
	Dry wt. (mg/Plant)	153.33×10^3	89.28×10^3	130.71×10^3	136×10^3	209.16×10^3	222.4×10^3	335×10^3

B: Brinjal; Sb: Soyabean; BG: Black Gram; T: Tomato.

However, stimulatory effect on seed germination and growth in terms of biomass production could not be recorded in soyabean crop as evident by Table 7.1. Behaviour of tomato plant in response to treated dose was found different from all other test crop plant. All concentrations used in the study and even in pure effluent, all parameters have been found an enhanced value.

Variabilities among the different test crops responding to the dose of concentration may also be due to genetic nature related with their genetic style. Biomass and yield under field condition of four above mentioned test crop plants showed similar trend in number of brinjal/plant and yield in terms of weight was found gradually increased from 5 to 25 per cent concentration and started declining beyond that.

Varied trend of dose was recorded between biomass at 21 days old black gram crop and yield at harvesting stage. Maximum value was found at 10 per cent at 21 days while maximum yield was found at lowest concentration *i.e.* 5 per cent in case of black gram. It may be due to regular irrigation, which could increase the different components in term of organic, inorganic compounds and salts after 21 days upto a level which was available at 10 per cent concentration.

Similar trend was found in soyabean crop, maximum yield has been found at 5 per cent concentrated effluent. It is totally different from the results of 21 days old crops where effluent showed adverse effect. Positive correlation with concentration of effluent and numbers of tomato per plant and yield in term of weight was found in *L. esculentum.*

Thus it can be concluded that albeit at very low concentration (5 per cent) effluent of star paper mill have not shown any adverse effect on germination, biomass and yield but probability of absorption of chemicals during seed germination was there. Moreover transportation and storing in the system of plants and their products could be avoided. Therefore irrigation of the paper effluent in *C. arietinum* and *G. max* at higher concentration should not be recommended. However paper mill indicate that these effluents contains some nutrients, which fulfill the requirement of manures to the germinating crops upto, certain extent. However before recommending these effluents for irrigation more scientific work is needed regarding deposition of chemical components in the products and also their effects on land surface and biota.

References

APHA, 1985. *Standard Methods for the Examination of Water and Wastewater*, 15th edn, New York, Washington D.C., pp. 1–1185

Bishop, A.W., 1983. Diposal of some industrial wastewater. *IAWPC Tech. Annual*, 10: 149–156.

Dutta, S.K. and Boissya, C.L., 1999. Effect of paper mill effluent on germination of rice and growth behaviour of its seedlings. *J. Ind. Poll. Cont.*, 13: 41.

Kumar, Prasanna P.G., Pandit, B.R. and Kumar Mahesh, R., 1997. Effect of dairy effluent on seed germination, seedling growth and pigment content of green gram (*Phaseolus aureus* L.) and black gram (*Phaseolus mungo* L.). *Adv. Plant Sci.*, 10: 129–136.

Levin and Rubin, 1995. *Statistics for Management*. Prentice Hall of India Private Limited, New Delhi.

Mishra, K. and Sahoo, S., 1989. Agropotentiality of paper mill wastewater. In: *Soil Pollution and Soil Organisms*, (Ed.) P.C. Mishra. Ashish Publishing House, New Delhi.

Rajanan, G. and Oblisami, G., 1979. Effects of paper factory effluent on soil and crop plant. *Ind. J. Environ. Hlth.*, 21: 120–130.

Sahai, R., Agarwal, N. and Khosla, N., 1986. Effects of fertilizer factory effluents on seed germination, seedling growth and chlorophyll content and biomass of *Phseolus radiatus* L. *Indian J. Ecol.*, 14: 21–25.

Snehlata, 1991. Phasic pretreatment effect of Cadmium on seedling growth and activity of hydrolyzing enzyme in *Phaseolus aureus* cultivars. *J. Envir. Biol.*, 12(B): 299–308.

Swaminathan, K., Arjunan, J. and Guruswamy, R., 1989. Effect of glucose factory effluents on seed germination and seedling development of ground nut (*Arachis hypogaea*). *Environ. Biol.*, p. 81–87.

Water and Wastewater Management in 2 Vols (2011) *Pages* **46–51**
Editors: **D.R. Khanna, R. Bhutiani & Gagan Matta**
Published by: **DAYA PUBLISHING HOUSE, NEW DELHI**

Chapter 8

Physico-chemical Characteristics of Doon Distillery, Doiwala Effluent and its *in vitro* Effect on Seed Germination of Four Agricultural Crops

Sangeeta Arora[1], A.K. Chopra[2], G. Prasad[3] and N. Joshi[1]
[1]Department of Environmental Sciences, Kanya Gurukul Mahavidyalaya, Haridwar
[2]Department of Zoology and Environmental Sciences,
[3]Department of Botany and Microbiology,
Gurukul Kangri University, Haridwar

ABSTRACT

The present chapter deals with the physico-chemical characteristics of Doon Distillery, Doiwala effluent and its *in vitro* effect on seed germination of four agricultural crops *i.e. Solanum melongena* (Brinjal), *Cicer arietium* (Black gram), *Glycine max* (Soyabean) and *Lycopersicon esculentum* (Tomato). The physico-chemical characteristics showed high degree of pollution load in the effuent as it is evident by high degree of BOD (35375mg/l), COD (81000 mg/l), MPN/100 ml ($30.56x10^5$) and SPC ($72.33x10^6$). The effect of effluent treatment on seed germination *in vitro* showed a significant negative correlation between different concentration of the effluent (5 to 100 per cent) and in percentage of seed gemination of *S. melongena, C. arietinum, G. max* and *L. esculentum*. In all these four crops maximum percentage of seed germination was observed at 5 per cent concentration of effluent. Germination of seeds treated with different doses was only recorded up to 25 per cent concentration. Effluent showed deleterious effect at 50 per cent and above that no seed germination was found.

Keywords: *Distillery, Effluent, Germination, Crops, Organic.*

Introduction

In the colonial period, it was not a governmental policy to industrialize the country. We have to depend almost entirely on other countries, especially on U.K. It is only after independence of the country in 1947 and after that large funds were made available for the industrialization of the country, but the planning commission that indeed started industrialization in India.

Albeit the role of industries in environmental pollution is well established but some of them are contributing high degree of pollution load especially in different aquatic system. Some of them may be considered as major polluting industries such as tanneries, chemical and fertilizer manufacturing units, refineries, paper and pulp mills, sugar and distilleries *etc*. Indiscriminate discharge of effluent from these industries in aquatic system or on land has altered the quality of both systems. Distilleries are primarily engaged in the production of alcoholic beverages *viz* ethyl alcohol by fermentation of molasses and subsequent distillation of the fermented wash. There are over 200 large distilleries in India and the spent wash of these are the largest effluent stream from distilleries. The spent wash is hot, highly acidic, coloured and contains high percentage of dissolved organic and inorganic matter. The effect of pollution caused, by the distillery spent wash on water and land are as lowering of pH value of the stream, increase in cropping period and increase in electrical conductivity of soil (Kaul *et al.*, 1995). The adverse effects of distillery waste are due to high BOD, COD, total solids, low values of dissolved oxygen and capability to impart colour. Stagnation of the effluent on land results in obnoxious condition in the region. If the soil is porous the effluent may also affect groundwater quality. Distilleries wastes in the form of spent wash are among the worst water pollutants produced by industries both in magnitude and strength.

Since the land surface water is major source of water supply for civic purposes in most of the cities, therefore this surface water (aquatic reservoirs, rivers *etc.*) should be prevented/minimized from pollutant in order to protect human health hazard. Infact a large no. of human diseases are water born and if civic water supply is contaminated it may cause severe outbreak of epidemic.Therefore, liquid waste treatment is very essential but the treatment plant is very costly affair in all the industries, which they cannot afford. Therefore, alternative method which can be useful and may minimize the harmful effect should be investigated and evaluated. Since our country is in underdeveloping stage, electricity supply is highly interrupted and labour problem is regular feature. Therefore, it is imperative to utilize these effluents in irrigation under liquid waste disposal programme. As a specific concentration of effluent will not be a general role for all agricultural crops and different parameter will not remain same for all the time in specific effluent. Therefore, it is highly needed that a regular monitoring of the effluent and concentration should be analyzed for each crop before taking the irrigation process in specific area. Disposal of liquid waste in irrigation helps as a solution of water for irrigation and as manure required for the crop. Keeping in view of the above facts efforts have been made to find out pollution load of Doon distillery effluent and its effect in different concentrations in seed germination of four agriculture crops *viz., Lycopersicon esculentum, Solanum melongena, Cicer arietinum* and *Glycine max*.

Materials and Method

Composite samples of the effluent were collected from inlet of Doon distillery in plastic container and brought to laboratory for the analysis of certain physico-chemical and microbiological parameters as per APHA (1985). The effluent was further used for the treatment of seeds of test crops. Various concentrations of the effluent *i.e.*5 per cent, 10 per cent, 25 per cent, 50 per cent, 75 per cent and 100 per cent were prepared using tap water. Tap water was used as such for control. Seeds of four crops *Lycopersicon esculentum, Solanum melongena, Cicer arietinum* and *Glycine max* were used as test plants.

The seeds of these test crops were surface sterilized with 0.1 per cent $HgCl_2$ solution for 20 second and were repeatedly washed with distilled water. These seeds were treated in six different concentrations. Seeds as such (without treatment) were used for control. The seeds were irrigated with equal quantity of the effluent. Four replicates were maintained for each concentration including control. Germination of seeds of two crops *i.e. Cicer arietinum* (black gram) and *Glycine max* (soyabean) were recorded on 7th day while that of *Lycopersicon esculentum* (tomato) and *Solanum melongena* (brinjal) on 15th day. The data was statistically analysed.

Results and Discussion

The physico-chemical and microbiological characteristics of Doon distillery, Doiwala effluent and its effect on the percentage of seed germination of four agricultural crops are presented in Tables 8.1 and 8.2.

Table 8.1: Physico-chemical and Microbiological Characteristics of Effluent of Doon Distillery, Doiwala (Values are mean±S.D. of six replicates)

Parameters	*Characteristics*
Colour	Deep Brownish
Odour	Smell of molasses
Turbidity (NTU)	11.53±2.07
Conductivity (at 20 m mhos cm)	9±2
Total Solids (mg/l)	65040±715.82
Dissolved solids (mg/l)	64221.66±697.07
Suspended solids (mg/l)	818.30±31.75
Dissolved oxygen (mg/l)	Nil
Biological oxygen demand (mg/l)	35375±305.98
Chemical oxygen demand (mg/l)	81000±0.00
Chloride (mg/l)	2153.66±106.89
pH	3.55±0.03
Hardness (mg/l)	10750±273 .86
Alkalinity (mg/l)	1666.66±258.19
Phosphate (mg/l)	1200+200
Sulphate (mg/l)	3886.83±6.79
Calcium (mg/l)	3660.64±41.39
Magnesium (mg/l)	1729.86±67.58
MPN/100 ml	$30.56\times10^5\pm0.47\times10^5$
SPC	$72.33\times10^6\pm0.60\times10^6$

The study revealed that the colour of Doon distillery effluent was deep brownish and yielded odour of molasses. Electrical conductivity was recorded 9 at 20 mhos. Conductivity of water and wastewater is the property caused by the presence of various ionic species. Conductivity of water and wastewater is due to the presence of total dissolved solids. Higher value of total solids *i.e.* 65,040 mg/l was recorded in the effluent. Total solids have profound effect on several parameter related with pollution load. Dissolved solids in the mill were recorded to be 64,221.6 mg/l. The mill contains considerable amount of dissolved solids because of the escape of viscose fibres and wood being used as raw materials in the factory. High values of these parameters signify the presence of both organic and inorganic compound in the effluent which increase the total pollution load of the effluent.

Alkalinity of the effluent was recorded to be 1666.66 mg/l. Alkaline value are important in calculating the dose of alum and biocides in water. Values of phosphate and sulphate were found to be 1200 mg/l and 3886.83 mg/l respectively. Phosphate is an important component which is being utilized as nutrient by majority of microorganisms especially algal species utilize the phosphate for their growth.

Table 8.2: *In vitro* Percentage of Seed Germination of Four Agricultural Crops (Values are mean±S.D. of 12 replicates)

Conc. of Effluent (Per cent)	*S. melongena (Brinjal)*	*C. arietinum (Black Gram)*	*G. max (Soyabean)*	*L. esculentum (Tomato)*
Control	81.1	73.3	77.1	58.8
5	85.5	70	81.1	66.6
	(+5.42)	(–4.50)	(+51.94)	(+13.26)
10	74.4	41.1	54.4	60.0
	(–8.26)	(–43.92)	(–29.44)	(+2.04)
25	45.5	34.4	22.2	42.2
	(–43.89)	(–53.06)	(–71.20)	(–16.6)
50	Nil	Nil	Nil	Nil
75	Nil	Nil	Nil	Nil
100	Nil	Nil	Nil	Nil
(r value)	–0.95	–0.88	–0.96	–0.85

Per cent increase or decrease in composition to control given in parenthesis.

r value: Coefficient of correlation.

Chi-Square calculated value in relation to percentage of effluent in the seed germination of all the crops=15.41.

Hardness in the effluent was recorded to be 10,750 mg/l. Hardness decreases after biological treatment due to the use of alum, lime *etc.* as coagulants which remove the salts of Ca and Mg, dissolved phosphates thereby reducing the hardness. Chloride in the effluent is recorded to be 2153.66 mg/l. High chloride content harm the metallic pipes and structure of these factories.

Albeit the dissolved oxygen is essential for aquatic living biota but DO was recorded nil in the distillery effluent (Table 8.1). Absence of dissolved oxygen in the effluent is directly linked with degradation of biodegradable organic components present in the effluent. Total oxygen of effluent was utilized by aerobic bacteria and other aerobic microorganism during the degradation process of biodegradable organic component and resulted in absence of oxygen. Manivaskam (1994) reported that decomposition of dissolved oxygen in receiving water give rise to odoriferous products of anaerobic decomposition. BOD content of the effluent was (35375 mg/l). High values of BOD are found before biological treatment, due to the presence of oxidizable organic matter, which by the addition of the alum and lime as coagulants and by continuous aeration gets considerably reduced. Bhargava (1977) has stated that the set of BOD and DO values taken together would not only show the pollution status, but sequential reaching of such sets would indicate the self purifying ability of the river. COD in the effluent was reported to be 81,000mg/l. The COD test gives a measure of the total oxidizable material present in the effluent.

The results of the effect of different concentrations of distillery waste showed varied percentage of seed germination of the crops like *S. melongena*, *C. arietinum*, *G. max* and *L.esculentum*. In case of *S. melongena* maximum percentage of seed germination (+5.42 per cent) was recorded at 5 per cent concentration of effluent in comparison to control. As the concentration of the effluent increased there was a decline in the percentage of seed germination. The declined trend was noted from 10 to 25 per cent (-8.26 to-43.89). At 50 per cent, 75 per cent and 100 per cent there was no germination. The gradual fall in percentage of germination at higher concentration might be due to increase in concentration of inhibitory compounds in the effluent. Since the effluent has acidic property, the effluent is treated with $Ca(OH)_2$ to enhance pH level before discharging out of the mill. The value of 'r' was–0.95 which represents highly negative correlation between the concentration of the effluent and seed germination.

Germination of seed of *C.arietinum* decreased from 4.5 to 53.6 in the range of 5 to 25 per cent concentration of effluent. Seed germination was completely inhibited above 25 per cent concentration. Negative relationship was established between effluent concentration and seed germination. This might be due to the seeds that take up water during germination to hydrolyze the stored food material in order to activate enzymatic action. As the absorption takes place by osmosis, the salt content outside the seed may act as a limiting factor which may be responsible for seed germination. Further high salt content may be another factor inhibiting seed germination. Our findings are in accordance to finding of Raja and Vijay Kumari (1982) who studied the impact of distillery effluent spent wash in seed germination, morphological characters yield and pigment concentration of *Trigonema foenum graceum* L. and observed that high concentrations of the effluent adversely affect plant growth and grain yield. Dilute distillery effluent treatment of 1 per cent and 0.1 per cent had a growth promoting effect and enhanced growth and yield of crop is more than control. Our results are also in agreement with findings of other workers (Ghosh and Kumar, 1998 and 2000) on the effect of distillery effluent on seed germination of *C. arietinum* and the effect of plywood industry effluent on seed germination and seedling growth of *C. arietinum* and observed that at higher concentration there was complete inhibition of seed germination and concentration of effluent.

In case of *G.max* maximum percentage of seed germination (+51.94) was observed at 5 per cent concentrated effluent than control. Decline trend of seed germination was recorded from 10 to 25 per cent concentration and above that no seed germination was observed. As the distillery effluent contains higher concentration of salts, their higher concentration inhibits germination process. Higher salt concentration and high BOD and COD in the effluent alter the hydrolytic enzyme activity, which are responsible for seed germination. Our results are in agreement with findings of other workers (Rajannan and Oblisami, 1979; Mishra and Sahoo, 1989 and Khan and Srivastava, 1996).

There was highly negative correlation ($p<0.05$, $r =-0.96$) between seed germination and concentration of effluent.

In case of *L. esculentum* maximum percentage of seed germination (+13.26 per cent) was oberved at 5 per cent concentrated effluent than control. Seed germination started decreasing from +2.04 to–16.6. Above 25 per cent no germination was seen. This might be due to mitotic irregularities of less concentration in these effluents. The cytodeformity induction by these effluents at higher concentration was due to the presence of toxic chemicals that damage cellular machinery in the cells of tomato.

Albeit effluent has adverse effect on seed germination of the different test crops but since at lower concentration *i.e.* 5 per cent effluent has shown a promotive tendency of seed germination therefore it can be utilized in irrigation at highly diluted condition. Irrigation at 5 per cent concentration may be useful for *S. melongena*, *C.arietinum*, *G.max* and *L. esculentum*. However, before recommending this effluent for irrigation careful scientific investigation regarding deposition of chemical components in the product and also their effect on land and biota should be taken.

Acknowledgement

Thanks are due to the Principal, Kanya Gurukul Mahavidyalaya Campus, Haridwar for providing necessary facilities.

References

APHA, 1985. *Standard Methods for the Examination of Water and Wastewater*, 15th edn. American Public Health Association (APHA), Washington, D.C, p. 1–1134.

Bhargava, D.S., 1977. Natural regeneration capacity of the natural water. *Water Poll. and Mgt.*, p. 228–237.

Ghosh, A.K. and Kumar, P., 1998. Effect of distillery effluent on seed germination of C. *arietinum* Linn. *Neobotanica*, 6: 21–22.

Ghosh, A.K. and Kumar, P., 2000. Effect of plywood industry effluent on seed germination and seedling growth of C. *arietinum* Linn. *Geobios*, 27: 77–80.

Kaul, S.A., Nandy, T. and Trivedy, R.K., 1995. *Pollution Control in Distilleries*. Environ. Media, Karad, India, p. 1–268.

Khan, S.A. and Srivastava, J., 1996. Distillery waste (effluent) utilization in agriculture. In: *Inter. Conf. on Plant and Environ. Pollu.*, 26–30 Nov. (ICPEP 96), Lucknow, India.

Manivaskan, N., 1994. *Physico-chemical Examination of Water Sewage and Industrial Effluents*. Pragati Prakashan, Meerut, India, p. 1–276.

Mishra, K. and Sahoo, S. 1989 : *Agro potentiality of paper mill waste. In: Soil Pollution and Soil Organisms*, Ed. P.C. Mishra, Ashish Publishing House, New Delhi.

Rajannan, G. and Oblisami, G. 1979. Effect of paper factory effluent on soil and crop plant. *Indian J. Environment Health*, 21(2): 120–130.

Ramgasam, V., Shanthmurthy, K.B. and Govindappa, D.A., 1980. Cytological effect of industrial effuents on somatic cells of *Allium cepa*. In: *Perspectives in Cytoloty and Genetics*, (Eds.) G.K. Manna and V. Sinha. Hindasia Publ., New Delhi, India.

Water and Wastewater Management in 2 Vols (2011) *Pages* **52–56**
Editors: **D.R. Khanna, R. Bhutiani & Gagan Matta**
Published by: **DAYA PUBLISHING HOUSE, NEW DELHI**

Chapter 9

Pollution Abatement and Environmental Preservation

Ashutosh Gautam[1] and D.R. Khanna[2]
[1]*India Glycols Ltd., A-1 Industrial Area, Bazpur Road, Kashipur – 244 713*
[2]*Department of Zoology and Environmental Sciences, Gurukul Kangri University, Haridwar*

ABSTRACT

In the recent years there has been a growing concern over the problem of municipal and industrial waste disposal. In India city sewage, distilleries, tanneries, breweries, sugar mill and food processing units are the major source of organic pollution in water resources. Our growing concern about the availability of clean environment/water makes us more conscious to develop effective and beneficial treatment technologies for the liquid wastes. Anaerobic treatment of liquid waste, having high concentration of biodegradable organic matter, is a competent method to overcome this problem. With the aid of anaerobic treatment technology such wastes can be used to extract some useful products/by products and energy, like-biogas. Biogas is a clean and effective substitute of other conventional energy sources, *i.e.* coal. The chapter highlights major issues of anaerobic treatment alongwith the cost economics.

Keywords: Wastewater, Anaerobic treatment, Biogas.

Introduction

Water is the elixir of life. In our day to day life we are using this vital natural resource for a variety of purposes, like drinking and other domestic uses, irrigation, power generation, industrial cooling and finally for waste disposal. Continuous practice of our society to use water resources as a natural dustbin is creating the greatest problem of this century-The water pollution. Water pollution may be of different types depending upon the constituents causing pollution, for example, organic pollution is mainly caused by the discharge of organic wastes or wastes containing organic material. Such wastes when enter a water body, dissolved oxygen is consumed in their oxidation or breakdown by microorganisms. The more the oxygen required for oxidation, the greater will be the deoxygenation, which can cause adverse impacts on the existing flora and fauna of the water body. Pollutants in the category of organic pollution generally comes from such sources as : domestic/city sewage, food processing units, distilleries, pulp and paper mill, sugar mill, breweries, dairy, slaughter house, *etc.* The water pollution problems in developing countries like India, are still at a stage where effects can be abate if action is taken in time. Always, it is easier and advantageous to control pollution at the early stages rather than investing huge capital to protect and purify the water resources at later stages like-Ganga Action Plan.

Waste

Waste is the disposal/discharge of material, residues and energy into the environment. Of these, some are unconvertible raw materials, some are un-recovered final products and some are by products. Thus discharge of waste in the environment creates a problem of environmental pollution. In other words it can be considered that we are loosing by two ways. Firstly, we are loosing our valuable material and secondly we are polluting our environment, thereby our natural resources.

Liquid waste containing biodegradable organic matter can be used to produce some useful products/energy like-methane rich fuel gas commonly termed as Biogas.

Water Pollution Control

On the discharge of untreated organic waste to aquatic environment:-

1. Dissolved oxygen level of the receiving aquatic body diminishes and results into subsequent damage to the aquatic organisms.
2. High turbidity hinders the penetration of sunlight which adversely affects the photosynthetic activity.
3. Floating impurities and foul smell due to the presence of organic matter makes water unsuitable for recreational purposes.
4. Water no longer remains suitable for domestic/irrigation purpose.

Bioaccumulation/Biomagnification/Bioamplification of toxic material takes place which ultimately attack the human population.

Therefore, the control of water pollution is essential and require instant attention and devotion. The most effective way to control water pollution is the treatment of wastewater in an efficient and suitable effluent treatment plant before its discharge. The main objective is to minimize harmful constituents to an acceptable level with or without by-product recovery. Various suggested methods are:

1. Physico-Chemical Methods.
2. Biological Treatment Methods.
3. Production of Green Manure (Composting).

Physico-Chemical methods are generally used in separating the heavy materials and to neutralize the acidity. These are-Screening, Sedimentation, Floculation and Neutralization *etc.*

The objective of biological treatment is to remove oxygen consuming organic matter by the help of microorganisms, capable in oxidizing the organic matter. Biological treatment is often the most competent method.

These are of two types:

1. *Aerobic*: microorganisms require oxygen to oxidize the organic matter and produce carbon dioxide as a by-product.
2. *Anaerobic*: microorganisms does not require oxygen and produce biogas as a by product.

Incineration is the drying of organic wastes and the dried material may be disposed off on land.

Liquid organic waste can be utilized to produce green manure by treating with solid wastes (composting). The production of green manure with this method is highly beneficial and the manure produced is always better than the chemical manure.

Anaerobic Treatment of Organic Wastes

Anaerobic digestion with methane recovery is indeed the most rational approach in present times of energy crisis. This treatment of organic wastes is also known as methanization or methane fermentation. Methane generation from the rotting or organic matter has been known since the 18th century. Volta identified this as marsh gas. The first use of this valuable gas as a fuel was recorded in 1875, when it was used for lightening the street light of a city in England. Due to various advantages (with few drawbacks) this method has gained a paramount interest and widespread acceptability. Several technologies are available in the market to produce biogas from organic wastes. Process design alternatives for anaerobic treatment of organic wastes may be identified as anaerobic filter process, anaerobic single stage process and acid methane segregation process.

Anaerobic Filter Process

Anaerobic filter process has been developed to treat strong wastes, having high concentration of organic matter. The reactor contains a solid support, usually plastic matrix or stones. In this process microorganisms grow attached to the support material or as free blocks in the interstices.

Single Stage Anaerobic Process

Single stage anaerobic process is a conventional anaerobic treatment process. This is a combined one step process of all activities (acid formation and methane formation) required to recover methane rich fuel gas from organic waste. In this process only a single reactor is required.

Acid Methane Segregation Process

Acid methane segregation process is a microbial segregation process where acid producing and methane producing microorganisms are allowed to grow in different reactors under different environmental conditions. This process enhanced the activity of microorganisms to utilize organic compounds more effectively.

Identification of Organic Wastes for Anaerobic Treatment

The liquid waste have high concentration of biodegradable organic material with adequate nutrients and micro-nutrients these are quite suitable for anaerobic treatment and can be utilized to produce biogas. These wastes should be free from toxic substances. In terms of BOD the wastes from distillery, starch, pharmaceutical, city, pulp and paper, slaughter house, tannery, brewery, sugar mill, *etc.* are highly suitable for the production of biogas with the help of anaerobic treatment. It is essential to know the status of organic waste before adopting the technology.

Process Parameters for Anaerobic Treatment

Every process depends upon several environmental conditions. For the anaerobic treatment temperature, pH, nutrients and the presence of toxic substances are important.

Temperature

The most suitable temperature range is 30-40°C. In the lower range of the temperature the process remain low due to low activity of the microorganisms, whereas in the higher range it initially touches the peak but after that is slows down due to the sudden fall in the activity and the number of microorganisms. Sudden variations in temperature results in temporary shut of the activity, but a long hold can cause degradation in the treatment process. On the basis of temperature range the process and the microorganisms are of three types *i.e.* Cryophilic (between 0-25 °C), conversion rates are slow. Mesophilic (between 30-40 °C) and Thermophilic (between 50-70 °C), the conversion rates are high due to high microbial activity. Normally the mesophilic range of temperature is most suitable and cost effective for the anaerobic treatment.

pH

The most suitable pH range for anaerobic treatment is 7.0-7.5, but it can be operated upto 8.0, the experimentation shows that in the higher pH range biogas have increased concentration of methane of the two bacteria, involved in anaerobic treatment, acidogens are capable of performing their activity between 5.0-6.0 pH range whereas 7.0-8.0 is suitable for methanogens.

Nutrients

Various studies showed that Organic waste must be rich in nutrients and micronutrients for better anaerobic treatment. The main nutrients are N and P. In different technologies the rate of nutrients varies as per the COD/BOD loading rate.

Toxic Substances

Like other biological processes anaerobic treatment may also be affected by the presence of toxic substances. For this process sulphide, ammonia and higher fatty acids are the most consequential.

These compounds always try to inhibit the microorganisms activity which ultimately end in the lowering of BOD/COD reduction and biogas production.

Advantages of Anaerobic Treatment

Anaerobic treatment process has several advantages from other biological processes. Few important are discussed hereunder.

1. This can accept high organic loading since oxygen requirement is nil.
2. This produces a useful by-product, *viz.*, Biogas.
3. This can be operated after a long non-operating gap.
4. Since the reactor is closed no offensive odour exists.
5. Sludge production is low.
6. Possible direct utilization of sludge as a manure.
7. Treated effluent can be used for aquaculture. The main advantage is the production of biogas, which can be used as a clean and environment friendly fuel.

Biogas

Biogas is a composite mixture of methane, carbon dioxide and hydrogen sulphide produced naturally by the decomposition of organic matter. The composition of all the three gases differ from time to time and technology to technology. The main composition is–Methane (CH_4)-50-70 per cent, Hydrogen sulphide (H_2S)–2-5 per cent and Carbon dioxide (CO_2)–rest. The calorific value of the biogas is approximately 5000 K.Cal/NM^3. Biogas is a good substitute of conventional energy sources like wood, coal, kerosene, *etc.* Biogas production from organic wastes varies from waste to waste. Generally, it ranges from 0.1 M^3 to 0.55 M^3/kg. of COD reduced.

Benefits from Biogas

There is an assumption to know exactly how much biogas is beneficial to us.

Let us consider a strong organic waste *i.e.* waste from distillery (Spentwash). 1 M^3 of distillery waste can produce 25-35 NM^3 of biogas.

(The variation is due to technology difference.)

- ☆ Methane percentage –70 per cent
- ☆ Calorific Value –5000 K.Cal/NM^3
- ☆ Calorific value of coal –4500 K.Cal/NM^3
- ☆ Combustion efficiency of coal –75 per cent
- ☆ Combustion efficiency of gas –90 per cent
- ☆ Equivalent coal –1.33 Kg/NM^3

therefore, 1M^3 of distillery waste can save 33-50 Kg of coal. Other than this, saving of transportation cost of coal, saving of energy in terms of diesel and conservation of environment are indirect benefits.

Conclusion

The foregoing discussion evidently revealed that biogas is highly beneficial. It can reduce the problem of energy crises and environmental pollution. Therefore, it is essential to popularize anaerobic treatment as an obligatory step to all, producing/discharging organic wastes into the water bodies.

References

Gautam, Ashutosh, 1992. Energy from waste and pollution control. *Chemical EngineeringWorld*, 28(9): 89–90.

Kaul, S.N., Nandy, T. and Trivedy, R.K., 1995. *Pollution Control in Distilleries*. Enviro Media, Karad, India.

Sastry, C.A., 1989. Water pollution control. *Industrial Safety Chronical*, 89: 48–61.

Water and Wastewater Management in 2 Vols (2011) *Pages* **57–62**
Editors: **D.R. Khanna, R. Bhutiani & Gagan Matta**
Published by: **DAYA PUBLISHING HOUSE, NEW DELHI**

Chapter 10

Nutrient Dynamics in Rhithron Zone of Shivalik Himalayan Stream Sahastradhara, Dehradun, Uttarakhand

D.S. Malik and Pawan K. Bharti
Department of Zoology and Environmental Science,
Gurukul Kangri University, Haridwar – 249 404, Uttarakhand

ABSTRACT

Sahastradhara hill-stream is a major tributary of river Song, 15 km far from Dehradun. It has a great importance as a picnic tourist spot of the region, whereas there is no more industrial and agricultural pollution. Hill streams of the area are the cradle of nutrients especially calcium and magnesium in the form of ions. Anthropogenic factors mainly tourists activities and catchments runoff may influence the index of nutrients in stream water. Nutrients play a vital role into niche restoration maintenance, self-regulation and water quality. Location variation in nutrients concentration of the stream was studied with special reference to correlation with physico-chemical parameters of the stream water. Calcium and Magnesium were observed in higher range in Sahastradhara stream in comparison to other hill-streams. Calcium and Magnesium were found maximum 104.21 mg/l and 68 mg/l respectively in summer season.

Keywords: *Nutrient dynamics, Sahastradhara stream, Rhithron zone, Water quality.*

Introduction

The Himalayas are the cradle of a large number of streams and mighty rivers. Lesser Himalaya consists of a huge part of Uttarakhand. Dehradun, Haridwar and Rishikesh are the main religious and beautiful cities, which are situated in the foothills of Garhwal Himalaya. A hill-stream

Sahastradhara, 15 km far from Dehradun city is the study site for the accounting of nutrients in natural water and fluctuations at the different locations along with the stream. Adequate understanding of the Himalayan proglacial streams is extremely important for the development of a realistic program for utilizing the potential of water that exist in the form of snow and ice in the area.

The present study reveals to characteristics of the water nutrient chemistry, influenced by tourist's activity and quarrying of the geologically sedimental environments and to determine the nature and degree of anthropogenic impacts on qualitative and quantitative variations occurred in nutrients in relation to physico-chemical parameters of stream.

Materials and Method

The water samples were collected monthly from sampling site Sahastradhara stream from different sampling stations, Site-I, Site-II, Site-III, Site-IV and Site-V during December, 2003 to May, 2004 in morning period 9:00 A. M. to 10:00 A. M.

The samples for nutrients and physico-chemical parameters were collected and analyzed by using rinsed borosil glassware, with the help of the procedure described by APHA (1995), Trivedi and Goel (1984), Santhanam *et al.* (1989).

Results and Discussion

Sahastradhara stream is flowing throughout a valley of Dehradun mountain chain in foothills of Shivalik Himalaya, enriched with limestone and lignite rocks, which affect the water quality of stream according to the locations. Nutrients concentration and related physico-chemical parameters from selected five sites are depicted in Tables 10.1 and 10.2.

Table 10.1: Nutrients Parameters (Mean values in mg/l) of Sahastradhara Stream

Parameters	*Site-I*	*Site-II*	*Site-III*	*Site-IV*	*Site-V*
Chloride (mg/l)	(11.36–15.62)	(11.36–16.58)	(12.78–18.46)	(9.94–18.46)	(8.52–16.58)
Mean	12.78±1.73	13.73±1.93	15.26±2.1	13.49±3.25	11.6±3.06
Calcium (mg/l)	(80.16–88.17)	(80.16–92.18)	(77.15–104.20)	(79.15–99.19)	(80.16–96.19)
Mean	84.67±3.58	85.67±4.28	91.68±10. 13	89.18±8.26	88.43±6.19
Magnesium (mg/l)	(45–53)	(55–62)	(63.0–68.0)	(52.0–28.0)	(45.0–55.0)
Mean	48.66±3.09	57.33±3.29	64.66±2.36	55.33±2.49	51.00±5.29
Sodium (mg/l)	(6–8.5)	(7.0–8.2)	(8.0–9.5)	(7.0–10.0)	(7.5–11.0)
Mean	7.375±0.96	7.725±0.506	8.75±0.75	8.87±1.14	9.25±1.35
Potassium (mg/l)	(14–16)	(14.5–17.0)	(16.0–17.5)	(15.5–17.5)	(16.0–18.0)
Mean	15.13±0.74	15.75±0.90	16.62±0.65	16.5±0.79	17.0±1.35

Sahastradhara stream has the spatio-temporal variations of water temperature, which plays a vital role in all physico-biochemical reactions and self-purification power of aquatic system (Badola and Singh, 1981). Higher value of temperature was found 18 °C in summer and minimum 12 °C in winter season. A direct relationship was established between the water temperature and free carbon dioxide (Hynes, 1970). Turbidity is striking characteristic of the physical status of the water bodies. Although in Sahastradhara stream water is clear because there is no more pollution, siltation was the main source of turbidity in tributaries. Detritus and other non-organic material being added to water mass due to rainfall and tourist activities (Camron, 1996). Maximum turbidity was recorded 7.0 JTU during rainy days and minimum 1.0 JTU in December. The maximum depth of photic zone provides the better biological production for all aquatic organisms.

Table 10.2: Physico-chemical Parameters (Mean values) of Sahastradhara Stream

Parameters	*Site-I*	*Site-II*	*Site-III*	*Site-IV*	*Site-V*
Temperature (°C)	(14–18)	(13–17.5)	(13–17)	(13.5–18)	(12–18)
Mean	15.75–1.14	15.37±1.63	15.0±1.41	15.75±1.6	15.75±2.277
Velocity (m/s)	(0.3–0.4)	(0.28–0.32)	(0.8–1.0)	(0.78–0.82)	(0.6–0.7)
Mean	0.3±0.14	0.9±0.1	0.35±0.05	0.8±0.02	0.65±0.05
Turbidity (JTU)	(1.0–5.0)	(2.0–5.0)	(3.0–6.0)	(2.0–6.0)	(3.0–7.0)
Mean	2.75±1.47	3.25± 1.30	4.5± 1.15	3.75±1.48	4.75± 1.48
Total solids (mg/l)	(955–1427)	(1090–1332)	(1115–1443)	(1135–1520)	(1148–1535)
Mean	1218±183.2	1234.2±89.8	1277.2± 122.6	1305.7±166.1	1290.7±146.1
TDS (mg/l)	(840–1278)	(895–1217)	(1010–1322)	(1039–1380)	(965–1420)
Mean	1093.25±174.9	1094.25±120.5	1140.5±134.6	1189±158.01	1134.25±176.1
TSS (mg/l)	(113–149)	(119–195)	(105–207)	(105–140)	(115–211)
Mean	124.75± 14.39	140.0±28.16	136.75±40.95	116.75±13.8	156.5±41.69
pH	(7.2–8.1)	(7.3–8.1)	(7.5–7.9)	(7.6–8.0)	(7.2–8.1)
Mean	7.625±0.43	7.6±0.28	7.75±0.15	7.75±1.66	7.6±0.339
Free CO_2 (mg/l)	(1.40–1.69)	(1.62–1.92)	(1.62–2.20)	(1.65–1.85)	(1.62–2.20)
Mean	1.6525±0.39	1.75±.017	1.825±0.25	1.74±0.09	1.87±0.21
DO (mg/l)	(7.04–11.08)	(6.84–10.07)	(7.04–9.47)	(6.04–9.06)	(6.04–9.47)
Mean	8.41±1.62	8.46±1.18	8.65±0.95	7.55±0.87	7.49± 1.36
BOD (mg/l)	(1.60–2.0)	(1.7–2.2)	(2.0–2.5)	(1.8–2.2)	(1.9–2.3)
Mean	1.875±0.162	1.98±0.23	2.25±.18	2.0±0.14	2.05±0.15
COD (mg/l)	(2.1–3.0)	(2.2–2.9)	(2.6–3.2)	(2.4–3.0)	(2.4–2.9)
Mean	2.55±0.32	2.6±0.25	2.9±0.22	2.68±0.22	2.7±0.18
Hardness (mg/l)	(190–210)	(185–220)	(200–250)	(180–230)	(185–235)
Mean	196.25±8.2	200.0±15.41	222.5±19.2	200.0±18.71	198.75±21.02
Alkalinity (mg/l)	(150–480)	(140–500)	(170–600)	(130–570)	(120–540)
Mean	250.0±134.7	262.5± 140.8	305±172.1	292.5±168.8	317.5±175.2

Total solids were found in the range of 955 mg/l to 1535 mg/l towards summer season, due to the gradual increases in velocity of river which favoured effective sedimentation (Subramanian, 1979). Chemical oxygen demand represents chemically oxidizable organic matter load in water, while biochemical oxygen demand is only biodegradable materials. In the present study high values observed during summer months may be attributed by maximum biological activities and high temperature that stimulate the growth of micro-organisms (William *et al.*, 1993).

The pH of natural water was controlled in a great extent by the interaction of hydroxyl ions arising from the hydrolysis of bicarbonate (Sharma, 1986). The pH of Sahastradhara stream was recorded slightly alkaline (7.2–8.1). Hardness is mainly due to percentage of calcium and magnesium salts of bicarbonates, carbonates, sulphates and chlorides, while the value of alkalinity occured due to presence of bicarbonates. The highest concentration of hardness was analyzed 250 mg/l during warm months. A positive relationship between hardness and alkalinity was recorded in river Ganga at Rishikesh (Chopra and Patrick, 1994).

Maximum chloride concentration was recorded (18.46 mg/l) in the month of May and minimum in December (11.36 mg/l). Chloride and hardness showed a positive relationship to one another (Chopra and Patric, 1994). Chloride was found in the form of chloride ion, and one of the major inorganic anion present in natural water.

Calcium and magnesium were the dominant cations, and these represent the main weathering products, but significant hydro-chemical differences between the five sampling sites associated with the bedrock geology exist (Jenkins *et al.*, 1995) Calcium is one of the essential nutrients, which plays an important role in biological system. Maximum calcium concentration was recorded (104.208 mg/l) in the month of May and minimum in January (77.154 mg/l). Calcium concentrations were found maximum about 7.5 per cent of the total solids present in the stream water. Positive relationship between, calcium and temperature was also reported by Khanna and Singh (2000) in river Suswa, Dehradun. Magnesium is also an essential element but it is toxic at higher concentration. The amount of magnesium in Sahastradhara was found maximum (68.0 mg/l) in the month of April and minimum (45.0 mg/l) in December.

During the summer season nutrients concentration in hill-streams became more. Miller *et al.*(1997) described the nutrients availability in selected environmental settings of the Potomac river and Cameron, E. M. (1996) showed the similar type of fluctuation in Fraser river of British Columbia. Sodium is one of the most common cations has no adverse effect on human health at lower concentration. Bond, H.B. (1979) described similar nutrients concentration pattern in a stream draining a mountain ecosystem in Utah. Maximum sodium concentration in Sahastradhara was recorded (11.0 mg/l) in the month of May and minimum in January (6.0 mg/l).

Potassium is naturally occurring element, released by the clay minerals, weathering and leaching from growing vegetation and decomposition of organic matter (Berndtsson, 1990). Maximum potassium concentration was recorded (18.0 mg/l) in the month of April and minimum in December (14.0 mg/l). Pande and Mishra (2000) also observed similar results on Sahastradhara hill-stream Dehradun. Relationship between nutrients and other physico-chemical parameters are given in the Table 10.3 and percentages of nutrients among the total solids are given in Table 10.4, during the study period. These high concentrations confirm the location of the calcareous formation associated with the Central Himalaya thrust. Similarly, samples from stream draining the slope of Doon valley, exhibit high concentrations of Mg relative to K and Na. Again this can presumably be attributed to local geological features. In general, the stream at site III with high Ca concentrations also exhibit relatively low K concentrations compared to the other sites.

The present results conclude that significant differences in stream water nutrient concentrations exist among different environmental settings within the five subunits. The environmental setting with the highest potential by more soluble nutrients, fluctuations in nutrient concentrations were the land use and carbonate bedrock that was predominated in the Doon valley especially, at main tourist spot and sulphur springs site III. Ca and Mg are the dominant cations in this alkaline tributary of river

Song. Rock weathering is the most important mechanism controlling the water chemistry. The high concentrations of alkali earth metals, high alkalinity and the high (Ca+Mg)/(Na+K) ratio indicate that the water chemistry of the stream is controlled largely by carbonate weathering and partially by run-off process and tourists activities. The spatial variations in TDS are attributed to climatic and lithological control over the ionic concentrations.

Table 10.3: Correlation (=CORREL) Between Nutrients (Mean values) and Physico-chemical Parameters (Mean values) of Sahastradhara Stream

Parameters	*Chloride*	*Calcium*	*Magnesium*	*Sodium*	*Potassium*
Temperature (°C)	–0.86097	–0.51833	–0.92056	–0.01186	–0.14275
Velocity (m/s)	–0.17595	–0.12077	0.007842	0.179873	0.225839
Turbidity (JTU)	0.012697	0.818834	0.400371	0.92284	0.959357
Total solids (mg/l)	–0.00448	0.788427	0.273292	0.949891	0.906666
TDS (mg/l)	0.133154	0.71132	0.25802	0.764443	0.682478
TSS (mg/l)	–0.35586	0.096981	0.004271	0.35689	0.46296
pH	0.681329	0.761113	0.623067	0.395679	0.345766
Free CO_2 (mg/l)	–0.03224	0.686443	0.396104	0.833271	0.90927
DO (mg/l)	0.679888	–0.12231	0.471533	–0.65112	–0.55633
BOD (mg/l)	0.607899	0.912138	0.845587	0.615583	0.694262
COD (mg/l)	0.595229	0.954861	0.801927	0.654354	0.704159
Hardness (mg/l)	0.814816	0.795954	0.897315	0.312648	0.386789
Alkalinity (mg/l)	–0.04111	0.829565	0.331085	0.97707	0.985243

Table 10.4: Percentage of Different Nutrients Among the Total Solids at Different Location of Sahastradhara Stream at Dehradun

Nutrients	*Site-I*	*Site-II*	*Site-III*	*Site-IV*	*Site-V*
Total solids	100 per cent	100 per cent	100 per cent	100 per cent	100 per cent
Chloride	1.0492 per cent	1.1125 per cent	1.1948 per cent	1.0331 per cent	0.8987 per cent
Calcium	6.951 per cent	6.941 per cent	7.178 per cent	6.830 per cent	6.851 per cent
Magnesium	3.995 per cent	4.645 per cent	5.062 per cent	4.238 per cent	3.951 per cent
Sodium	0.6055 per cent	0.6263 per cent	0.6852 per cent	0.6793 per cent	0.7166 per cent
Potassium	1.242 per cent	1.276 per cent	1.301 per cent	1.264 per cent	1.317 per cent

References

APHA, 1995. *Standard Methods for Examination of Water and Wastewater*, 19th edn. American Public Health Association Inc., New York, pp. 1150.

Badola, S.P. and Singh, H.R., 1981. Hydrobiology of the river Alaknanda of Garhwal Himalaya. *Indian J. Ecol.*, 8(2): 269–276.

Berndtsson, R., 1990. Transport and sedimentation of pollutants in a river reach: A chemical mass balance approach. *Wat. Resour. Res.*, 26(7): 1549–1558.

Bond, H.B., 1979. Nutrient concentrations patterns in a stream draining a montane ecosystem in Utah. *Ecology*, 60(6): 1184–1196.

Cameron, E.M., 1996. Hydrogeochemistry of the Fraser River British Columbia: Seasonal variation in major and minor components. *J. Hydrol.*, 182(1–4): 209–255.

Chopra, A.K. and Patrick, N.J., 1994. Effect of domestic sewage on self-purification of Ganga water at Rishikesh. *A. Bio. Science*, 13(11): 75–82.

Hussain, S.I., Subramanian, V. and Dhanpal, K., 1989. Chemical characteristics of meltwaters from a Himalaya glacier, India. *J. Hydrol.*, 106: 98–106.

Hynes, H.B.N., 1970. *The Ecology of Running Waters*. Liverpool University Press, Liverpool, p. 1–555.

Jenkins, Alan, Wilkinson, Jeremy and Collins, Robert, 1995. Hydrochemical characteristics of headwater stream in the middle hills and high mountains of the Himalayas, Nepal: Atmospheric hydrologic and ecological interaction–Part C. In: *International Conference on Ecohydrology of High Mountain Areas*, pp. 255–263.

Jenkins, A., Sloan W.T. and Cosby, B.J., 1995. Stream chemistry is the middle hills and high mountain of the Himalaya, Nepal. *Journal of Hydrology*, 166(1–4): 61–79.

Khanna, D.R. and Singh, R.K., 2000. Seasonal fluctuations in the plankton of Suswa River at Raiwala Dehradun. *Env. Conservation J.*, 1(2 and 3): 89–92.

Miller, C.V., Denis, J.M., Ator, S.W. and Brakebill, J.W., 1997. Nutrients in stream during baseflow in selected environmental settings of the Potomac river basin. *J. American Wat. Resources Association*, 33(6): 1155–1171.

Odum, E.P., 1971. *Fundamentals of Ecology*. W.B. Saunders and Co., New York, pp. 1–344.

Pande, R.K. and Mishra, Asha, 2000. Water quality study of freshwaters of Dehradun (Sahastradhara stream and Mussoorie Lake). *Aquacult.*, 1: 57–62.

Santhanam, R., Velaynatham, P. and Jegatheesan, G., 1989. *A Manual for Freshwater Ecology*. Daya Publishing House, Delhi, p. 1–92.

Sharma, R.C., 1986. Effect of physico-chemical factors on benthic fauna of Bhagirathi River Garhwal Himalaya. *Indian. J. Ecol.*, 13(1): 133–137.

Subramanian, V., 1979. Chemical and suspended sediment characteristics of river of India. *J. Hydrol.*, 44: 37–55.

Trivedi, R.K. and Goel, P.K., 1984. *Chemical and Biological Methods for Water Pollution Studies*. Environmental Publication, Karad, pp. 1–25.

William, M.W., Brown, A. and Melack, J.M., 1993. Geochemical and hydrologic controls on the composition of surface water in the high elevation basin, *Sierra Navada. Limnol. Oceanogr.*, 38: 775–797.

Water and Wastewater Management in 2 Vols (2011) *Pages* **63–70**
Editors: **D.R. Khanna, R. Bhutiani & Gagan Matta**
Published by: **DAYA PUBLISHING HOUSE, NEW DELHI**

Chapter 11

Aquatic Birds Diversity at Yashwant Nagar Talab, Mhow, M.P.

Gopal Sahay Saxena[1], M.M. Prakash[2] and B. Arora[1]
[1]*Department of Zoology, Govt. P.G. College, Mhow*
[2]*Department of Biotechnology and Bioinformatics, Holkar Science College, Indore, M.P.*

ABSTRACT

Present study describes the seasonal variation of aquatic birds. Total 19 species of water birds were reported from this water body including 7 species of winter visitor and 7 species of local migratory. Species wise monthly count was also discussed in this chapter.

Keywords: *Water fowl, Monthly count, Migratory birds, Winter visitors.*

Introduction

Aquatic birds play an important role in wetland ecosystem because they act as a consumer in trophic levels of such system. They are also considered as an indicator of the changes occurring in the aquatic environment. Various workers such as (Bezzel, 1974; Karlson *et al*, 1976; Nilsson and Nilsson, 1976; Erikson, 1984; Koskimies, 1987 and Shukla *et al*, 2004) also expressed similar view and stated that almost all types of aquatic organisms can serve as an indicator. Literature available stated that very few scientists worked on the diversity of aquatic birds in M.P. (Prakash and Saxena, 2005; Malhotra *et al.*, 2005 and Mishra, 2006). Looking to the importance of aquatic birds in the management of local small water bodies, the present study was undertaken to study the aquatic avian diversity at Yashwant Nagar Talab, Mhow.

Materials and Method

Yashwant Nagar Talab, Mhow (M.P.) is a perennial water body, situated on the A.B. Road and 23 km away from Mhow, Indore. This Talab was constructed in 1934 by Holkar State on river Karam. It is used by local people to irrigate agricultural land and for bathing and washing purposes.

The water birds were identified with the help of binoculars, consulting Wood Cooks (1983) and Ali and Ripley (1983). The counting was carried out during morning hours strictly throughout the year.

Table 11.1: Water Birds along with Classification and Status in Yashwant Nagar Talab, Mhow (2005–06)

Sl.No.	Groups	Common Name	Order	Family	Status
I.	**Greebs**				
1.	*Podicepus ruficollis*	Little grebe	Podicipediformes	Podicipedidae	RLM
II.	**Cormorants and Darters**				
2.	*Phalacrocorax niger*	Little cormorant	Phalacrocoraciformes	Phalacrocoracirae	RLM
3.	*Anhingra rufa*	Darters	Phalacrocoraciformes	Phalacrocoracirae	RLM
III.	**Herons and Egrets**				
4.	*Ardeola grayii*	Pond Heron	Ciconiformes	Ardeidae	R
5.	*Bubulcus ibis*	Cattle egret	Ciconiformes	Ardeidae	R
6.	*Egretta garzetta*	Little egret	Ciconiformes	Ardeidae	R
IV.	**Storks**				
7.	*Ibis leucocephala*	Painted Stork	Ciconiformes	Ciconiidae	RLM
V.	**Geese and Ducks**				
8.	*Dendrocygana javanica*	Lesser whistling teal	Anatiformes	Anatidae	RLM
9.	*Anas acuta*	Pintail	Anatiformes	Anatidae	WV
10.	*Anas platyrhynchos*	Mallard	Anatiformes	Anatidae	WV
11.	*Anas clypeata*	Shoveller	Anatiformes	Anatidae	WV
12.	*Aythya ferina*	Pochard	Anatiformes	Anatidae	WV
13.	*Aythya fuligula*	Tufted duck	Anatiformes	Anatidae	WV
VI.	**Cranes**				
14.	*Grus antigone*	Sarus Crane	Gruiformes	Gruidae	RLM
VII.	**Coot**				
15.	*Fulica atra*	Coot	Gruiformes	Rallidae	WV
VIII.	**Waders**				
16.	*Himantopus himantopus*	Blackwinged Stilt	Charadriiformes	Chariidae	RLM
17.	*Tringa hyoleucos*	Common sandpiper	Charadriiformes	Chariidae	WV
18.	*Vanellus indicus*	Redwattled Lapwing	Charadriiformes	Chariidae	R
IX.	**Terns**				
19.	*Sterna auranita*	Indian river Tern	Charadriiformes	Laridae	R
		19	**06**	**09**	

Results and Discussion

Results obtained in the present investigation are summarized in Tables 11.1–11.4. Total 19 species of waterfowl were observed. They were belonging to 7 orders and 9 families. Out of 19 species, 7 species were migratory, 7 species were local migratory and 5 species were resident (Table 11.1). Prakash (1999) described 12 species of aquatic birds from Bahadur sagar (Jhabua, M.P.) However, Malhotra *et al.* (2005) reported 32 species of waterfowl from Sirpur tank (Indore, M.P.) belonging to 10 families and 7 orders. Out side from M.P. Pandey (1993) reported 54 species of water birds from Pongdom reservoir (Himachal Pradesh). Joyti *el al.* (2001) studied Gharana wetland reservoir (Jammu) and reported 24 species of waterfowls. Barman *et al.* (1995) described 62 species of water birds from Deepar Beal wildlife sanctuary (Assam). Kumar and Bohra (2002) recorded 103 species of birds belonging to 43 families and 13 orders from Udhuwa Lake (Jharkand).

Waterfowl observed in the present investigation were kept in 9 grouped *viz.*, Grebes, Cormorants and Darters, Herons and Egrets, Storks, Geese and Duks, Cranes, Coot, Waders and Terns. Similar grouping was observed by Prakash and Saxena (2005) from Krishnapura Talab, Indore. On the basis of count and per cent contribution (Table 11.2) the most dominated group in the present investigation was of Geese and Ducks 7 months followed by Waders dominated in three months, while Herons and Egrets dominated in two months. In most of the month's, species wise Geese and Ducks dominated this water body, followed by Herons and Egrets (2nd rank). On third place Cormorants was placed. Rest came under the 4th place as they were supported by only one species. In most of the months Tern was lowest except in November where Cranes took this position. The order of sequence of dominance of various groups in descending orders can be represents as under:

Geese and Ducks > Coot > Waders > Cormorant and Darter > Grebes >
Herons and Egrets > Crane > Tern

Species wise study of water fowl at Yashwant Nagar Talab showed that species number were highest from November to March (19) followed by October (18), and April to June (14). From the species point of view rainy months were very poor as they included only 9 species (Tables 11.2 and 11.3). Similar trends were followed by total monthly count. The presence of maximum number of species during winter was may be due to the inclusion of some migratory species. In rainy season when present water body was full of water and forcely overflow (which washed all the vegetation *etc.*) the count and species number reduces in these months because of absence of their food materials and flood condition. The month wise order of dominance of various species in descending order can be presented as:

April

Waders > Coot > Geese and Ducks > Grebes > Herons and Egrets > Cormorant and Darter > Tern > Storks > Crane.

May

Waders > Geese and Ducks > Coot > Herons and Egrets > Grebes > Cormorant and Darter > Storks > Tern > Crane.

June

Waders > Geese and Ducks> Herons and Egrets> Coot > Cormorant and Darter > Grebes > Tern > Storks > Crane.

Table 11.2: Species-wise Monthly Count (2005–06) of Water Birds Present at Yashwant Nagar Talaab, Mhow

Sl.No.	Species Groups	Feb	Mar	Apr	May	Jun	Jul	Aug	Sep	Oct	Nov	Dec	Jan
I.	**Greebs**												
1.	*Podicepus ruficollis*	84	80	82	78	80	76	30	16	50	50	75	78
II.	**Cormorants and Darters**												
2.	*P. nigar*	40	46	60	56	80	82	30	20	20	30	36	38
3.	*Anhingra rufa*	10	15	15	18	15	10	10	10	14	16	10	10
III.	**Herons and Egrets**												
4.	*Ardeola grayii*	12	15	15	18	15	10	12	15	16	18	16	16
5.	*Bubulcus ibis*	30	32	40	40	36	34	30	10	16	20	22	22
6.	*Egretta garzetta*	28	28	26	62	62	26	24	20	20	26	28	30
IV.	**Storks**												
7.	*Ibis leucocephalus*	6	10	10	12	10	0	0	0	0	3	10	10
V.	**Geese and Ducks**												
8.	*Dendrocygana jvana*	230	220	205	200	186	100	10	10	20	200	222	220
9.	*Anas acuta*	180	80	0	0	0	0	0	0	100	160	180	180
10.	*Anas platyrhyncha*	30	20	0	0	0	0	0	0	5	10	30	30
11.	*Anas clypeata*	240	200	0	0	0	0	0	0	216	222	240	236
12.	*Aythya ferina*	100	70	0	0	0	0	0	0	20	90	100	100
13.	*Aythya fuligula*	16	14	0	0	0	0	0	0	14	20	20	20
VI.	**Cranes**												
14.	*Grus antigone*	2	2	2	4	4	0	0	0	2	2	2	2
VII.	**Coot**												
15.	*Fulica atra*	400	200	200	150	100	0	0	0	200	410	400	400

Table 11.3: Species-wise Monthly Percentage (2005–2006) of Water Birds Presenting Yashwant Nagar Talab, Mhow

Sl.No.	Species Groups	Feb	Mar	Apr	May	Jun	Jul	Aug	Sep	Oct	Nov	Dec	Jan
I.	**Greebs**												
1.	*Podicepus ruficollis*	4.99	6.05	8.78	8.42	9.14	21.46	18.29	13.67	5.63	3.32	4.60	4.85
II.	**Cormorants and Darters**												
2.	*P. nigar*	2.37	3.48	6.43	6.04	9.14	23.16	18.29	17.09	2.25	1.99	2.20	2.36
3.	*Anhingra rufa*	0.59	1.13	1.60	1.94	1.71	2.82	6.09	8.54	1.57	1.06	0.61	0.62
III.	**Herons and Egrets**												
4.	*Ardeola grayii*	0.71	1.13	1.60	1.94	1.71	2.82	7.31	12.54	1.80	1.19	0.98	0.99
5.	*Bubulcus ibis*	1.78	2.42	4.28	4.31	4.11	9.60	18.29	8.54	1.80	1.32	1.35	1.36
6.	*Egretta garzetta*	1.66	2.11	1.78	6.69	7.08	7.34	14.63	17.09	2.25	1.72	1.71	1.86
IV.	**Storks**												
7.	*Ibis leucocephalus*	0.35	0.75	1.07	1.29	1.14	0	0	0	0	0.39	0.64	0.62
V.	**Geese and Ducks**												
8.	*Dendrocygana jvana*	13.67	15.14	21.97	21.59	20.57	28.24	7.31	8.54	2.25	13.28	13.62	13.68
9.	*Anas acuta*	10.70	6.05	0	0	0	0	0	0	11.27	10.63	11.04	11.19
10.	*Anas platyrhyncha*	1.78	1.51	0	0	0	0	0	0	0.56	0.66	1.84	1.86
11.	*Anas clypeata*	14.26	15.14	0	0	0	0	0	0	24.25	14.75	14.73	14.67
12.	*Aythya ferina*	5.94	5.29	0	0	0	0	0	0	2.25	5.98	6.13	6.21
13.	*Aythya fuligula*	0.94	1.05	0	0	0	0	0	0	1.57	1.32	1.22	1.24
VI.	**Cranes**												
14.	*Grus antigone*	0.11	0.15	0.21	0.43	0.45	0	0	0	0.22	0.13	0.12	0.12
VII.	**Coot**												
15.	*Fulica atra*	23.78	15.14	21.43	10.79	11.42	0	0	0	22.54	27.24	24.55	24.87

Table 11.4: Group-wise Monthly Counts (2005–06) of Water Birds of Yashwant Nagar Talab, Mhow

Sl.No.	*Species Groups*	*Feb*	*Mar*	*Apr*	*May*	*Jun*	*Jul*	*Aug*	*Sep*	*Oct*	*Nov*	*Dec*	*Jan*
1.	Greebs	84	80	82	78	80	76	30	16	50	50	75	78
2.	Cormorants and Darters	50	61	75	74	95	92	40	30	34	46	46	48
3.	Herons and Egrets	70	75	81	120	113	70	66	45	52	64	66	76
4.	Storks	6	10	10	12	10	0	0	0	0	3	10	10
5.	Geese and Ducks	796	604	205	200	186	100	10	10	375	702	792	786
6.	Cranes	2	2	2	4	4	0	0	0	2	2	2	2
7.	Coot	400	200	200	150	100	0	0	0	200	410	400	400
8.	Waders	260	275	266	276	275	10	10	10	168	225	232	202
9.	Terns	14	14	12	12	12	6	8	6	6	6	6	6
	Total	1682	1321	933	926	875	354	164	117	887	1508	1629	1608

July

Geese and Ducks> Cormorant and Darter > Grebe> Herons and Egrets> Waders > Tern.

August

Herons and Egrets> Cormorant and Darter> Grebe> Geese and Ducks= Waders > Tern.

September

Herons and Egrets> Geese and Ducks > Waders > Coot > Grebe>Cormorant and Darter > Storks> Tern. > Crane.

October

Geese and Ducks> Coot > Waders > Herons and Egrets> Grebe>Cormorant and Darter > Tern.> Crane.

November

Geese and Ducks> Coot > Waders > Herons and Egrets> Grebe>Cormorant and Darter > Tern > Storks> Crane.

December

Geese and Ducks> Coot > Waders > Grebe> Herons and Egrets> Cormorant and Darter > Storks> Tern > Crane.

January

Waders > Geese and Ducks> Coot > Waders > Grebe> Herons and Egrets> Cormorant and Darter > Storks> Tern.> Crane.

February

Geese and Ducks> Coot > Waders > Grebe> Herons and Egrets> Cormorant and Darter > Tern > Storks> Crane.

March

Geese and Ducks> Waders > Coot > Grebe> Herons and Egrets> Cormorant and Darter > Tern > Storks> Crane.

Thus present study suggest that this water body can be a good house of aquatic birds specially in winter and summer, if properly managed and keep free from hunting and anthropogenic activities. The interesting point of this water body is the presence of three sided cover of good vegetation which provides good resting and nesting place for various birds.

Acknowledgement

Authors are thankful to Dr.N.K.Dhakad, Principal, Holkar Science College, Indore and Prof. P.C. Chinchore, Principal, Govt. P.G. College, Mhow for facilities and encouragement.

References

Ali, S. and Ripley, S.D., 1983. *Handbook of the Birds of India and Pakistan.* Oxford Univ. Press, New York.

Bergman, R., Saikia, P., Singh, H.J., Talukdar, B.K. and Bhattacharjee, P.C., 1995. Study of the population trends of water bird in deeper or heal wildlife sanctuary. *Assam Pavo.*, 33 (1 and 2): 25–40.

Bezzel, E., 1974. On the evaluation of waterfowl hiotopes. In: *Int. Conf. Conser. Wetland and Waterfowl,* Heligenhanfen FRG, 2–6, 12: 294–299.

Erikson, M.O.G., 1984. Acidification of lakes: Effect of water birds in Sweden. *Ambio,* 115: 260–262.

Green, A., 1993. Threatened waterfowl recovery plans of Asia. In: *Proc. of Int. Symp.,* Karachi, Pakistan, 14–20 Dec., 1991. IWRB Special Publication, 25: 80–84.

Jyoti, M.K., Sharma, K.K., Sharma, P.S. and Singh, P., 2001. Biodiversity of Gharana wetland reserve with special reference to water birds. In: *Abstract, Nat. Sem. on Biodiversity Conservation,* Jiwaji Univ., Gwalior, Nov. 26–28, pp. 16.

Karlson, J., Ladergreen, A. and Rudebeck, G., 1976. *Drastiska for andringer Och flagelfauna; krankesjon Och Bi-orkesakrasion,* 1973–1976 (In Swedish) Anser, IS 165–184.

Koskimies, P., 1987. Monitoring finish bird fauna: Bird as environmental indicator (In *Finish* with English summary). Pub. Ministry Environment, A., 49: 1–255.

Kumar, A. and Bohra, C.J., 2002. Faunistic composition, communities structure, population dynamics and ethology of the birds of Udhuwa lake (Santhal Pargana) and the strategics for the conservation in new millennium. In: *Ecology and Conservation of Lakes, Reservoir and River,* Vol. 2, (Ed.) A. Kumar. ABD Publisher, Jaipur, p. 411–434.

Malhotra, Manjeet, Prakash, M.M. and Pawar, K., 2005. Diet variation in waterfowl during winter at Sirpur Tank, Indore. *Indian J. Environ. and Ecoplan.,* 10(1): 129–138.

Mishra, P.K., 2006. Conservation of water birds in M.P. (India). *Env. Cons. Journal,* 7(3): 59–62.

Nilsson, S.G. and Nilsson, I.N., 1976. Valuation of South Swedish Wetland for conservation with the proposal of a new method for valuation of wetland as breeding habitat for birds (In Swedish with English summary) fauna. *Ack Flora,* 71: 136–144.

Pandey. Sanjeev, 1993 Changes in water birds diversity due to the construction of Pong Dam reservoir. Himachal Pradesh, India. *Ecological and Conservation,* 66 : 125–136.

Prakash, M.M., 1999. Impact of anthropogenic activities on the aquatic birds population at Bahadur Sagar (Jhabua. M.P.). In: *Advances in Fish and Wildlife Ecology and Biology,* (Ed.) B.L. Kaul. Daya Publishing House, New Delhi, 2: 174–178.

Kash, M.M. and Saxena, Gopal, 2005. Population dynamics of water fowl at Kishanpura Talab, Indore (M.P.). *National Journal of Life Science,* 2(122): 45–48.

Shukla, R.N., Shrivastava, R.K., Dubey, K.K., Shukla, Alok and Shrivastava, A.K., 2004. Biological monitoring: An ideal tool for water pollution studies. *Nat. J. Life Sciences,* 1(1): 169–171.

Woodcocks, M., 1983. *Collins Hand Guide to the Birds of the Indian Sub-continent.* William Collins Sons and Co. Ltd., London.

Water and Wastewater Management in 2 Vols (2011) *Pages* **71–75**
Editors: **D.R. Khanna, R. Bhutiani & Gagan Matta**
Published by: **DAYA PUBLISHING HOUSE, NEW DELHI**

Chapter 12

Biodiversity of Fish Fauna of Kishanpura Lake, Indore, M.P.

Shailendra Sharma[1], L.K. Mudgal[2], Praveen Sharma[2] and D.K. Belsare[3]

[1]Department of Zoology, P.N.B. Gujarati Science College, Indore
[2]Department of Zoology, Govt. Girls P.G. College, Moti Tabela, Indore
[3]Institute of Natural Resources and Human Environment, Bhopal, M.P.

ABSTRACT

The water quality of Kishanpura lake is assessed in order to estimate its potency for fish culture. The present study deals with fish species diversity. During the study period 23 species of fishes were recorded, which belong to five orders ten families and 17 genera. The present study was carried out at Kishanpura lake during Apr 2005–Jan 2006.

Keywords: *Water quality, Fishes, Biodiversity, Kishanpura Lake.*

Introduction

Indian fishes constitutes an important sector of national economy for various reasons. India has vast potential for development of inland fisheries. The fish population of our aquatic ecosystem plays a significant role in the human economy. Approximately 21, 723 fishes are known to science of which 40 per cent live in freshwater, that too majority of them in tropics. Potentiality of inland resources of India are the richest in the world. Fishes of Inland water of the Indian subcontinent have been also studied since a century by Hemilton-Buchaan (1822), Day (1878), Hora (1920-59), Dutta (1970), Jayram (1991), Jhingran (1982) and Nanda and Tiwari (2001). State of Madhaya Pradesh is neglected in this regard even though gross work of this aspect were made by Dubey and Mehra (1959), Gupta and Rao (1978), Rao *et al.* (1988, 1991, 1993) and Sharma and Mudgal (2003).

No attempt has been made on Kishanpura lake in this regard, so far this is the first limnological study conducted in this lake conducted and details of fish biodiversity were recorded which will helpful in the development, management and conservation of fish species diversity in the lake. In our present study an attempt is made to document the fish fauna in the Kishanpura lake during May 2005-Jan 2006. There is no earlier study on fish species diversity of this lake.

Materials and Method

Sampling of fish has been made for every month through out the study period, collection of fish was made directly from the fishermen during the time of fishing.Two types of fishes nets were used like Gillnet, Cost net. The fish were brought to the laboratory and preserved in 5 per cent Formaldehyde solution. After noting the coloures and general pigmentation of fishes. The identification of the fishes was done with the help of standard keys and books (Jayaram, 1994; Shrivastava, 1998; Jhingran, 1991 and Day, 1958).

Description of Study Area

The Kishanpura lake is situated at near Chhota Betma on Indore-Dhar road at 22 km from Indore. The lake is located at latitude 22°40′30″ and longitude 75°39′00″. The neighboring areas of the lake are agricultural fields the main source of water of this lake is rainy water. Suitability of water body for fish culture depends upon its physico-chemical characters, hence the important physico-chmeical characters of water from the lake was analyzed (Table 12.1).

Table 12.1: Physico-chemical Parameters of Kishanpura Lake Indore

Parameters	*Value*
Temperature	27°C
pH	6.89
Transparency	68 cm.
Dissolved Oxygen	7.06 mg/l
Total Alkalinity	2.56 mg/l
Hardness	120 mg/l
Chloride	0.0031 mg/l
Calcium	4.49 mg/l
Phosphate	0.49 mg/l

Results and Discussion

The result of our study confirms the occurrence of 23 species belonging to the five orders, out of the five orders Cypriniformes was dominated with 13 species followed by Siluriformeis with 4 species, order Perciformes with 4 species and order Osteoglossiformes and Beloniformes represented by one species only. The change in the composition of fish assemblage often indicated a variation in the water quality parameters such as pH, Temperature, Dissolves Oxygen, Hardness *etc.* (Table 12.2).

Table 12.2: Fish Species Diversity of Kishanpura Lake Indore

Order – Cypriniformes		
Family – Cyprinidae		
Genus and species–	1.	*Puntus ticto*
	2.	*Punctus sarana*
	3.	*Catla catla*
	4.	*Cirrhinus mrigla*
	5.	*Labeo rohita*
	6.	*Labeo calbasu*

Contd...

Table 12.2–Contd...

	7.	*Labeo goniua*
	8.	*Cyprius carpio*
	9.	*Danio malavaricus*
	10	*Rasvaoradaniconlus*
Family – Cobitinae		
Genus and species–	1.	*Noemachilus aureus*
	2.	*Lepidocephalus guntea*
Family – Siluridae		
Genus and species –	1.	*Wallago attu*
Order – Osteoglossiformes		
Family – Notopteridae		
Genus and species –	1.	*Notopterus notopterus*
Order – Silurifornes		
Family – Bagridae		
Genus and species –	1.	*Mystus seengnala*
	2.	*Mystus bleekeri*
Family – Clariidae		
Genus and species –	1.	*Clarias batrachus*
Family – Hetropneustidae		
Genus and species –	1.	*Hetropenustes fossilus*
Order – Perciformes		
Family – Chammidae		
Genus and species –	1.	*Channa punctus*
	2.	*Channa striatus*
	3.	*Channa gachua*
Family – Mastacenbelidae		
Genus and species –	1.	*Mastacembalus armatus.*
Order – Beloniformes		
Family – Belonidae		
Genus and species –	1.	*Xenentodon concila*

The change in the composition of a fish assemblage often indicate a variation in the water quality parameters. Such as pH, Temp., D.O. and Nutrient (Jhingran, 1982; Kumar and Paul, 1990). Due to more fecundity of major carp and suitable environmental condition there exists a relatively higher no. of Cypriniforimes. Such type of observation was reported by Talwar and Jhingran (1991); Das and Chand (2003); Pathak and Mudgal (2005) in Indian fishes.

Table 12.3: Fish Species Diversity of Kishanpura Lake Indore

Sl.No.	Scientific Name of Fish	Local Name or Vernacular Name	Status
1.	*Rasbora daniconius*	Ajara	++
2.	*Puntius ticto*	Puthf	+++
3.	*Puntius sarana*	Sherni	++
4.	*Catla catla*	Catla	+++
5.	*Cirrhinus mrigala*	Mrigal	+++
6.	*Labeo rohita*	Ronu	+++
7.	*Labeo calbasu*	Kalot	+++
8.	*Labeo gonius*	Sarsi	++
9.	*Cyprinus carpio*	Kangi	+
10.	*Danio Malabanicus*	–	+
11.	*Notopterus notopterus*	Patola	++
12.	*Mystus seenghala*	Singhara	+++
13.	*Mystus bleekeri*	–	++
14.	*Heteropneustes fossilis*	Singhi	+++
15.	*Clarias batrachus*	Mangur	++
16.	*Channa punctatus*	Kabra	++
17.	*Channa striatus*	Giral	++
18.	*Channa gachua*	Giral	+
19.	*Mastacembalus armatus*	Bam	+
20.	*Noemachilus aureus*	–	+
21.	*Leidocephalus guntea*	–	+
22.	*Wallago attu*	Padhni	++
23.	*Xenentodon concila*	Sooja	+

Acknowledgements

The authors wish to thank Dr. O.P.Joshi, Principal of P.M.B. Gujarati Science College, Indore for encouragement to undertake the present investigation and to Dr. Rakesh Trivedi, Head of Department of Zoology, P.M.B. Gujarati Science College, for laboratory facilities. Authors are thankful to the Regional Office of the University Grants Commission, Bhopal (M.P.) for financial assistance.

References

APHA, 1998. *Standard Method for the Examination of Water and Wastewater*, 20th edn. American Public Health Association, New York.

Dubey, G.P. and Mehra, R.K., 1993. Fish and fisheries of Chambal river. In: *Proc. All India Zool. Congr.*, 1(2): 647–665.

Das, S.K. and Chand, B.K., 2003. Limnology and biodiversity of ichthiyofauna in a pond of southern Orissa, India. *J. Ecotox. Env. Monit.*, 13(2): 97–102.

Datta, A.K., 1970. *Fauna of Rajasthan*. India Part–7 Fishes Rec. Zoo. Survey, India, p. 62.

Day, F., 1958. *The Fishes of India*. William Dawson and Sons .td., London, Vol. 1: p. 777 and 2: p. 198.

Gupta, D.R. and Rao, K.S., 1978. Experimental studies on tagging as an effective technique in ageing and fish growth on Gandhisagar reservoir. *Bioresearch*, 2(182): 51–18.

Hamilton-Bachanan, 1822. *An Account of the Fishes Found in the River Ganges and its Branches*. Edinburg and London. Viii # 405 PP 39 P.

Jhingran, V.G., 1982. *Fish and Fisheries of India*. Hindustan Pub. Corporation, India.

Jayaran, K.C., 1994. *The Freshwater Fishes of India Pakistan, Bangladesh, Burma and Sri Lanka*. Zoological Survey of India, Calcutta, p. 475.

Kumar, Vijay K. and Paul, R., 1990. Physico-chemical studies on the Bhosga reservoir in Gulberga, Karnataka. *Ecol. Biol.*, 2: 332–335.

Lagler, K.F., 1956. *Freshwater Fishery Biology*, 2nd edn. WMc Brown Co., Durbungue.

Natrajan, A.V., 1985. Potential and prospects of inland fisheries in India. In: *Harvest and Postharvest Technology of Fish*. Society of Fisheries Technologists (India) Cochin, p. 16–18.

Nanda, S.N. and Tiwari, J.N., 2001. A survey of fish fauna in the Sambalpur–Hirakund–Burta region, Orissa. *Environment and Pollution*, 8(1): 43–44.

Pathak, S.K. and Mudgal, L.K., 2005. Limnology and biodiversity of fish fauna in Virla reservoir, M.P. (India). *Environment Conservation Journal*, 6(1): 41–45.

Rema Devi, K. Indra, *et al.*, 1999. On collection of fish fauna from Chennai, Chenglepur and Thiruvallur districts of Tamil Nadu. *Rec. Zool. Survey of India*, 97(4):151–166.

Srivastava, G.J., 1980. *Fishes of Eastern Uttar Pradesh*. Vishwavidyalya Prakashan, Varanasi, p. 207.

Sharma, A. and Mudgal, L.K., 2004. Fish diversity of Yashwant Sagar reservoir, Indore, M.P. *Him. J. Evn. Zool.*, 18(2): 117–119.

Singh, D.F., 1993. *Fish Diversity of the Western Ghats*. Draft Report, SACON.

Talwar, P.K. and Jhingran, A.G., 1991. *Inland Fishes of India and Adjacent Countries*. Oxford and IBH Publishing Co. Pvt. Ltd., New Delhi, p. 1158.

Figure 13.2: Close Up View of the BAF and its Filtering Material

Figure 13.3: The Bench Scale BAF Setup in the Lab. Sedimentation tank is situated at the right hand casher.

Experiment

A bench scale BAF was setup in the early stage of the experiment. The schematic diagram of BAF was shown in Figure 13.1. After the setup, the experiment was conducted in three phases. The first phase was the isolation and cultivation of dominant biofilm forming bacteria, which were capable of degrading or removing pollutants. These bacteria were further selected and used as seed culture (or inoculums) for the BAF in a process known as bioaugmentation. The second phase involved the analysis and characterization of wastewater samples before and after being treated by the BAF. The third phase involves the microbiological analysis whereby dominant types of bacteria isolated from the biofilm formed in the BAF were characterized and identified by their genera.

Biofilter

The bench scale BAF was made of glass. The system consisted mainly of 2 tanks, the sedimentation tank and the compartmentalized Biofilter. The BAF measures at 50 cm X 120 cm X 40 cm and was made up of 12 compartments containing the filtering materials. Schematic diagram of the BAF and its filtering material was shown in Figure 13.1. The filtering materials were chosen based on their ability to adsorb certain pollutants and their large water holding capacity; with no clogging risk and minimal pressure drop. A glass lid was used to cover the compartmentalized filter tank to prevent direct exposure to atmospheric air, thus creating a partial anaerobic condition. In the experiment that followed, wastewater was circulated into the BAF continuously for 7 days. The first 7 days was to propagate biofilm growth and from that onwards the BAF was used in the treatment of polluted river water. The wastewater used in this experiment was taken from a drain flowing into Sungai Melana, Taman Universiti. The flow rate of the BAF was set at 250ml/min.

Selection of Bacteria for Bioaugmentation

Isolation and screening of dominant biofilm forming bacteria from raw wastewater was carried out. The isolated bacteria were cultured on nitrate, sulfate and phosphate containing specialized media to select bacteria capable of removing these nutrients. Since the increased concentration of these nutrients may be used as a pollution indicator.

Water Quality Parameter Analysis

Wastewater qualities were analyzed based on physical and chemical parameters. Analysis of BOD,COD, nitrate, phosphate, sulphate, turbidity and ammonia were determined using HACH DR 4000 SPECTROPHOTOMETER. pH was measured using Orion Model 420A pH meter. Total suspended solids were analyzed using Whatman 0.45μm glass fiber filter paper.

Microbiological Analysis

Biofilm was collected from the fiber sponges in every compartment of the BAF. The samples were plated out to determine bacterial population and their dominancy. Dominant bacteria were isolated and grown as pure colonies. Biochemical tests were carried out to determine the types of bacteria present in the BAF. Biofilm dry weight analysis was also carried out.

Results and Discussion

Parameters

The BAF operated under partial anaerobic condition generally improved the influent condition from murky colored with foul odour to a clear and odorless effluent. Formation of slimy layers that may be associated with biofilm was examined microscopically. Besides that, black precipitation found in the BAF was indicative of the presence of sulphate reducing bacteria (SRB). The BAF was also proved to be effective in removing a maximum of 93.6 per cent and 80.0 per cent of TSS and turbidity respecively. A significant decrease in TSS may be implied to organic substances and pollutants being adsorbed onto filtering materials. This in turn will create an environment suitable for the propagation of indigenous microorganisms. These microorganisms will then biodegrade the pollutants as part of their metabolic activities. Consequently, the COD level; which is associated with the organic content of wastewater also showed a maximum removal percentage of 70.5 per cent. The pH of the effluent was monitored and found to be in the range of pH 7 to 7.5.

Nutrient Removal

Apart from the physical parameters, chemical parameters were also evaluated based on the removal percentage of nitrate, phosphate, sulphate and ammonia. The BAF was able to remove nitrates may be facilitated by denitrification process due to the partially anoxic condition created in the filter. This indicates that denitrifying bacteria present in the BAF had reduced nitrate to nitrite and finally to gaseous nitrogen. As for ammonia removal, zeolite inserted in the BAF was capable of adsorbing ammonia (Matsumoto 1997). When the concentration of ammonia in the bulk solution is reduced, the adsorbed ammonia is then released from the zeolite and will then be oxidized by denitrifying bacteria present. The maximum removal of phosphate was only 40.7 per cent. This poor removal suggested that under anaerobic conditions, the presence of polyphosphate bacteria may release phosphate into their surrounding, resulting in the increase of phosphate. In this study, the role of limestones as a phosphorus-adsorbing agent was insignificant in reducing the phosphorus content. As for sulfate removal, an average removal of 63.4 per cent was reported. The removal was mostly due to the BAF being operated under anaerobic condition, which favoured the reduction of sulfates by the SRB.

Table 13.1: Nutrients Removal

Nutrient (mg/L)	*Raw Wastewater*	*Biofilm*	*Suspension*
Nitrate	35	15.3	19
Phosphate	3	1.83	2.45
Sulphate	80	40.5	78

Role of Bacteria in Wastewater Treatment

To determine whether the microorganisms immobilzed in the BAF or the microorganisms in the bulk solution were responsible for the degradation of pollutants in the wastewater, 2 parallel studies were carried out in which similar water quality parameters were analyzed using the untreated wastewater and wastewater treated using the BAF.

Table 13.2: Comparison of Final Effluent Concentration during Treatment of Wastewater with and without the BAF

Parameter	*Without*	*BAF*
BOD (mg/l)	3.4	8.2
COD (mg/l)	580.5	215.5
Turbidity (FAU)	74	22
TSS (mg/l)	79	21
Nitrate (mg/l)	18.9	0.6
Ammonia (mg/l)	5.11	2.97
Sulphate (mg/l)	24.7	15

Table 13.3: Comparison Between BAF with Standard B (DOE)

Parameter	*Influent*	*Effluent*	*Standard B*
COD (ppm)	1075.5	215.5	100
TSS (ppm)	93	21	100
Co (ppm)	0.019	0.011	1
Pb (ppm)	0.012	0.001	0.5
pH	7.25–7.39	7.24–7.48	5.0–9.0

The results proves that the microorganisms immobilized in the BAF are responsible for the major removal of pollutants in the wastewater compared to the indigenous microorganisms naturally present in the bulk solution.

Microbiological Analysis

For the microbiological analysis, the dominant bacteria in each compartment were determined using biochemical tests. The results showed that the dominant bacteria consisted of *Pseudomonas sp*, *Bacillus* sp, *Aeromonas* sp, *Actinobacillus* sp and *Moraxella* sp. These results also demonstrated that the bioaugmentation process has been proved successful in increasing selected bacteria such as those of *Pseudomonas* sp and *Moraxella* sp.

Conclusion

It may be concluded that the BAF, is capable of removing COD, BOD, turbidity, TSS and nutrients such as nitrates, phosphates and sulphates from the drainage system studied. Although it is impossible to remove the pollutants totally, this wastewater treatment system was successful in minimizing the level of pollutants released into the river system. This system is also capable of functioning for a long period of time with minimal maintenance due to its biological regeneration. The BAF can be very useful equipment in various industries due to its compactness and efficiency for biological purification, their good integration in the environment, their low energy consumption and sludge production and the fact they do not need secondary clarifiers (Jacob *et al.*, 1996). The BAF is indeed a sustainable technology ; as building a sustainable future has been a great challenge today, this new innovation may improve the quality of the enviorment by means of practicing efficient and quality waste management system.

References

Corbitt, R.A., 1999. *Standard Handbook of Environment Engineering*, 2nd edn. McGraw Hill, 5: 89–593.

Gavrilescu, M. and Macoveanu, M., 2000. Attached-growth process engineering in wastewater treatment. *Bioprocess Engineering*, pp. 95–106.

Gavrilescu, M., 2002. *Engineering concerns and new developements.* In: *Anaerobic Wastewater Treatment.* Springer-Verlag.

Ismail, R., Mohd Azaai Kassim and Ng, T.H., 2001. *The Performance of Biologically Active Filter in Treating Municipal Landfill Leachate.* Universiti Teknologi, Malaysia.

Jacob, J. *et al.*, 1996 . Dynamic simulation of biofilters. *Simulation Practice and Theory*, 4: 335–348.

Matsumoto, S. 1997. The conception and execution of improved river water quality using a newly developed purification method: The Simanto-gawa system. *UNEPIETC Newslett.*

Water and Wastewater Management in 2 Vols (2011) *Pages 82–86*
Editors: **D.R. Khanna, R. Bhutiani & Gagan Matta**
Published by: **DAYA PUBLISHING HOUSE, NEW DELHI**

Chapter 14

Limnological Study of Sapna Reservoir at Betul, M.P.

R.C. Banatwala[1], S.D. Dongre[2] P.K. Mishra[2] and R.G. Verma[2]
[1]*Department of Zoology, J.H. Govt. P.G. College, Betul, M.P.*
[2]*M.G.M., Itarsi, M.P.*

ABSTRACT

Preliminary limnological studies were conducted to find out quality of water in a permanent reservoir of Sapna Dist. Betul. Dissolved oxygen and pH of the water was normal while the alkalinity or the form of bicarbonate were quite high. Temperature fluctuation was between 13°C to 30°C. Flora and fauna of the reservoir was quite rich.

Keywords: *Reservoir, Limnological, Water, Flora, Fauna.*

Introduction

In Indian low line areas including temporary ponds, reservoir and pools have been ecologically investigated by few workers. Unni (1983) studied the water chemistry in Chhindwara, Rao and Gupta (1986) studied the ecology of Gandhisagar, Singhal (1985) studied the relationship among physical and chemical factor and plankton characteristics of ponds in Haryana. Ali(1993), cultured *Cyprinus carpio* in domestic wastewater ponds. The present preliminary investigation was conducted to study the physico-chemical condition of Sapna Dam of Betul District (M.P.).

Materials and Method

The present study was conducted on Sapna Dam reservoir at Betul during July 2003 to December 2003. Sapna Dam is a water reservoir which fullfills, the requirements of surrounding 400 hectare agricultural land on the other hand drinking water supply is also done. The water sample were collected during July 2003 to December 2003. Some of the physico chemical parameters like temperature, CO_2, pH, alkalinity, DO *etc.* were studied..The water sample, were collected at an interval of 12 hrs. schedule. The physico-chemical and biological parameters were determined by following Standard methods of APHA (1998) and Mathur (1982).

Table 14.1: List of Plankton Available in the Reservoir of Sapna Dam, Betul during July 2003 to December 2003

Plankton	*Abundant*	*Common*	*Rare*	*Accidental*
Chlorophyceae				
Chlamydomonas sp.	x			
Pandurina morum	x			
Eudorina elegans		x		
Chlorococoum sp.	x			
Chlorococum humicola			x	
Chlorella vulgaris	x			
Characium limenticum		x		
Pediastrum tetras				x
Tetradron tirqonum				x
Actinestrum hantzschil				
Coelastrum microporum			x	
Oocystis sp.				
Scendesmus quadricauda			x	
S. abundans		x		
S. alterans				x
Ankistrodesmus falcatus			x	
Cosmarium hammeri				x
C. scabrum			x	
C. quinarium			x	
Clostenim sp.			x	
C. littorale				x
Uronema gigas			x	
Euglenophyceae				
Euglena acus		x		
E. oxyuris			x	
E. sp.				x
E tripteris				x
E. gracilis				x
E. munuta				
Phacus tortus			x	
P. Meson	x			
P. helikoides				x
P. longicauda			x	

Contd...

Table 14.1–Contd...

Plankton	Abundant	Common	Rare	Accidental
P. acuminatus	x			
P. brachykentron			x	
P. orbicularis		x		
P. pyrum		x		
Trachelomonas varians			x	
T. similis		x		
T. armata			x	
T. superba			x	
T. hispida		x		
Lepocinclis acuta		x		
L. fusiformis				x
L. ovum			x	
L. textra				x
L. ovalis	x			
Cyanophyceae				
Merismopedia glauca				x
Oscillatoria princeps		x		
O. curviceps			x	
O. tenuis			x	
O. chalybea				x
Spirulina laxissima				x
Anabaena sp.	x			
Bacillariophyceae				
Navicula cuspidata			x	
Nitzchia palaea				x
Gomphonema sp.	x			
Frustulia sp.				x
Protozoa				
Amoeba proteus		x		
Arcella discoides	x			
A. vulgaris		x		
Centropyxis aculeata				x
C. hemispherica		x		
Diffugia lebes		x		
Euglypha brachiata				x
E. mucronata				x
Euplotes sp.				

Contd...

Table 14.1–Contd...

Plankton	*Abundant*	*Common*	*Rare*	*Accidental*
Rotifera				
Brachinus calyciflorus	x			
B. caudatus	x			
B. budapensis				x
B. falcatus		x		
B. forficula		x		
B. patula				x
Platyas quadricornis				x
Lepadella ovalis	x			
L. rhomboides	x			
Lecana (monostyla) bulla			x	
Lecana curvicornis	x			
Filinia Longiseta		x		
F. opoliensis				x
Trichocera sp.	x			
Synachaeta sp.		x		
Haxarthra mira				x
Ostracoda				
Cypris sp.		x		
Stenocypris sp.	x			
Cladocera				
Moina brachiata	x	x		
Alona sp.				
Copepoda				
Mesocyclops sp.	x			

Note: x: Present.

Results and Discussion

Dissolved Oxygen (DO)

Dissloved Oxygen shows fluctuation from 7.5 to 10.1 mg/l. The minimum dissolved oxygen was of 7.5 at the high temperature of 30°C and increased according to the decreasing temperature.

Free CO_2

Free CO_2 content of the reservoir was nil at good favourable condition of Sapna reservoir.

pH

The pH show very low fluctuation from 7 pH to 7.5 pH in the reservoir.

Alkalinity

The Alkalinity of the reservoir was found to be increased from 78 to 249 mg/l. Decreasing temperature and the increasing density of the plants was the reason.

The Carbonate alkalinity increased from 7 to 17 mg/l and the low fluctuation is due to lack of phytoplankton population.

Temperature

The temperature showed very high fluctuation from 30°C (in July) to 13°C (in December) the total fall of temperature is 17°C.

Plankton

Plankton were found rich in the study pond.

Variation in phytoplankton population was noted that the phytoplankton were maximum in the evening time. Phytoplankton is directly correlated with free CO_2 the similar result were noted in the zooplankton population it means zooplankton avoids light. The dissolved oxygen was found minimum in the month of October it was due to the activity of plankton in the water. The pH of water is found alkaline during study period. A positive correlation was found between pH and total plankton.

Acknowledgements

Authors are thankful to Principal J.H. Govt. College Betul and M. G. M. Itarsi for providing laboratory facilities.

References

Ali, S., 1993. Histopathological evaluation of Carpio cultured in domestic wastewater pond.

APHA, 1980. *Standard Methods for Examination of Water and Wastewater*. American Public Health Association, 1015 Fifteen Street N.W. Washington, 15: 1–1134.

Mathur, R.P., 1982. Diurnal Variation in a shallow tropical freshwater pond Bareilly (U.P.). *Ad. Bios.* 13(1): 51–56.

Rao, K.S. and Gupta, D.R., 1986. Benthic studies in Gandhi sagar reservoir with special reference to macro-zoobenthos. In: *Nat. Sem. as Prespective in Hydrobiology*.

Singhal, R.N., 1985. Relationship among physical, chemical and plankton. *Trop. Ecol.*, 25: 43–53.

Unni, K.S., 1983. Comparative water chemistry dominated and a macrophyte dominated taken in Chhindwara M.P. *Proct. Nov. Acal. Sci.*, 52: 81–88.

Water and Wastewater Management in 2 Vols (2011)
Editors: D.R. Khanna, R. Bhutiani & Gagan Matta
Published by: DAYA PUBLISHING HOUSE, NEW DELHI

Chapter 15

Impact of IDPL Effluent on Water Quality of River Ganga at Shyampur Khadir, Rishikesh, Uttarakhand

G. Prasad[1], V. Shanker[1] and A.K. Chopra[2]
[1]*Department of Botany and Microbiology,*
[2]*Department of Zoology and Environmental Science,*
Gurukula Kangri University, Haridwar

ABSTRACT

Present chapter deals with the impact of IDPL effluent on water quality of river Ganga at Shyampur Khadir, Rishikesh. A significant deterioration of water quality of Ganga was recorded at Shyampur Khadir. Water quality of river Ganga at Pashulok Barrage (1.5 km before discharge of IDPL effluent) was better in term of various physicochemical characteristics, valued mg/l, chloride-5.6, DO-8.8, BOD- 7.14 and COD- 20.95 was observed. Bacterial load of water in term of MPN and SPC was found 11.6×10^2/100 ml and 102×10^3/ml respectively. But a high degree of deterioration of water quality was recorded at Shyampur Khadir located about 1.5 km away from discharge point in down stream of the river. Various parameters were found highly enhanced *i.e.* Hardness- 137.8, chloride- 14.3, BOD- 58.8, COD-162.45, MPN–64×10^2 and SPC- 290×10^3. Main cause of deterioration of water quality of Ganga at Shyampur Khadir was found due to the discharge of IDPL effluent which was highly polluted as evident by low level of DO- 2.8 and high degree of BOD- 181.0, COD- 261.8 mg/l and MPN- 209×10^2 and SPC- 380×10^3.

Keywords: Effluent, Water quality, Pollution, Drain.

Introduction

Humans use water in the home, in industry, in agriculture and for recreation. These applications differ widely in quantity and quality of the water which they require. In general consumption of water per capita daily is more higher in urban areas than rural areas all over the world. Since last two decades regular demand of water has much enhanced due to over growth of population and rapid increase in urbanisation which had a direct effect on demand of water quantity in terms of liter/ capita.

More food grains production is highly needed to fulfill nutritional requirement of the society and has forced to take up additional agricultural land for the purpose. Maximum consumption of water occurs in agriculture followed by domestic use and industry. Since a very little percentage of Freshwater, hardly 3 per cent is available for civic use and out of which only 0.2 per cent is in land water. The land surface water is a major source of civic supply in a large number of cities because these are located on the bank of the various rivers.

Inland surface water receives different kinds of pollutant from different sources in varying amount which has a direct effect on water quality of that aquatic system in terms of various physico-chemical and bacteriological counts. Practice of indiscriminate discharge of untreated or partially treated municipal and industrial wastewater has become a major cause of water pollution in India and abroad. Most of the rivers of country are being used for disposal of wastewater (domestic and industrial sewage) as raw sewage/or treated sewage. Since the maximum number of the cities including both Municipality corporation and Industries, either they do not possess treatment plant or they are not fit for adequate treament or not in working condition, resulting pouring of raw sewage directly at different points in the several rivers and Ganga is one of them.

Ganga receives contaminants form of raw sewage at different places throughout its course. Rishikesh which seems an entry point of river Ganga in the plains, contains large number of drains carrying city sewage and industrial effluent and pouring it in the river at different points. Albeit a large number of industries are found in the Rishikesh region but Indian drug pharmaceutical limited (IDPL) is a major source of pollution of river Ganga. A lot of information is available regarding impact of drains on water quality of receiving system. Some important contribution in this area in the recent past can be mentioned (Shanker *et al.*, 1986, Chopra and Rehman 1992, Chopra and Patric 2000, Prasad and Shankar 1999, Prasad *et al.*, 2003 and Khanna and Chugh 2004). To find out intensity of pollution load of IDPL effluent and their impact on the water quality of Ganga at Shyampur Khadir, the present investigation was carried out.

Materials and Method

Haridwar is located in North India at foot hills of Shivalik hill range at altitude of 294.15 meter from sea level. The area has a tropical climate with seasons winter, summer and rains in a year. Following sampling points were selected to find out the real impact of IDPL effluent on water quality of River Ganga at Shyampur Khadir about 1.5 km in the down stream.

Sampling Sites

1. Pashulok Barrage (up stream, without any out falls)
2. IDPL drain outfalls (carrying-IDPL effluent)
3. Shyampur Khadir (A point- showing impact of effluent on Ganga water after proper mixing dilutions)

Sample Collection Analysis

Water samples were collected from above mentioned sampling stations once in a month for estimation of various physical, chemical and bacteriological parameters. Samples were collected according to APHA (1985). DO was analysed at the spot by Winkler-idiometric method. Samples were collected in sterilized BOD bottles for bacteriological examination. These samples were brought to the laboratory and were analysed within 8 to 10 hours. Mac Conkey broth was used for estimation of most probable number(MPN) of total coliform bacteria and Nutrient agar media (NAM) for standard plate count of bacteria. pH of the medium was adjusted to 7.4. After inoculation and incubation at 35 °C MPN and SPC were recorded after 24 hrs. COD was estimated by autoclave method (1985). BOD was measured by seeding method. Chloride and sulphates were measured by titration method.

Results and Discussion

An average value of Physico-chemical and bacteriological characteristics studied for the period of three years at different sampling stations *i.e.* Pashulok Barrage, IDPL drain and Shyampur Khadir has been presented in the Table 15.1. Water quality in terms of average value was found better at Pashulok Barrage 8.8mg/l but most polluted water was recorded in IDPL drain, where it (DO) was found very low 2.8 mg/l (Table 15.1). Similarly other parameters of organic contaminants *i.e.*, BOD and COD were found very low at Pashulok Barrage 7.13 mg/l and 20.5 mg/l respectively, whereas these values were maximum recorded in IDPL effluent *i.e.* 181 mg/l and 261.8 mg/l. MPN and SPC were also found very low at Pashulok Barrage compared with IDPL drain. Low values of pollution indicator parameters at Pashulok Barrage were probably due to absence of any kind of discharge as well as minimum chances of soil erosion which contributes lot in raising of organic contaminates in the Ganga water. Besides this low temperature *i.e.*, 17.96 has also played the significant role in minimizing the bacterial multiplication and lowering the consumption of dissolved oxygen by bacterial population.

Table 15.1: Mean Values of Physico-Chemical and Bacteriological Characteristics for the Period of Three Years of IDPL Effluent and their Impact on Water Quality of River Ganges at Shyampur Khadir, Rishikesh

Parameters	*Sampling Stations*		
	Pashulok Barrage	*IDPL Drain*	*Shyampur Khadir*
Temperature (°C)	17.96	25.25	21.80
Turbidity (NTU)	7.00	95.00	7.00
pH	7.50	6.77	7.60
Conductivity (m mhos)	0.122	0.358	0.257
Hardness ($CaCO_3$) (mg/l)	103.43	191.00	137.80
Chloride (mg/l)	05.6	45.00	14.33
Sulphate (mg/l)	2.73	20.42	3.98
DO (mg/l)	8.80	2.80	5.33
BOD (mg/l)	7.13	181.00	58.80
COD (mg/l)	20.95	261.80	162.45
MPN (per 100ml)	11.16×10^2	209.3×10^3	64×10^2
SPC (per ml)	102.6×10^3	380×10^3	290×10^3

Albeit enhanced value of BOD,COD,MPN, SPC and low level of DO in IDPL effluent, was directly related with organic contaminants but it seems that temperature has also played the role of a governing factor for all above parameters as it was found more about 7°C from Pashulok Barrage. Higher temperature towards 30°C facilitates for the multiplication of most genera and species of bacteria on one hand and also enhances the consumption of dissolved oxygen. Therefore a high level of reduction of dissolved oxygen occurs which is evident by the recorded date *i.e.*, 8.80 > 2.80. Findings of different parameters related with organic pollution load of Shyampur Khadir indicated a high degree of deterioration of water quality at this sampling site compared with Pashulok Barrage. Since there was no any other discharge between IDPL factory and Shyampur Khadir except IDPL drain therefore it was only IDPL effluent which caused a serious loss of Ganges water in terms of water quality and it is evident by the recorded data at both sampling sites *i.e.*, IDPL drain and Shyampur Khadir (Table 15.1).

References

APHA, 1985. *Standard Methods for Examination of Water and Wastewater*, 15th edn. Amercian Public Health Association, Inc., New York, p. 1134.

Chopra, A.K. and Rehman, Abdul, 1992a. Effect of domestic sewage on physico-chemical qualities of the Ganga canal water at Jwalapur, Haridwar, India. *International J. Environmental Studies*, 41:183.

Chopra, A.K. and Patick, Nirmal, 2000. Effect of domestic sewage on self purification of Ganga water at Rishikesh. II. Microbiological parameters. *J. Natcon.*, 8: 81–86.

Khanna, D.R. and Chugh, T., 2004. *Microbial Ecology: A Study of River Ganga*. Discovery Publication House, New Delhi, pp. 285.

Prasad, G. and Shanker, V., 1999. Impact of Industrial and municipal wastewater discharging in to Ganga on water quality and algal population in Rishikesh and Haridwar. In: *Proceeding of Sustainable Ecosystem and Environment*, p. 133–146.

Prasad, G., Shanker, V. and Chopra, A.K., 2003. Environmental state of river Ganges in Rishikesh and Haridwar and its management. *Environment Conservation J.*, 4(1–3): 51–58.

Shanker, V., Sangu, R.P.S. and Joshi, G.C., 1986. Impact of distillery effluent on water quality and ecosystem of River Reh in Doon valley. *Poll. Res. J.*, 5(34): 137–142.

Water and Wastewater Management in 2 Vols (2011) *Pages 91–95*
Editors: **D.R. Khanna, R. Bhutiani & Gagan Matta**
Published by: **DAYA PUBLISHING HOUSE, NEW DELHI**

Chapter 16

Ganges Water Pollution and its Management through Religion

A.K. Chopra[1], G. Prasad[2] and D.R. Khanna[1]
[1]Department of Zoology and Environmental Sciences
[2]Department of Botany and Microbiology,
Gurukula Kangri University, Haridwar

ABSTRACT

The Ganges, a major river of the Indian subcontinent, associated with myth and reality of people, is becoming polluted on its way down to the Indian Ocean. The pollution starts from the beginning itself but the prominent cities which are adding substantially to the woes of the Ganges are Kanpur, Allahabad, Varanasi, Patna and Kolkata. The major polluting industries are the leather industries that use large amounts of Chromium and other chemicals, and much of it finds its way into the meager flow of the Ganges. Besides, a large volume of waste-estimated at nearly 1 billion liters of mostly untreated raw sewage is dumped in the river per day. Also inadequate cremation procedures contribute to a large of partially burnt or unburnt, corpse floating down the Ganga, not to mention livestock corpses. The 'Ganges Action Plan' was set up which formed a number of waste treatment plants and most of the work was limited to monitoring of the various parameters of the water from time to time. But today, there is an urgent need of arousing the religious sentiments of the people of the region who have a forages revered the Ganges not just as a river but as a Goddess. The self-purification of the people will help in maintaining the long-standing self-purification property of the river Ganges intact.

Keywords: Water, Pollution, Religion, Industries, River.

Introduction

The river Ganges is held as sacred and the symbol of India's age-old civilization and culture. It is a major river of the Indian subcontinent and is the lifeline of the people of India. The Ganges is unique because it is considered holy by the Hindu people of India. No river in the world plays a more important economic, social and cultural role in the lives of people than the Ganges. The river is also known as Maa Ganga or Mother Ganges and millions of people take a holy dip in the river each day. Millions of people rely on the river for their physical and spiritual sustenance. Also they believe that the water heals you from sin, and if you get cremated in the Ganges you are sure to get a better after-life.

The Ganges has an exalted position in the Hindus ethos. It is repeatedly invoked in the Vedas, the Puranas and the two Indian epics, the Ramayana and the Mahabhartha. It is believed that Bhagiratha, who after server austerities, propitiated the Ganga Devi, a Goddess and she agreed to come down to earth. Seeing the severity of impact of her fall that could be borne by none less than Shiva himself, Bhagirtha went into meditation again and obtained Shiva's consent after many more austerities. Finally, the river came down and fell into Shiva's matted hair, and hence to earth. This is the presumed site of the present-day temple at Gangotri.

Bhagiratha led the way on horse back and the river followed and an ocean formed from the waters there.This is the Sagar Island of today, where the Ganges flows into the Bay of Bengal ("Sagara" is also Sanskrit for ocean). Hindus use the water of the Ganga to cleanse any place or object for ritual purposes. Bathing in the Ganges is still the lifelong ambition of many of India's believing masses, and they congregate on its banks for the tremendously overcrowded Sangam, Sagar Mela or Kumbh Mela which are held on auspicious dates every few years.

The river emerges in spectacular fashion from an ice cave under the Gangotri glacier, a vast expanse of ice five miles by fifteen, at the foothills of the Himalayas (14000 ft) in North Uttarakhand. It is the source of Bhagirathi, which joins Alaknanda (originating nearby) to form Ganges at the craggy canyon-carved town of Devprayag. From Devprayag it goes to Rishikesh, Haridwar and then passing some of the most populous cities of India, including Kanpur, Allahabad, Varanasi, Patna and Kolkata and then to the Bay of Bengal flowing a distance of about 2507 Km. The largest tributary to the Ganges is the Ghaghara, which meets it before Patna, in Bihar, bearing much of the Himalayan glacier melt from Northern Nepal. The Gandak, which comes from near Kathmandu, is another big Himalayan tributary. Other important rivers that merge with the Ganges are the Son, which originates in the hills of Madhya Pradesh, the Gomti which flows past Lucknow, Yamuna, Chambal and Karvi.These tributaries are very useful in maintaining the volume of water in the river Ganga throughout the year.

There are two major dams on the Ganges,one at Haridwar diverts much of the Himalayan snowmelt into the Upper Ganges Canal, built by the British in 1854 to irrigate the surrounding land. This causes severe deterioration to the waterflow in the Ganges, and is a major cause for the decay of Ganges as an inland waterway. The other dam is a serious hydroelectric affair at Farakka, close to the point where the main flow of the river enters Bangladesh, and the tributary Hooghly (also known as Bhagirathi) continues in West Bengal past Kolkata. This barrage feeds the branch of the river by a 26-mile long feeder canal. On its way, the river is also fed by a dwindling supply of subsoil streams. These streams are drying up because much of the forest has been cut down.

Life-line

400 million people live in the Ganges river basin and depend on the river for their lives. The water is used extensively for irrigation, agriculture production, industries and domestic purpose.It also helps in maintaining water table of undergroundwater by recharging. Floods of the Ganges have enriched the soil. This makes the Ganges Valley a great agricultural region in India. Fishing also is an important use of the Ganges. The electricity is generated at the waterfalls of upper part of the Ganges.The river is also important for transportation, many cities get their drinking water from the Ganga. There

are many major industries along the Ganges. The important ones are food processing, clothes and other textiles, leather, paper and pulp, manufacturing of tools and chemical processing. In the lower part of the river there is metal and other mineral production.

Pollution

It is quite unfortunate that Ganges the lifeline is becoming polluted. The pollution has become a huge problem on the Ganges. The river is polluted by enormous quantities of human and industrial waste. The Disease-causing organisms and poisonous chemicals abound in her ever-sacred waters.

Many of the cities have steps called Ghats along the riverbank so the people can bathe in the water. They also wash their clothes in the river. At many places there are times where massive bathing (millions of people) takes place and millions of people take bathe, during a few days. People are cremated and dumped in the river because they believe that the river will purify them (sometimes cremated people are not burned well because they can't afford much kindle). It is common Hindu practice to scatter the ashes of the dead in the Ganges. The people along it dump sewage into the river. Factories dump chemicals and waste into the river. Narora nuclear power plant discharges its water into the lower Ganges canal. Agricultural chemicals wash into it. There's too much grazing of animals that make waste, which runs into the river.

The pollution begins soon after the Ganges begins its slow journey through North India planes, some165 miles downstream at Rishikesh,most of its dry-season flow is diverted to Canals, first at Haridwar and then near Aligarh. At the same time, town and industries discharge a large amount of waste in the sacred waters. The major polluting industries on the Ganges are the leather industries, especially near Kanpur,which use large amounts of Chromium and other chemicals and much of it finds its way into the meager flow of the Ganges. By the time the river leaves Kanpur, one of the big industrial centers along the river, the load of human, animals and industrial waste becomes overwhelming, threatening the rare species of fish, dolphins and soft-shell turtles. Allahabad is the sacred meeting point for the Ganges and the Yamuna, a river that becomes heavily contaminated as it oozes past the country's capital, Delhi. In ancient city of Varanasi, it is considered particularly auspicious to die here, where hundreds of corpses are burnt on the banks every day. Many bodies are pushed into the water without being burnt. However, industry is not the only source of pollution but a large volume of waste-estimated at nearly 1 billion liters of mostly untreated raw domestic sewage is dumped into the river per day. Also, inadequate cremation procedures contributes to a large number of partially burnt or unburnt corpses floating down the Ganges, not to mention livestock corpses. Thus pollution multiplies as the river moves downstream because very little freshwater is generated between towns. The global warming is also causing a decline in the snowfall on glacier, needed to replenish it.

The pollution of water can be a host of diseases-hepatitis, amoebic dysentery, typhoid, cholera and cancer-among the roughly 400 million people who live in the vast Gangetic basin.Waterborne diseases such as diarrhoea and cholera sap the lifeblood from the people causing death malnutrition, especially for children. Hindus believe that a dip in the holy Ganges will cleanse their souls of sin. But the pollution that bedevils the river could do untold damage to the bodies of the faithful who bathe in it. The people cannot escape the physical on/slaught of raw sewage, rotting carcasses, industrial effluent, fertilizers and pesticides that infect the river from the Himalayan foothills to the Bay of Bengal.

Ganga Action Plan

Alarmed by the rising level of pollution, an ambitious Ganga Action Plan was launched with great fanfare in 1986 by the Govt. of India to clean the river. But little has been achieved. Many sewage treatment plants were set up in major cities like Haridwar, Kanpur, Allahabad, Varanasi and Kolkata. Unfortunately treatment plants are not doing their job for which they were designed. The sewage is not being treated to their capacity. The plants are very energy intensive and expensive to run. The main plants remain off during the frequent energy blackouts, do not remove pathogenic bacteria and cause a build up of pollution alongside several villages. The plants need constant power supplies that are not available. It has met only 39 per cent of its primary target for sewage treatment. This results in sewage to back up throughout the city flowing out into the religious bathing areas, out through manhole covers and into the streets. The health problems are devastating.

Besides, less than half of the grossly polluting industrial units lining the 1,560 mile river had installed effluent treatment plants, and over 18 per cent of them did not function properly. " The authorities claim that almost all industrial units are diverting their effluent to treatment plants but the ground reality is that their claim is exaggerated. The experts say the level of pollution in the river has been contained, even if it has not been eliminated. Electric crematoria were built but not in use due to religious sentiments. Meanwhile, the government is planning to build more than 50 dams and barrages to regulate the river-flow, supply water and generate power. The largest of them near Tehri has already attracted controversy amid concerns about safety and the environment.

Fighting Pollution with Religion

Religion is the most powerful tool of the Indian civilization. Not from today but from time immemorial religion has remained one of the most important aspects of the people of this land. Only an individual, that is,one not divided against one self, can imbibe this humanizing state and claim to be religious. Without the stars in the sky and the sun rising behind the hill, without that woman with a child in her lap, and the man in an ancient turban passing by, without the glint of cheer in children's eyes, there can be no religion worth the name. Religion is the symbiotic relationship between man and his environment, promising a homogenized congenial ambience. The religious sentiments of the people have been responsible for each and every revolution that has ever taken place in this land of Gods. Today also there is need of a revolution not in the outer world but in the hearts and minds of the people who have for long worshipped the Ganges not only as their lifeline but as a Goddess who looks after her children just like a benevolent mother. Today this sentiment has to be aroused in each and every Hindu that they have now troubled their Mother a bit too much and that she is in a lot of pain.

Religion may succeed where the government has not. It is imperative to revive the religion feelings among the society so that besides that Government efforts to enforce using the scientific application for prevention and control of Ganges pollution, the guidelines given by ancient Rishis/Sages for prevention and control of pollution are self-generated among the people. A large number of norms have been suggested in the Hindu's sacred literature. It has been suggested that there should not be any bank-side construction on river Ganges. A few examples of the instructions given in sacred literature are quoted here.

न दन्तधावन कुयादि गंगा गर्भे विचक्षणः
परिधेयाम्बराम्बूनि गंगा प्रेतसिनत्यजेत्।
पदमपुपाण'

i.e. Neither should one brush his teeth nor should one wash his clothes in the holy river Ganges.

मलं प्रक्षालयेत्तीरे ततः स्नान समाचरेत ।
'मेधानिति'

i.e. One should take bath in the river only after defecating elsewhere.

Thus, taking the example from such religious scriptures, the religious teachers should awaken the people about maintaining the sanctity of the river Ganges.Further, the religious heads of different organizations should come forward in cleaning and management of Ganga as they can catch the sentiments of people. People have great faith in them and follow whatever they preach. One of the most gallant examples of such a revolution spearheaded by religion was the effort made by the people of Shanti Kunj, Haridwar way back in 2000 when they lead their volunteers to clean the Ganges strip of Haridwar without any Govt. aid. Their efforts of 2000 were really commendable. Today there is a need of not only initializing such programs but also to see to it that these programs continue till each and every patch of this sacred river is cleared from dirt and also each and every national starts feeling it as his or her responsibility to keep their Goddess pure. The religious organization can facilitate in running environmental education and training centers/programs and can promote active participation of pilgrims, riverside local villages, schools, community-based organizations in the cleaning and protection of the river. The religious sentiments of purifying the self should be aroused. The self-purification of the people will help not only in cleaning the Ganges water but also in maintaining the long-standing self-purification of the river Ganges intact.

References

Anderson, J.W., 1992. The great clean up of the holy Ganges. *The Washington Post*, September 25.

CD-ROM, 1995. *Ganges River*. Compton's Interactive Encyclopedia.

CD-ROM, 1995. *Ganges River*. Grolier Electronic Encyclopedia.

CD-ROM, 1996. *India*. World Book Electronic Encyclopedia.

Chaphekar, S.B. and Mhatre, G.N., 1986. *Human Impact on Ganges River Ecosystem*. Concept Publishing Company, New Delhi.

Mago, C., 2001. All plan: No action on Ganga. *The Hindustan Times*, New Delhi.

Singh, R., 1992. *The Ganges.* Aperture, New York.

Water and Wastewater Management in 2 Vols (2011) *Pages* ***96–106***
Editors: **D.R. Khanna, R. Bhutiani & Gagan Matta**
Published by: **DAYA PUBLISHING HOUSE, NEW DELHI**

Chapter 17
Rainwater Harvesting

Mukesh Ruhela[1], R. Bhutiani[2] and Aradhana Anand[1]
[1]Department of Environment Sciences, Guru Jambeshwar University, Hissar
[2]Department of Zoology and Environmental Science, Gurukul Kangri Vishawavidyalaya, Haridwar

Introduction

As we all know that about 70 per cent of the earth is water but most part of the earth is ocean. Only 3 per cent of total water available on earth is freshwater. Out of the three percent, only 1 per cent is accessible as surface freshwater where as the rest 2 per cent is locked away in the form of ice caps and glaciers in the polar regions. 1 per cent water is regularly renewed by rain fall and other means and thus available on a sustainable basis and easily considered accessible for human use.

Increasing population together with the intensifying urbanization and industrialization affects availability of freshwater both qualitatively and quantitatively. Water is clearly the single largest problem facing India today, though the country was once upon a time categorized as a water rich society because of non-uniform distribution of rainfall. Over the years about 90 per cent of its annual rainfall occurs during the summer monsoon *i.e.* from July to September whereas the rest of the months the country get less rainfall. Even India is wettest but still there is water scarcity. Water scarcity will be one of the major threats to humankind during this century (Prinz, 2000). The problem of water shortage in arid and semi-arid regions is low rainfall and uneven distribution through out the season, which makes rain fed agriculture a risky enterprise.

There were lot of methods which were applied by ancient people, as the available water resources taken from streams, rivers and groundwater will not be sufficient in most dry areas to cover the needs of agriculture and urban areas, we have to reassess the value of certain traditional irrigation methods, to find out their value to ease future water scarcity (Prinz and Singh, 2000).These traditional methods played a much greater role in the past and were the backbone of ancient civilization in arid and semi-arid areas around the world (Agarwal and Narain, 1997; Prinz 1996). Therefore new interest came up in recent decades to evaluate traditional water management techniques (Prinz *et al.*1999) most of them being simple, sure to implement and of low capital investment. The classical sources of irrigation water are often at the break of overuse and therfore untapped sources of (irrigation) water have to be sought for increasing agricultural productivity and providing sustained economic base. Water harvesting for dry-land agriculture is a traditional water management technology to ease future water scarcity in many arid and semi-arid regions of world.

Older Concept

The Indus Valley Civilization, that flourished along the banks of the River Indus and other parts of Western and Northen India about 5,000 years ago, had one of the most sophisticated urban water supply and sewage systems in the world. The fact that the people were well acquainted with hygiene can be seen from the covered drains running beneath the streets of the ruins at both Mohenjodaro and Harappa. Another very good example is the well-planned city of Dholavira, on Khadir Bet, a low plateau in the Rann in Gujarat. One of the oldest water harvesting systems is found about 130 km from Pune along Naneghat in the Western Ghats. A large number of tanks were cut in the rocks to provide drinking water to tradesman who used to travel along this ancient trade route. Each fort in the area had its own water harvesting and storage system in the form of rockcut cisterns, ponds, tanks and wells that are still in use today. A large number of forts like Raigarh had tanks that supplied water.

In ancient times, houses in parts of western Rajasthan were built so that each had a rooftop water harvesting system. Rainwater from these rooftops was directed into underground tanks. This system can be seen even today in all forts, palaces and houses of the region.

Underground baked earthen pipes and tunnels to maintain the flow of water and to transport it to distant places, are still functional at Burhanpur in Madhya Pradesh, Golkunda and Bijapur in Karnataka, and Aurangabad in Maharashtra.

Traditional Method

Traditional rainwater harvesting, which is still prevalent in rural areas, was done in surface storage bodies like lakes, ponds, irrigation tanks, temple tanks *etc.* In urban areas, due to shrinking of open spaces, rainwater will have to necessarily be harvested as groundwater, hence harvesting in such places will depend very much on the nature of the soil *viz.* clayey, sandy *etc.* The below listed are the various kinds of traditional rainwater harvesting methods.

Kunds

In the sandier tracts, the villagers of the Thar Desert had evolved an ingenious system of rainwater harvesting known as kunds or kundis. The first known construction of a kund in Western Rajasthan was during 1607 AD by one Raja Sursingh in village Vadi-ka-Melan, in the Mehrangarh Fort in Jodhpur. Kund, the local name given to a covered underground tank, was developed primarily for tackling drinking water problems. Usually constructed with local materials or cement. Kunds were more prevalant in the western arid regions of Rajasthan, and in areas where the limited groundwater available was moderate to highly saline. Groundwater in Barmer, for instance, nearly 76 per cent of the district's area, has total dissolved salts (TDS) ranging from 1,500-10,000 parts per millon (ppm).Under such conditions kunds provided convenient, clean and sweet water for drinking. Kunds were owned by communities or privately, which the rich having one or more kunds of their own. Community kunds were built through village co-operation or by a rich man for the entire community. The kund consists of a saucer-shaped catchments area with a gentle slope towards the center where a tank is situated. Openings or inlets for water to go into the tank are usually guarded by a wire mesh to prevent the entry of floating debris, birds and reptiles. The top is usually covered with a lid from where water can be drawn out with a bucket. Kunds are large circular in shape with little variation between the depth and diameter which ranges from 3.0-4.5 m. Lime plaster or cement is typically used for the construction of the tank, since stone as a building material is not always available and is relatively more expensive. Either of these material can be used to plaster the horizontal and vertical soil surfaces, although cement ensures a longer life span. The success of a kund depends on the selection of the site, particularly its catchments characteristics. An adequately large catchments area has to be selected or artificially prepared to produce adequate runoff to meet the storage requirements of the kund.

Kul Irrigation Method

Spiti is an important trading post on the route connecting Ladakh and the plains of Himachal Pradesh. Villages in the Spiti subdivision are located between 3,000 m and 4,000 m, which means they are snowbound six months a year. Rainfall is negligible in spiti because it is a rain shadow area. Spiti's lunar-like terrain was transformed into an agrarian success story by an ingenious system, devised centuries ago to tap distant glaciers for water. But shortsighted developmental policies, though well-intentioned, now threaten both this unique irrigation system and the social consciousness that spawned it. The soil is dry and lacks organic matter. But, despite these handicaps, the Spiti valley has been made habitable and productive by human ingenuity. But Spiti's unique contribution to farming is kul irrigation, which utilises kuls (diversion channels) to carry water from glacier to village. The kuls often span long distances, running down precipitous mountain slopes and across crags and crevices. Some kuls are 10 km long, and have existed for centuries. The crucial portion of a kul is its head at the glacier, which is to be tapped. The head must be kept free of debris, and so the kul is lined with stones to prevent clogging and seepage. In village,the kul leads to a circular tank from which the flow of water can be regulated. For example, when there is need to irrigate, water is let out of the tank in a trickle. Water from the kul is collected through the night and released into the exit channel in the morning. By evening, the tank is pratically empty, and the exit is closed. This cycle is repeated daily. The culture also is instrumental in maintaining the carrying capacity of the surrounding cultivable land. However, this system, carefully nurtured through the centuries, now runs the risk of being upset through government intervention.

Bamboo Rainwater Harvesting

In Meghalaya an ingenious system of tapping of stream and spring water by using bamboo pipes to irrigate plantations is widely prevalent. It is so perfected that about 18-20 litres of water entering the bamboo pipe system per minute gets transported over several hundred meters and finally gets reduced to 20-80 drops per minute at the site of the plant. The tribal farmers of Khasi and Jaintia hills use the 200-year-old system. The bamboo drip irrigation system is normally used to irrigate the betel leaf or black pepper crops planted in areca nut orchards or in mixed orchards. Bamboo pipes are used to divert perennial springs on the hilltops to the lower reaches by gravity. The channel sections, made of bamboo, divert and convey water to the plot site where it is distributed without leakage into branches, again made and laid out with different forms of bamboo pipes. Manipulating the intake pipe positions also controls the flow of water into the lateral pipes. Reduced channel sections and diversion units are used at the last stage of water application. The last channel section enables the water to be dropped near the roots of the plant.Bamboos of varying diameters are used for laying the channels. About a third of the outer casing in length and internodes of bamboo pieces have to be removed while fabricating the system. Later, the bamboo channel is smoothened by using a dao, a type of local axe which is a round chisel fitted with a long handle. Other components are small pipes and channels of varying sizes used for diversion and distribution of water from the main channel. About four to five stages of distribution are involved from the point of the water diversion to the application point.

Temple Tank

People in ancient time used to store water in temples and other religious places, this was another method for storing water.

As the appropriate choice of technique depends on the amount of rainfall and its distribution, land topography, soil type and soil depth and local socio-economic factors, these systems tend to be very site specific. The water harvesting methods applied strongly depend on local conditions and include such widely differing practices as bunding, pitting, micro catchments water harvesting, flood water and groundwater harvesting (Prinz, 1996; Critchley and Siegert, 1991).

"the 9th IRCSC (International Rainwater Catchement System Conference), held in Brazil in July 1999, Dr. A. Appan said:

"The concept of rainwater catchment system technology is as old as the mountains. The standard adage- as in all water supply schemes is- store water (in a tank reservoir) during the rainy season so that you can use it when you need it most during the summer. In other words 'Save for the dry day!' The principle, methods of construction, usage and maintenance are all available. And, most important of all, there are many financial models to suit developing and developed countries. What is most needed is the moral acceptance of the technology and the political will to implement the system."

A non-conventional approach that is gaining significance as an effective long term strategy for supplementing traditional sources of freshwater supplies is the harvesting and utilization of rainwater.

Water Harvesting

Suface water is inadequate to meet our demand and we have to depend on groundwater.Due to rapid urbanization, infiltration of rainwater into the sub-soil has decreased drastically and recharging of groundwater has diminished. Under such circumstances conservation of water by improved technology based on traditional concept has become the need of the hour.

Types of Water Harvesting

A brief description of these water-harvesting techniques along with sub-types is given below:

Rainwater Harvesting

Rainwater harvesting is defined as a method for inducing, collecting, storing and conserving local surface runoff for agriculture in arid and semi-arid regions (Boers and Ben-Asher, 1982). Rainwater harvesting covers three types of water harvesting.

1. Water collected from roof tops, courtyards and similar compacted or treated surfaces is used for domestic purpose or garden crops.
2. Microcatchment water harvesting is a method of collecting surface runoff (sheet or rill flow) from a small catchment area and storing it in the root zone of an adjacent infiltration basin. The basin is planted with a single tree or bush or with annual crops (Prinz, 2001).
3. Macrocatchment water harvesting is also called " Water harvesting from long slopes" or "harvesting from external catchment systems" (Pacey and Cullis, 1988). In this case, the runoff from hillslope catchments is conveyed to the cropping area, which is located below the hill foot on flat terrain.

Flood Water Harvesting

Can be defined as the collection and storage of creek flow for irrigation use. Flood water harvesting, also known as 'large catchment water harvesting' or 'Spate Irrigation', may be classified into following two forms:

1. *Floodwater harvesting within stream bed*: In this case the water flow is dammed and as a result, inundates the valley bottom of the flood plain. The water is forced to infiltrate and the wetted area can be used for agriculture or pasture improvement.
2. *Floodwater diversion*: In this case the wadi water is forced to leave its natural course and conveyed to nearby cropping fields.

It is difficult to give exact figure on the present area under various forms of floodwater harvesting system. Pakistan has more than 1.5 millon ha under floodwater harvesting. Such method of floodwater harvesting is adopted in countries of North Africa and Middle East regions.

Groundwater Harvesting

It is a rather new term and employed to cover traditional as well as unconventional ways of groundwater extraction. Qanat system, underground dams and special types of wells are few examples of the groundwater harvesting techniques.

Qanats, widely used in Iran, Pakistan, North Africa and Spain consists of a horizontal tunnel that taps undergroundwater in an alluvial fan, brings it to the surface due to gravitational effect. Qanat tunnels have an inclination of 1-2 per cent and a length of up to 30 km. Many are still maintained and deliver steadily water to fields for agriculture production and villages for drinking water supply.

Groundwater dams like 'Subsurface Dams' and 'Sand Storage Dams' are other fine example of groundwater harvesting. They obstruct the flow of streams in a river bed there by water is stored in the sediment below ground surface and can be used for aquifer recharge. Sand filled reservoirs have the following advantages:

1. Evaporation losses are reduced.
2. No reduction in storage volume due to siltation.
3. Health hazards due to mosquito breeding are avoided.

Groundwater harvesting does not play the same role globally as rain and floodwater harvesting.

Underground Storage

As several disadvantages are connected with surface storage of water-large evaporation losses, loss of storage caused by siltation, pollution problems and loss of agricultural land, underground storage of water may be an interesting alternative. This storage can be done in near surface aquifers (*e.g.* in wadi beds), calling for a conjunctive management of water resources, or in cisterns. Cisterns are man-made caves or underground constructions to store water. Often the walls of these cistern are plastered, their water losses by deep percolation or by evaporation can be minimal. The construction of cisterns was already practiced several thousand years ago.

Traditionally, in Mediterranean houses, one cellar room was specifically designed to store rainwater. Similar in-house cisterns are known from Rajasthan, NW India. In the same region, 'Kunds', covered underground tanks with a plastered catchment, are found (Agarwal and Narain, 1997). Now a days cisterns are often constructed using concrete.

Modern Methods for Rainwater Harvesting

The Modern methods of rainwater harvesting are categorised under two, they are Artifical Recharging and Rainwater Harvesting.The former is classified into Absorption Pit Method, Absorption Well Method, Well cum Bore Method and Recharge trench cum injection well. The later is categorised into Individual Houses and Grouped Houses which are further classified into Percolation Pit Method, Bore Well with Settlement Tank, Open Well Method with filter bed Sump and Percolation Pit with Bore Method. In the Indian region over the last few millennia, such climate fluctuations may have given rise to traditional village tanks, ponds and earthern embankments numbering more than 1.5 million, that still harvest rainwater in 660,000 villages in India (Pandey, 2001).

Artificial Recharging

1. Absorption Pit Method
2. Absorption Well Method
3. Well cum Bore Method
4. Recharge trench cum injection well

Absorption Pit Method

A percolation/absorption pit is a hand bore made in the soil with the help of an augur and filled up with pebbles and river sand on top. The depth of these pits will be anywhere between 4 and 8 meters depending on the nature of the soil. If the soil is clayey, the pit has to be dug to a depth till a reasonable sandy stratum is reached. The diameter of these pits will be 25 cm (10 inches). A square/circular collection chamber with silt arrester is provided at the top.

Absorption Well Method

These wells are constructed using cement rings readily available in the market. The diameter of these rings range from 2ft. to 6 ft. The depth up to which these wells are dug depends on the nature of the soil and the diameter depends on the number of roof top pipes that are likely to be connected to each one of these wells. These wells are left unfilled and are covered with RCC slabs of suitable thickness to facilitate free pedestrian and vehicular movement on the ground.

Well-cum-Bore Method

In areas where the soil is likely to be clayey up to say 15ft. and more, it is advisable to go in for a percolation well up to 10 ft. or 15 ft. and a hand bore pit within this well up to a depth of 10 ft. to 15 ft. from its bottom. A PVC pipe of 5 inch diameter is inserted into the bore for the entire length.

Recharge Trench-cum-Injection Well

This technique is ideally suited for areas where permeable sandy horizons within 3 to 5m. below ground level and continues up to the water level under unconfined conditions, by which copious water available can be easily recharged.

In this technique, 1 to 2m wide and 2 to 3m deep trench is excavated, the length of which depends on the site availability and volume of water to be handled. An injection,well of 100 to 150 mm diameter is constructed, piercing through the layers of impermeable horizons to the potential aquifer reaching about 3 to 5 meters below water levels (1 to 10m) from the bottom of the trenches. Depending upon the volume of water to be injected, the number of injection wells can be increased to enhance the recharging rate.

Rainwater Harvesting

Rainwater can be harvested from the following surfaces:

Rooftops

If buildings with impervious roofs are already in place, the catchment area is effectively available free of charge and they provide a supply at the point of consumption. Paved and unpaved area *i.e.*, landscapes, open fields, parks,storm water drains, roads and pavements and other open areas can be effectively used to harvest the runoff. The main advantage in using ground as collecting surface is that water can be collected from a larger area. This is particularly advantageous in area of low rainfall.

Waterbodies

The potential of lakes, tanks and ponds to store rainwater is immense. The harvested rainwater can not only be used to meet water requirements of the city, it also recharges groundwater aquifers.

Stormwater Drains

Most of the residential colonies have proper network of stormwater drains. If maintained neatly, these offer a simple and cost effective means for harvesting rainwater.

Recharge Structures

Rainwater may be charged into the groundwater aquifers through any suitable structures like dugwells, borewells, recharge trenches and recharge pits.

Various recharge structures are possible which promotes the percolation of water through soil strata at shallower depth (*e.g.* recharge trenches, permeable pavements) whereas others conduct water to greater depths from where it joins the groundwater (*e.g.* recharge well). At many locations, existing structures like wells, pits and tanks can be modified as recharge structures, eliminating the need to construct any structures afresh. Here are a few commonly used recharging methods.

Recharging of Dugwells and Abandoned Tubewells

In alluvial and hard rock areas, there are thousands of wells which have either gone dry or whose water levels have declined considerably. These can be recharged directly with rooftop run-off. Rainwater that is collected on the rooftop of the building is diverted by drainpipes to a settlement or filtration tank from which it flows into the recharge well (borewell or dugwell).

If a dugwell is used for recharge, the well lining should have openings (weep-holes) at regular intervals to allow seepage of water through the sides. Dugwells should be covered to prevent mosquito breeding and entry of leaves and debris. The bottom of recharge wells should be desilted annually to maintain the intake capacity.

Providing the following elements in the system can ensure the quality of water entering the recharge wells:

1. Filter mesh at entrance point of rooftop drains
2. Settlement chamber
3. Filter bed

Settlement Tank

Settlement tanks are used to remove silt and other floating impurities from rainwater. A settlement tank is like an ordinary storage container having provisions for inflow (bringing water from the catchment), outflow (carrying water to the recharge well) and overflow. A settlement tank can have an unpaved bottom surface to allow standing water to percolate into the soil.

In case of excess rainfall, the rate of recharge, especially of borewells, may not match the rate of rainfall. In such situations, the desilting chamber holds the excess amount of water till it is soaked up by the recharge structure. Thus, the settlement chamber acts like a buffer in the system.

Any container (masonry or concrete underground tanks, old unused tanks, pre-fabricated PVC or ferro-cement tanks) with adequate capacity of storage can be used as a settlement tank.

Recharging of Service Tubewells

In this case the rooftop runoff is not directly led into the service tubewells, to avoid chances of contamination of groundwater. Instead rainwater is collected in a recharge well, which is a temporary storage tank (located near the service tubewell), with a borehold,which is shallower than the water table. This borehole has to be provided with a casing pipe to prevent the caving of soil, if the strata is loose. A filter chamber comprising of sand, gravel and boulders is provided to arrest the impurities.

Recharge Pits

A recharge pit is 1.5m to 3m wide and 2m to 3m deep. The excavated pit is lined with a brick/ stone wall with openings (weep-holes) at regular intervals. The top area of the pit can be covered with a perforated cover. Design procedure is the same as that of a settlement tank.

Soakaways/Percolation Pit

Percolation pits, one of the easiest and most effective means of harvesting rainwater, are generally not more than 60x 60x 60 cm pits, (designed on the basis of expected runoff as described for settlement tanks), filled with pebbles or brick jelly and river sand, covered with perforated concrete slabs wherever necessary.

Recharge Trenches

A recharge trench is a continuous trench excavated in the ground and refilled with porous media like pebbles, boulders or broken bricks. A recharge trench can be 0.5 m to 1 m wide and 1 m to 1.5 m deep. The length of the recharge trench is decided as per the amount of runoff expected. The recharge trench should be periodically cleaned of accumulated debris to maintain the intake capacity. In terms of recharge rates, recharge trenches are relatively less effective since the soil strata at depth of about 1.5 meters is generally less permeable. For recharging through recharge trenches, fewer precautions have to be taken to maintain the quality of the rainfall runoff. Runoff from both paved and unpaved catchments can be tapped.

Recharge Troughs

To collect the runoff from paved or unpaved areas draining out of a compound, recharge troughs are commonly placed at the entrance of a Residential/Instutional complex. These structure are similar to recharge trenches except for the fact that the excavated portion is not filled with filter materials. In order to facilitate speedy recharge, boreholes are drilled at regular intervals in this trench. In design part, there is no need of incorporating the influence of filter materials.

This structure is capable of harvesting only a limited amount of runoff because of the limitation with regard to size.

Modified Injection Well

In this method water is not pumped into the aquifer but allowed to percolate through a filter bed, which comprises sand and gravel. A modified injection well is generally a borehole, 500 mm diameter, which is drilled to the desired depth depending upon the geological conditions, perferably 2 to 3 m below the water table in the area. Inside this hole a slotted casing pipe of 200 mm diameter is inserted. The annular space between the borehole and the pipe is filled with gravel and developed with a compressor till it gives clear water. To stop the suspended solids from entering the recharge tubewell, a filter mechanism is provided at the top.

Diversion of Run Off into Existing Surface Water Bodies

Construction activity in and around the city is resulting in the drying up of water bodies and reclamation of those tanks for conversion into plots for houses. Free flow of storm run off into these tanks and water bodies must be ensured. The storm run off may be diverted into nearest tanks or depression, which will create additional recharge.

Formula Used for Potential of Harvested Water

Water harvesting potential = Rainfall (mm) x Area of catchment x Runoff coefficient

Benefits of Rainwater Harvesting

Development and augmentation of freshwater supplies through installation of Rainwater harvesting systems will offer following benefits:

1. It provides relatively high quality water, soft and low in minerals at low cost
2. Direct capturing of rainwater significantly reduces our reliance on water from dams/ reservoirs and canal systems. This will exert less pressure on national storage capacity at marco level and can reduce the need to expand dams or build new ones
3. Rainwater harvesting promotes self-sufficiency and fosters an appreciation for water as a resource. It also promotes water conservation.
4. Local erosion and flooding from impervious cover associated with buildings is lessened, as a portion of local rainfall is diverted into collection tanks.
5. Encourages households and institutions to be well equipped with an onsite and decentralized water supply of reliable quality.

Disadvantages

The technical progress of the 19th and 20th century occurred mostly in the so called developed countries in moisture moderate climate zones without a need for rainwater harvesting. Further more emphasis was put on big dams, groundwater development and piped irrigation projects with high input of fossil energy and electricity, this was another reason that rainwater-harvesting techniques have been set aside or totally forgotten. The technical progress of the 19th and 20th century occurred mostly in the so called developed countries in moisture moderate climate zones without a need for rainwater harvesting. As a consequence of colonialization agricultural practices from moderate climate zones were implanted in drier climate zones. Short comings of rainwater harvesting technologies are mainly due to the limited supply and uncertainly of rainfall. Adoption of this technology requires a bottom up approach rather than the more usual top dowm approach employed in other water resources development projects. This may make rainwater harvesting less attractive to some governmental agencies tasked with providing water supplies in developing countries, but the mobilization of local government and NGO resources can serve the same basic role in the development of rainwater-based schemes as water resources development agencies in the larger, more traditional public water supply schemes.

Conclusion

Rainwater harvesting has the potential to increase the productivity of arable and grazing land by increasing the yields and by reducing the risk of crop failure. They also facilitate re/or afforestation, fruit tree planting or agroforestry. With regard to tree establishment, rainwater harvesting can contribute to the fight against desertification. Water harvesting technology is especially relevant to the semi-arid and arid areas where the problems of environmental degradation, drought and population pressures are most evident. It is an important component of the package of remedies for these problem zones. Most of techniques are relatively cheap and can therefore be a viable alternative where irrigation water from other sources is not readily available or too costly. Unlike pumping water, water harvesting saves energy and maintenance costs. Using harvested rainwater helps in decreasing the use of other valuable water sources like groundwater, there is no doubt that implementation of water harvesting techniques will expand. Remote sensing and Geographical Information Systems can help in the determination of areas suitable for water harvesting (Prinz *et al.*, 1998).

References

Agarwal, A. and Narain, S., 1997. During wisdom: Rise, fall and potential of India's traditional water harvesting systems. *State of India's Environment Report* 4. Center for Science and Environment, New Delhi.

Appan, Adhityan, 1999. Opening Address at the 9th International Rainwater Catchment System Conference at Petrolina, Brazil, 6 to 9 July 1999.

Critchley, W.R.S. and Reij, C., 1989. Water harvesting for plant production. Part 2. Case studies and conclusions from Sub-Saharan Africa. (Draft)

Evanari, M., Shanan, L. and Tadmor, N.H., 1971. *The Negev: The Challenge of a Desert*. Harvard University Press, Cambridge, Mass.

http:/www. rainwaterharvesting.org/Urban/Urban. htm

http:/www.rainwaterharvesting.org/Urban/Components.htm

http:/www.eng.warwick.ac.uk/ircsa/7th.html#10

Pandey, D.N. 2001 A bountiful harvest of rainwater. *Science*, 293, 1763.

Prinz, D., 1996. *Water Harvesting: History, Techniques, Trends*. Z.f. Bewaessrungswirtschaft, 31(1): 64–105.

Prinz, D., 2000. Global and european water challenges in the 21st century. Keynote Speech. In: *Proceedings, 3rd Inter-Regional Conference on Environment, Water "Water Resources Management in the 21st Century"*, 1–3 June, Budapest/Hungary, pp. 247–254.

Prinz, D., 2001. Water harvesting for afforestation in dry areas. In: *Proceedings of 10th International Conference on Rainwater Catchment Systems*, Mannheim, 10–14 Sept., pp. 195–198.

Prinz, D. and Wolfer, S., 1999. Traditional techniques of water management to cover future irrigation water demand. Z.f. Bewasserungswirtschaft, 34 (1): 41–60. ISSN 0049–8602. *A Water Harvesting Manual for Urban Areas.*

Pacey, A. and Cullis, A., 1986. *Rainwater Harvesting*. The Collection of Rainfall and Runoff in Rural Areas. Intermediate Technology Publ., London.

Rao, M. Mohan, 1999. District collector of Dewas, Madhya Pradesh, while inaugurating the Bhoojal Samvardhan Mission, Dewas, on May 28, 1999. Source: Jeyakumar; Rainwater Harvesting Manual P–21

Srivastava, Karun C. Keynote Address on *Rainwater Harvesting*, IAS. Municipal Commissioner, Municipal Corporation of Greater Mumbai.

Water and Wastewater Management in 2 Vols (2011) *Pages* **107–110**
Editors: D.R. Khanna, R. Bhutiani & Gagan Matta
Published by: DAYA PUBLISHING HOUSE, NEW DELHI

Chapter 18

Epilithic Periphyton and Detritus Ecology of the Spring-fed Stream Eastern Nayar in Garhwal Himalaya

V.P. Balodi, Anoop K. Dobriyal, H.K. Joshi, S.P. Uniyal and Anju Thapliyal
Department of Zoology, HNB Garhwal University Campus, Pauri Garhwal – 246 001, Uttarakhand

ABSTRACT

The chapter deals with the study of epilithic periphyton and detritus standing stock in the river Eastern Nayar, which forms an important part of the food for macrozoobenthos and the fish, thus contributing greatly to the productivity of the stream.

Keywords: Stream, Lotic, Algal, Biomass, Detritus.

Introduction

Epilithic periphyton and detritus plays an important role in the lotic environments where the current does'nt allow plankton to stay on in a considerable quantity. The fish inhabiting these streams mostly feed on the macrozoobenthos and the phytobenthos along with the detritus. It is the riparian vegetation, which usually contributes the detritus. Eastern Nayar is the parent stream of the Nayar that originates in the Doodhatoli peak at an elevation of about 3116 masl and confluences with other stream, the Western Nayar at Nagun Kamand near Satpuli at an elevation of about 650 masl.It is rich in carni-omnivore, omnivore and detritivore fishes. The Nayar basin is one of the important drainage basin in the Ganges river system and consists of 1997 km^2 area extending from 29° 45' to 30°15' latitude and 78°34' to 79°12' longitude (Survey of India toposheet No. 53 N/4, 0/1, J/12, J/16, K/13, K/14 and

K/9).The authors while conducting studies on the food and feeding biology of fishes inhabiting the Nayar river observed that the epilithic periphyton and the detritus is main food of many fishes, hence it was decided to conduct some ecological studies on these two neglected aspects.

Materials and Method

The physico-chemical parameters were analysed by the standard methods (Welch, 1948, APHA,1975). For the quantitative estimation of periphytic algal biomass, the sample was taken from a known surface area, in centimeter square from each sampling spot. After scrapping, the periphytic algal was preserved in 5 per cent formalin solution and was brought to the laboratory for further analysis. In laboratory, the algae collected from known area (*i.e.* 1cm^2 or otherwise calculated accordingly) was concentrated in either 50 ml or 100 ml depending upon its concentration. The counting was made in Sedgwick-Rafter counting chamber. Calculation was made as follows:

$$n = (a \times 1000) \times b$$

where,

n: Number of units of periphyton/cm^2

a: Average number of periphyton in 1 chamber of 1 mm^3 capacity

b: Concentration prepared in ml.

For dry biomass, another sample of periphytic algae from 1cm^2 area was collected and dried at 105°C in oven.For Ash free dry weight biomass, the same sample was further dried at 550°C in a muffle furnace and biomass was measured. The estimation of the standing detritus associated with one meter square substratum up to a depth of approximately 15cm. was done. In the laboratory, detritus was oven dried to constant weight (at 70° C). The final values were expressed in gram Dry weight/m^2.

Results and Discussion

The seasonal values of physico-chemical parameters in the Eastern Nayar are presented in the Table 18.1 and the data relating to epilithic periphyton and detritus is presented in the Table 18.2.

Table 18.1: Seasonal Variations in Physico-chemical Parameters of the Eastern Nayar during 1996 [Average of spot no.1 (Thalisain) and spot No. 2 (Satpuli)]

Parameters	*Season*			
	Winter	*Summer*	*Monsoon*	*Autumn*
Water Temp. (°C)	11.45±2.43	19.85±4.67	24.07±3.89	16.3±5.44
Velocity (m/sec)	0.33±0.059	0.40±0.112	0.563±0.194	0.33±0.056
Turbidity (NTU)	7.7±1.79	9.90±1.14	59.87±42.96	22.4±14.91
pH	8.0±0.209	7.75±0.104	7.43±0.175	7.77±0.233
Total Hardness (ppm)	75.23±1.62	69.48±4.36	72.25±9.99	73.32±2.57
DO (ppm)	11.35±0.33	10.55±0.24	9.15±0.72	10.62 ±0.28
Free CO_2 (ppm)	Nil	Nil	1.43±0.79	0.085±0.134
Total Alkalinity (ppm)	36.45±3.49	31.8±5.07	25.52±4.39	35.75±4.08
Nitrates (ppm)	0.037±0.004	0.048±0.015	0.073±0.010	0.055 ±0.008
Phosphate (ppm)	0.024±0.005	0.030±0.007	0.041±0.009	0.037±0.004

Table 18.2: Seasonal Values of Periphytic Algae and its Biomass, and Detrius (Standing stock) in g DW/m² in the Upper Stretches of Eastern Nayar during 1996

Season	*Periphytic Algae*	*Biomass*		*Seasonal Value of Detritus Standing Stock*
	Units/cm²	*Dry-weight mg/cm²*	*AFDW mgC/cm²*	*gDW/m²*
Winter	124.00±11.30	0.20±0.03	0.126±0.002	13.96±1.38
Summer	92.67±11.5	0.19±0.02	0.096±0.011	7.29±2.3
Monsoon	24.00±41.57	0.053±0.092	0.026±0.046	1.06±1.84
Autumn	62.00±33.87	0.133±0.068	0.067±0.038	11.27±2.20

Water temperature and current velocity are two important physical factors that affect the entire ecology of a lotic system. River Eastern Nayar had a moderate water temperature round the year due to its spring-fed nature of origin. It was maximum 24. 0±73.89°C during the monsoon months (June-August) and was minimum 11.45±2.43° C during winter. The velocity or water current was also not too high if we compare it with the other hillstreams like Bhagirathi, Alaknanda and the Mandakini. It was due to less gradient in the stream during its course downwards. A maximum velocity was recorded as 0.563±0.194 m/sec during monsoon and the minimum was 0.33±0.05 m/sec in the Winter-Autumn. Due to moderateness in these two important physical parameters, the stream becomes conducive for the growth of periphyton as well as the detritus. And because the marcozoobenthos and fish feed on them, the entire productivity of the stream is positively affected. This observation lends support to Dobriyal and Singh (1988) and Dobriyal *et al.* (2002). River Mandakini, a glacier-fed river was studied by Negi (1998) who reported that due to low water temperature and high velocity of water current the river was less productive in comparison to the spring-fed streams.

The dissolved oxygen contents were quite high in the stream due to primary productivity activities of the periphyton (Dobriyal, 2003)and clarity of water due to less anthropogenic activities. It was recorded as high as 11.35±0.33 in the winters when the respiratory activity of the biota also ceases due to less bacterial activities. It was lowest 9.15±0.72 in the monsoon due to influx of organic and inorganic ion into the stream along with the flash flood. The alkalinity and pH values of the stream were high during winter (36.45±3.49 ppm and 8.0±0.21 respectively), which can be correlated with the high growth of algae during this period. The lowest values were recorded obviously in monsoon (25.2±54.39 ppm and 7.43±0.18) due to rains being normally acidic in nature.

The quantitative study of periphytic algae indicated that it was observed seasonally maximum (124.0±11.30 units/cm²) in winter and minimum (24.0±41.57 units/cm²)in monsoon. The algal biomass was observed maximum 0.26±0.03 mg/cm² (DW) and 0.126±0.015 mgC/cm² (APDW) in winter and minimum 0.053±0.092 mg/cm² (DW) and 0.026±0.046 mgC/cm² (AFDW) in the monsoon. The detritus (standing stock) was observed maximum 13.96±1.38 g DW/m² in winter and minimum 1.06±1.84 g DW/m² in monsoon 1996.

The algal biomass and detritus forms are important component of the stream ecology as the macrozoobenthos as well as fish feed on it. According to Welcomme (1985) in rhithronic waters where phytoplankton is virtually absent, the production of periphytic algae and submerged vegetation is very important. The longitudinal study of river indicated that the algal biomass was high at the lower spot (Dangal) of the stream (142 units/cm², dry weight 0.3 mg/cm², and AFDW-0.14 mgC/cm²) in relation to the upper spot (Thalisain-126 units/cm^2dry weight 0.26 mg/cm^2 and AFDW 0.14 mgC/cm^2) which can be correlated to the heterogeneity and stability of the substrarum. The detritus standing stock was also observed high at lower spot. *i.e.* (10.546.33 g DW/m^2) at Dangal in comparison to the upper one *i.e.* (6.25±4.39 gDW/m^2) Thalisain.

Qualitative study on the periphytic algae indicated that the algal carpet of periphytic algae was mostly of the *Cladophora,Spirogyra, Ulothrix, Hydrodictyon, Zygnema* (all chlorophyceae) and the diatoms like *Synedra, Diatoma, Navicula, Cymbella, Tabellaria, Gomphonema, Fragilaria* and *Nitzschia, etc.* The detritus standing stock was contributed by the dried twigs and leaves of the riperian vegetation at the bank of stream, mainly the herbs, shrubs and trees of Quercus and Pinus *etc.* The fish observed in the stream were mainly the omnivore (*Tor tor, Tor putitora, Nemacheilus* sps., *Barilius* sps. *etc.*) detritovore (*Tor chilinoides*) and the herbivore (*Garra* sps., *Crossocheilus* sps.).

References

APHA, 1976. *Standard Methods for the Examination of Water and Wastewater*, 14th edn. American Public Health Assoc. Washington, D.C.

Dobriyal, A.K., 2003. Structure and function of an ecosystem with special reference to River Nayar. In: *National Symposium on Ecology and Aquatic Biodiverstiy* (SEBAE), 15–17 February, 2003 extended Abstract No. 86: 1–4.

Dobriyal, A.K., Balodi, V.P., Kumar, N. and Bisht, M.S., 2002. An appraisal of the fishery potential of river Nayar with a note on its improvement possibilities: Coldwater fish genetic resources and their conservation. *Nature Conservator*, 7: 235–242.

Dobriyal, A.K. and Singh, H.R., 1988. Ecological basis for the ichthyofaunal variation in two hillstreams of Garhwal Himalaya. In: *The First Indian Fisheries Forum Proceedings*, (Ed.) M. Mohan Joseph. Asian Fishery Society, Indian Branch, pp. 313–317.

Negi, K.S., 1998. Some aspects of fish biology of *Crossocheilus latius latius* (Ham.) correlated with its riverine Environment.

Welch, P.S., 1948. *Limnological Methods*, McGraw Hill, New York.

Welcomme, R.L., 1985. *River Fisheries*. FAO Fisheries Technical Paper 262, Rome, pp. 325.

Water and Wastewater Management in 2 Vols (2011) *Pages* **111–115**
Editors: D.R. Khanna, R. Bhutiani & Gagan Matta
Published by: DAYA PUBLISHING HOUSE, NEW DELHI

Chapter 19

Water Quality Assessment of a Polluted Urban Lake with respect to Zooplankton

Suresh B. Zade[1] and Shashikant R. Sitre[2]
[1]P.G. Department of Zoology, M.B. Patel College, Sakoli District, Bhandara
[2]Department of Zoology, Govt. Institute of Science, Nagpur

ABSTRACT

The plankton community is a heterogeneous group of tiny plants and animals adapted to suspension in the water bodies of the entire globe. The movements of plankton are so feeble that they essentially remain at the mercy of water currents *i.e.*, waves and tides. The producers of the aquatic ecosystems are the plant plankton *i.e.*, phytoplanktons while consumers are the animal plankton *i.e.* zooplankton. The phyto and zooplankton constitute a dynamic chain on which fish fauna sustains and grows to be useful for the benefit of mankind. The pollution of water bodies is increasing day by day and a large number of water bodies are progressing towards eutrophication. The status of a water body is known from the presence of various kinds of species in it. The organic enrichment favours growth of a large number of species in a water body which results in its growth and development. Keeping these points in view assessment of water quality of a urban water body (Sonegaon lake) situated in the Nagpur city is done during monsoon and post monsoon months to know the exact status of zooplankton species prevailing in lake water. The lake water is polluted due to Ganesh idol immersion, garland immersion as well as load of detergents from washing and other activities. The lake is slowly progressing towards eutrophication as evident from the present studies on zooplankton.

Keywords: Zooplankton, Water quality, Lake, Pollution.

Introduction

Today the freshwater lakes of the world are undergoing fast degradation subsequently leading to eutrophication (Rao and Durve, 1989). The anthropogenic sources *viz.* activities of man are mainly responsible for polluting the freshwater resources all over India. The freshwater lakes are closed ecosystems in which zooplankton hold a key position in the metabolism of water bodies, trophic levels, food chains and energy flow. The occurrence and abundance of zooplankton in freshwater lakes depends on its productivity, which in turn is influenced by physico-chemical parameters and available nutrients. The organically enriched water bodies sustain a large variety of zooplankton species throughout the year. Keeping these points in view a freshwater lake of Nagpur city is investigated with respect to biodiversity of zooplankton during monsoon and post monsoon months.

In India studies on freshwater zooplankton were carried out by many prominent investigators like Babu Rao (1997), Chandrasekhar (1996), Rai (1982), Sharma and Hussain (2001), Verma and Dutta Munshi (1987), Dhanpathi (2000), Rao and Durve (1989), Dutta *et al.* (1987), Somani and Pejaver (2004) and Kodarkar (1994).

Materials and Method

The Sonegaon lake is an old lake situated on Southern side of Nagpur city. Previously the habitation was very sparsle, near the lake. But due to rapid progress of Nagpur city the localities expand and now at present the lake is covered from all the sides by residential colonies. Some part on Western bank of lake is encroached by localities and built houses therein. Previously the lake was surrounded by fields and the water was abundantly available but now due to encroachment on banks the water seeping into the lake basin has reduced to a large extent, the effect is that during extreme Summer months the volume of water shrinks to a size of very small pond in the center. Two old temples are situated on eastern bank of this beautiful lake which is visited by many devotees. During "Ganesh idol immersion programme" a lot of garlands and other pooja material are dumped into the basin thereby enriching the lake waters. Some local residents also utilize the lake for washing and bathing activities daily thereby adding detergent load to waters polluting it. The lake base is shallowed down and water table is hardly four feet in the lake due to siltation of banks. During winter months lots of migratory birds are frequently observed in the lake water. In order to study the zooplankton biodiversity in this beautiful lake. Zooplankton samples were collected from littoral zone of surface water during monsoon and post monsoon months by filtering 50 litres of Sonegaon lake water through plankton net made of silk bolting cloth (mesh size 45 µm) in early morning hours twice a month. The zooplankton samples were immediately preserved in 4 per cent formalin solution for further analysis and kept in laboratory till further analysis. The samples were observed under "Sedgwick Rafter Counting Cell" (SR Cell) having dimensions 50 mm x 20 mm x 1 mm, under light microscope. The identification of zooplankton was done using standard literature (Battish, 1992; Edmondson, 1992; Dhanpathi, 2000; Ward and Whipple, 1958). The different forms were observed and recorded individually.

Results and Discussion

The Zooplankton of Sonegaon Lake is represented by five different groups *viz.* protozoa, cladocera, copepoda, ostracoda and rotifera (Table 19.1). The ostracoda group is represented by 2 species rotifers by 13 species cladocera by 8 species, copepoda by 3 species and protozoa by 2 species during monsoon and post monsoon season in general. During monsoon months the ostracoda group is represented by 2 species while in post monsoon it is represented by *Cypris species* alone. Tonapi (1980) has reported higher population of ostracoda during monsoon months due to abundance of fine detritus available during this period of the year on which ostracods feed vigorously and develop.

The most abundant species during monsoon months were rotifers represented by about 12 different forms; while in post monsoon it is represented by about 8 different species. The pollution indicator sps. *Rotaria rotatoria* was observed in post monsoon months indicating that after rain, pollution level of nutrients and organics has slowly increased in lake water. The most abundant species observed in lake water was *Brachionus calyciflorus* from rotifera group. The occurrence of indicator species such as *Brachionus forficula* (Rao and Durve, 1989) and *Filinia longiseta* (Schindler and Noven, 1971; Mishra and Saksena, 1998) indicates that slowly the lake water is progressing towards eutrophication stage.

Table 19.1: Zooplankton Species Observed in Sonegaon Tank

Sl.No.	*Group*		*Species*	*Time Monsoon*	*Post Monsoon*
1.	Ostracoda	1.	*Cypris* sp.	+	+
		2.	*Stenocypris*	+	–
2.	Rotifera	1.	*Asplanchna* sp.	+	+
		2.	*A. intermedia*	+	–
		3.	*Brachionus forficula*	+	+
		4.	*B. caudatus*	+	+
		5.	*Keratella tropica*	+	+
		6.	*Philodina*	+	–
		7.	*Testudinella*	+	–
		8.	*Filinia longiseta*	+	–
		9.	*Brachionus rubens*	+	+
		10.	*Epiphanes*	+	–
		11.	*Monostyla bulla*	+	+
		12.	*Rotaria rotatoria*		+
		13.	*Brachionus calyciflorus*	+	+
3.	Cladocera	1.	*Simocephalus*	+	–
		2.	*Macrothrix*	+	+
		3.	*Chydorus*	+	+
		4.	*Alona* sp.	+	+
		5.	*Bosmina longirostris*	+	+
		6.	*Diphanosoma* sp.	+	–
		7.	*Pleuroxus* sp.	+	–
		8.	*Sida* sp.	+	–
4.	Copepoda	1.	*Cyclops* sp.	+	+
		2.	*Diaptomus* sp.	+	+
		3.	*Copepod nauplius*	+	–
5.	Protozoa	1.	*Difflugia* sp.	+	–
		2.	*Arcella*	+	–

The freshwater zooplankton form an important group which in turn is consumed by variety of secondary consumers including commercially important groups of crustaceans such as prawns and fishes.Thus zooplankton community constitute an important component of aquatic ecosystem and many species are suitable for aquaculture practices.

Rotifers play an important role as grazers and suspension feeders within the zooplankton community. The difference in periodicity and population density of different rotifer species is due to biotic interactions and nutritional content of the lakes. Rotifers exhibit marked differences in their tolerance and adaptability to changes in physico-chemical and biological parameters. Such changes are dramatic and sudden in the case of urban ecosystems. Chandrasekhar (1996) observed that in summer and monsoon months the factors like water temperature, turbidity, transparency and dissolved oxygen play an important role in controlling the diversity and density of rotifers.

Most of the cladocera species are primary consumers and feed on microscopic algae and fine particulate matter in the detritus thus influencing cycling of matter and energy in benthic food chain. The factors like turbidity, materials in suspension, transparency, dissolved oxygen of the water play a key role in controlling the diversity and density of cladocera. According to Datta and Munshi (1995) abundance of cladocera can be attributed to thick deposit of organic matter in an aquatic ecosystem. In the present investigation it is noticed that during monsoon months due to turbulence in water a large amount of organic detritus is made available to cladocera on which about 8 different forms thrived which were reduced to about 4 types in post monsoon months.

So it can be inferred from the present study that the aquatic ecosystem of the urban polluted lake has abundant biodiversity represented by various forms during monsoon and post monsoon seasons. The lake water is slowly progressing towards eutrophication as indicated by the presence of indicator species.

References

Babu Rao, M., 1997. Studies on the ecology and fish fauna of an oligotrophic lake, Himayatsagar, Hyderabad, A.P. In: *Recent Advances in Freshwater Biology,* (Ed.) K.S. Rao, 3: 73–97.

Baker, R.L., 1979. Specific status of *Keratella cochiearis* (Gosse) and *K. earlinare* Ahlastrom (Rotifera, Brachionidae) Morphological and ecological considerations. *Can. J. Zool.,* 57(9): 1719–1722.

Battish, S.K., 1992. *Freshwater Zooplankton of India.* Oxford and IBH Publishing Co., p. 223.

Chandrasekhar, S.V.A., 1996. Ecological studies on Saroornagar Lake, Hyderabad with special reference to zooplankton communities. *Ph.D. Thesis,* Osmania University, Hyderabad.

Datta Munshi, J. and Datta Munshi, J.S., 1995. *Fundamentals of Freshwater Biology.* Narendra Publishing House, New Delhi, p. 222.

Dhanpathi, M.V.S.S., 2000. *Taxonomic Noters on the Rotifers from India (from 1889–2000).* IAAB, Hyderabad.

Dutta, N.C., Mandal, N. and Bandyopadhyay, B.K., 1987. Seasonal abundance of Rotifers in a perennial freshwater pond in Calcutta. *J. Environ. Biol.,* 8(1): 63–71.

Kodarkar, M.S., 1994. Biodiversity of zooplankton in Saroornagar lake, Hyderabad. *J. Aqua. Biol.,* 9(1 and 2): 30–33.

Mishra, S.R. and Saksena, D.N., 1998. Rotifers and their Seasonal Variation in a Sewage Collecting Morar (Kalpi) river, Gwalior. *J. Environ. Biol.,* 19: 363–374.

Rao, I.S., 1982. Ecology of the Manjira reservoir Sangareddy, A.P. *Ph.D. Thesis,* Osmania University, Hyderabad, p. 294.

Rao, N.G. and Durve, V.S., 1989. Cultural eutrophication of the lake Rangasagar, Udaipur, Rajasthan. *J. Env. Biol.,* 10: 127–134.

Schindler, D.W. and Noven, B., 1971. Vertical distribution and seasonal abundance of zooplankton in two shallow Ontario lakes. *J. Fish. Res. Bd. Can.*, 28: 245–256.

Sharma, B.K. and Hussain, M.D., 2001. Abundence and ecology of zooplankton in a tropical flood plain lake, Assam (N.E. India). *Ecology Environment and Conservation,* 7(4): 397–403.

Somani, V. and Prejaver, M., 2004. Crustacean zooplankton of lake Mkasunda, Thane, Maharashtra. *J. of Aqua. Biol.*, 19(1): 57–60.

Verma, P.K. and Munshi, J.S. Dutta, 1987. Plankton community structure of Bada reservoir of Bhagalpur (Bihar). *Trop. Ecol.,* 28: 200–207.

Yousuf, A.R., Balhi, M.H. and Quadri, M.Y., 1986. Limnological features of aforest lake of Kashmir. *J. Zool. Soc.*, 58(1–2): 29–42.

Water and Wastewater Management in 2 Vols (2011) *Pages* **116–121**
Editors: **D.R. Khanna, R. Bhutiani & Gagan Matta**
Published by: **DAYA PUBLISHING HOUSE, NEW DELHI**

Chapter 20

Environmental State of River Ganges in Rishikesh-Haridwar and its Management

G. Prasad[1], V. Shankar[1] and A.K. Chopra[2]
[2]Department of Botany and Microbiology
[2]Department of Zoology and Environmental Sciences,
Gurukul Kangri University, Haridwar

ABSTRACT

The present study deals with the influence of discharge of various drains carrying domestic and industrial wastes on the water quality of river Ganga at different locations of Rishikesh and Haridwar. It is indicated that though the quality of river water deteriorates at these locations, the river still has an excellence purifying ability. The condition of water quality is not much alarming. There is a need of monitoring of water quality from different depths at 5-6 different points of water current across the river width throughout the year. Various methods for the proper management of river water have been suggested.

Keywords: Ganga, Rishikesh, Quality, Management.

Introduction

The river Ganga is most important of holy river of India. More than 80 per cent people of our country worship it since time immemorial. Like other Northern Indian rivers, Ganga also originates from Himalayan glaciers, *i.e.*, Gomukha, before Devprayag, it is known as Bhagirathi. A large number of tributaries join the river and loose their identity during its course. Important tributaries such as Alaknanda in hills and even Yamuna in plains lose their identity at Devprayag and Allahabad respectively. Tributaries have important role in maintaining the quantity of water. Ganga has an important impact on the economy of Northern India and has been considered as lifeline of the country especially for the states of Uttar Pradesh, Bihar and West Bengal.

Ganga covers about 225 km from Gomukha to Bay of Bengal. A number of important cities are located on the bank of river and are getting water supply not only for domestic purpose but for Industrial uses also. Besides, it has been a major source of irrigation for the agricultural land and for recharging of the groundwater at certain areas. A very limited amount of freshwater is available on the earth at global level for human activity. Of the total available water, 93.96 per cent is confined to Ocean, 4.2 per cent to underground, 0.2 per cent to total inland surface including rivers, lakes and other reservoirs. A very small percentage of water *i.e.*, 0.0001 per cent is available in rivers.

Albeit large restrictions have been made by ancient Rishis to protect the Ganga from contaminants and pollutants but there is a sorry state. It is pardox that the people who worship it are polluting in a river, which is a part of cultural heritage of the country, to ecological assault. Among all the fourteen major rivers of the country, Ganga is being considered as most polluted river. Environmental health of river has not been deteriorated suddenly. There is information that about 3,70000 people died by consumption of Ganga water due to water born diseases such as diarrhoea, dysentery, cholera, typhoid and hepatitis from 1898 to 1907 (Jonthan *et al.*, 1984).

Keeping in view of religious and economic significance of the river, it becomes more imperative to maintain the environmental health of Ganga. Therefore, it is necessary to find out major sources of pollution and their impact on water quality and also to provide remedial measures in Indian context of Rishikesh-Haridwar region.

Major Resources of Pollution at Rishikesh-Haridwar and their Characteristics

Major drains carrying domestic sewage and industrial waste and evacuating them at different locations at Rishikesh and Haridwar were as follows:

	Domestic drains		*Industrial drains*
1.	Swargashram drains	1.	Doon Distillary drain
2.	Sarswati nala drain	2.	IDPL drain
3.	Kangra mandir drain	3.	BHEL drain
4.	Naisotadrain		
5.	Lalta Rao drain		

Various physico-chemical and microbiological parameters of the discharge of different drains entering river Ganga and different locations of the river water were analysed using standard techniques (APHA, 1986). Annual average data of characteristic features of various wastewater based on the present study has been presented in Table 20.1. The study showed that wastewater contains high amount of organic load that ultimately results in higher BOD, COD and bacterial population. Maximum BOD was found to be 5244.00 mg/l in Doon distillery while high profile of COD was recorded to be 170 mg/l at Kangra mandir drain. Bacterial population was found very high in IDPL and Sarswati drain followed by BHEL drain. Both BHEL and Sarswati drains carry high amount of faecal matter, which is evident by disposed faecal waste and excreta at the point of discharge of drain into Ganga. DO in the effluent was negligible and some time it was nil. All the above parameters indicated that these drains caused severe deterioration of water quality of Ganga wherever they discharged their waste.

Table 20.1: Annual Average Values of Physico-chemical and Bacteriological Parameters of Some Domestic and Industrial Drains Responsible for Ganga Pollution

Parameters	*Drains*						
	SN	*ID*	*DD*	*KM*	*NS*	*LR*	*BH*
Temp.(°C)	27.5	25.5	38	21.8	22.0	20.0	35.0
Turbidity (NTU)	109.6	102.31	380.0	52.3	62.26	62.37	160.11
pH	7.26	6.59	3.8	7.82	7.63	7.36	7.62
Conductivity (mv)	0.580	0.467	0.858	0.315	0.288	0.322	0.620
Hardness (mg/l)	189.0	180.0	352	190.0	188.0	250.0	298.0
Alkalinity (mg/l)	252.5	104.0	251.0	108.0	140.0	195.0	212.0
DO (mg/l)	1.20	2.36	Nil	4.61	1.50	3.96	Nil
BOD (mg/l)	142.63	176.0	5244.0	36.21	82.61	93.8	193.21
COD (mg/l)	267.15	278.6	15252.0	170.61	212.11	198.27	242.16
Chlorides (mg/l)	38.5	48.5	210.0	21.8	28.0	28.0	34.64
Sulphates (mg/l)	14.6	21.58	23.5	12.0	8.0	12.0	12.5
SPC/ml x 10^3	446.0	380.0	350.0	42.0	160.0	210.0	580.0
MPN/100 ml x 10^2	240.0	223.0	22.0	93.0	193.0	110.0	210.0

SN: Sarswati Nullah; ID: IDPL; DD: Doon Distillary; KM: Kangra Mandir; NS: Nai Sota; LR: Lalta Rao; BH: BHEL.

Further it was observed that all organic pollution load parameters *viz.* BOD, COD, MPN and SPC were tremendously enhanced in the rainy season. Increased values of these parameters are mainly due to addition of garbage. This garbage reaches into the river through stream water and soil erosion. Stream water also contains high amount of surface dust which is rich source of both organic nutrients and aerobic microbial flora which caused depletion of DO due to utilization of oxygen during decomposition of organic matter and enhances the value of BOD, MPN and SPC. This trend was recorded in all drains uniformly. Besides this, during summer and rains, large number of pilgrims visit Haridwar and Rishikesh which results peak of human activity and generation of large quantity of solid waste, besides addition of detergent, human excreta discharge which ultimately reach into Ganga. A number of sewers and a few Nalas open into Ganga from various Ashrams from Har Ki Pauri to Avadhoot mandal.

Impact of Domestic and Industrial Drains on Water Quality of Ganga

It was found that in main stream of Ganga as well as in its tributaries from Gangotri to Rishikesh, the DO was better in winter and there was a slight depletion in summer. BOD, MPN and SPC also showed the same pattern. There was no significant difference in these parameters in Ganga and its tributaries. But the picture of water quality from Rishikesh to Ganga was different due to large number of out falls starting from Rishikesh to Haridwar. The discharge of domestic waste as well as industrial waste in Ganga results in deterioration of water quality upto certain extent downstream depending on the volume of waste and organic matter, volume and flow rate of the river as well as the season and other factors.

The data showed maximum deterioration of water quality during rainy season and an evidence of influence of drains on water quality has also been well established under seasonal variation. It is obvious that DO, BOD, COD, MPN and SPC during rains were 8.0 mg/l, 5.7 mg/l, 16.77 mg/l and 20.5 x 10^2, 26 x 10^3 respectively at Muni Ki Reti. Decreased level of DO *i.e.* 6.4 mg/l and enhanced values of BOD (42.25 mg/l), COD (120 mg/l), MPN (82.5 X 10^2 and SPC (266 x 10^3) are evidence of the influence of Sarswati drains at Trivenighat. Most deterioration of water quality was found at IDPL mixing zone at Shyampur Khadri in Rishikesh. It was perhaps due to high volume of effluent discharge into river Ganga and due to large surface area of river current which results in comparatively slower flow rate, shallow depth that could not dilute the discharge of effluents rapidly.

It was established that these drains were deteriorating the Ganga water quality but as the sampling was done from nearer to mixing point and few meter down stream and not from the different depths and opposite to the bank of river and 1 to 2 km down stream, therefore the exact impact on water quality as whole spectrum cannot be ruled out. Further is was evident that river purifies the water downstream up to certain extent at Haridwar which is around 25 km down from Rishikesh. Annual average values of various parameters of the water samples collected from Har Ki Pauri indicated that there was excellence recovery in DO (8.1 mg/l), BOD(4.0 mg/l), COD(20.0 mg/l), MPN(10 x 10^2) and SPC (30 x 10^3) as compared to Triveni ghat and IDPL mixing zone. Water quality improved up to level of site at Muni Ke Reti (Table 20.2).

Table 20.2: Annual Average Values of Physico-chemical and Bacteriological Parameters at Few Sites of River Ganga

Parameters	*Sites*					
	Muni ki Reti	*Trivenighat*	*Pashulok Barrage*	*IDPL*	*Neal Dhara*	*Hari ki Pauri*
Temp. (°C)	16.84	19.80	18.5	22.0	20.0	21.5
pH	7.7	8.13	7.5	7.6	7.4	7.5
Total solids (mg/l)	462.8	950.6	695.0	596.0	284.0	470.0
CO_2 (mg/l)	0.20	0.29	0.62	0.37	0.78	0.56
Chloride (mg/l)	3.5	18.8	8.07	15.6	10.08	3.50
DO (mg/l)	8.5	6.10	9.0	4.7	8.0	8.1
BOD (mg/l)	7.3	47.10	8.60	116	8.60	4.0
COD (mg/l)	19.50	100.0	17.0	195	38.0	20.0
SPC/ml x 10^3	20.0	280.0	135	280.0	70.0	30.0
MPN/100 ml x 10^2	8.4	16.6	7.0	60.0	11.0	10.0

It seems that during the course of flow dilution of both effluents and river water, occurred. Simultaneously there was a degradation of organic component and enhancement of DO due to mixing and absorbance of O_2 from atmosphere and by release of O_2 by various aquatic plants including algal genera and others. Further it appears that river quality of Ganga would be definitely better towards bottom of the river and opposite bank of the river as it is very difficult for the pollutants to move to opposite bank and to settle at bottom in any stream-current. However, exceptionally high current of wind may facilitate the movement of pollutants toward opposite bank on the river but their impact will be negligible.

Various physico-chemical and bacteriological features of different sites of river Ganga and Haridwar have also been reported by Shankar, 1987; Khanna, 1993; Chopra and Rehman, 1992 a, b, and Chopra and Patrick, 1994, 2000. The efforts to make a suitable management programme of domestic sewage as well as industrial effluent with the treatment by aquatic plants under laboratory condition was made by Shankar *et al.*, 1989 and Sharma *et al.*, 1999. They have found a drastic improvement in various parameters of sewage after treatment. Shankar *et al.* (1989) observed that certain industrial effluents are a better source for seed germination and plant growth of some agricultural crops and it may provide another source of wastewater management in Indian context after doing more detailed study.

Conclusion

With these findings, it is concluded that water quality as far as bathing is concerned, was not solubility deteriorated. But it does not mean that precaution should not be taken for protection of river Ganga.

The population explosion, change in mode of life, urbanization, industrialization and human activities of course these will cause severe risk on ecological system of river water. To protect the people and to prevent any kind of outbreak of epidemics, there should-be publicity that people should avoid consuming the river water as it may contain traces of raw sewage and chances of presence of human pathogen can also not be ignored.

Management Programme

1. It is obvious that direct discharge of raw sewage or industrial effluents is cause of deterioration of water quality at discharging point. Though purification occurs after a distance of 10-15 km up to certain extent but the chances of presence of human pathogens remain there. Therefore at any cost, direct discharge of wastes should be prevented.
2. Both domestic sewage and industrial effluents should be properly monitored and treated primarily and secondarily for important parameters. Further the waste should be treated in oxidation ponds. This will facilitate for decomposition of remaining organic components and will allow growing large number of algal genera and other aquatic plants and will generate the organic manure.
3. Due to large volume and load of water, percolation of water may occur which may cause another serious threat for underground water especially if wastewater contains heavy metals or toxic substances, which are not degradable. Besides this, the movement of underground water occurs as columns march very slowly. Once contamination takes place, it would not be only difficult but impossible to eradicate contaminants. Therefore other different designing of oxidation ponds should be done to protect the groundwater.
4. Before discharge of treated sewage into the river, volume of sewage discharged per cubic meter/second should also be considered on basis of flow rate and volume of river. During winter season discharge of treated sewage should not be permitted as water volume and flow rate of Ganga are highly reduced. Recommendation regarding discharge of treated sewage indicates that for safe dilution, river water should be eight times more than sewage and this criterion should be strictly followed.
5. The treated water should also be tested for its effects on germination of various agricultural crops, biomass production and ultimately on yield. Both biomasses as well as yield should be properly monitored for deposition of heavy metals and chemicals to prevent any kind of human health hazard.

Acknowledgements

The authors Prof. (Retd.) V.Shankar and Dr.G.Prasad are thankful to Ministry of Environment, Govt. of India, New Delhi for providing financial assistance for the research programme.

References

APHA, 1986. *Standard Methods for Examination of Water and Wastewater*, 15th edn. American Public Health Association (APHA), Inc., New York.

Chopra, A.K. and Rehman, Abdul, 1992a. Effect of domestic sewage on physico-chemical qualities of the Ganga canal water at Jwalapur, Haridwar (India). *Intern. J. Environmental Studies, (U.K.)*, 41: 183.

Chopra, A.K. and Rehman, Abdul, 1999b. Effect of domestic sewage on microbiological charateristics of Ganga canal at Jwalapur, Haridwar. *Him. J. Env. Zool.*, 6: 183–187.

Chopra, A.K. and Patrick, Nirmal J., 1994. Effect of domestic sewage on self-purification of Ganga water at Rishikesh. I. Physico-chemical parameters. *Advances in Biosciences*, 13: 75–82.

Chopra, A.K. and Patrick, Nirmal J., 2000. Effect of domestic sewage on self-purification of Ganga water at Rishikesh. II. Microbiological parameters. *J. Natcon.*, 8: 81–86.

Turk, J., Turk, A., and Arms, K., 1984. *Environmental Science*, 3rd edn., pp. 1–502.

Khanna, D.R., 1993. *Ecology and Pollution of Ganga River*. Ashish Publishing House, Delhi, p. 1–241

Shankar, V., 1987. *Ganga (Rishikesh to Garhmukteshawar)*. Technical Report, Submitted to Ministry of Environment, Govt. of India, pp 1–100

Shankar ,V. Prasad, G. and Sharma, S.K., 1989. Influence of industrial effluent on seed germination and seedling growth on some agricultural crops. J. Nat. and Physic. Sci. 3:6–11

Shankar ,V. Prasad, G. and Shankar, V., 1999. Screening of pollution reducing potential of some free floating hydrophytes. *Sustainable Ecosystem and Environment*, p. 147–154

Water and Wastewater Management in 2 Vols (2011) *Pages 122–130*
Editors: **D.R. Khanna, R. Bhutiani & Gagan Matta**
Published by: **DAYA PUBLISHING HOUSE, NEW DELHI**

Chapter 21

Study of Leachability Characteristics of Fly Ash and its Effect on Soil and Groundwater Contamination

Sweta Bhardwaj
Department of Zoology and Environmental Science, Gurukul Kangri University, Haridwar

ABSTRACT

In the present study the leachability characteristics of fly ash and its effect on soil and groundwater contamination. During the course of study it was found that despite the presence of the only trace amount of heavy metal in coal and coal ash, these materials have the potential to accumulate in soil and increase in concentration by food chain biomagnification.

Keywords: Thermal power plant, Fly ash, Leachate.

Introduction

The use of coal for electricity generation was resulted in increasing environmental problems associated with gaseous and particulate emissions from coal fired power plants. Ash management is one of the key aspects of environmental management in coal fired thermal power stations. The ash production in thermal power plant result from the combustion of coal, consisting roughly 20 per cent bottom ash and 80 per cent fly ash. India is trying to expand the electric power generation capacity, as current generation is seriously below peak demand. About 110 million tones of coal consumed in these power stations during a year produce about 30-40 million tones of fly ash *i.e.*, at the rate of one lakhs per day. Only about 10 per cent is utilized in the manufacture of bricks and cement. For the disposal of rest, nearly 28000 hectares of land has already been converted into ash dumps. Chemical constitutents of fly ash mainly depend on the chemical composition of the coal. Fly ash is the finer size particle (Patil *et al.*, 1991).

Leachate contamination may severely pollute neighboring ground and surface waters. The chemical quality of land fill leachate differs greatly from one landfill to another and fluctuates seasonally within individual landfill.Leachate composition is waste and site specific depending on waste type, landfill age, amount of infiltering water. Thus land disposal of fly ash creates the potential for groundwater pollution *via* the migration of heavy metals from the fly ash. The disposal ponds are in interaction with surface and groundwater systems and there is a risk of contamination of these water systems. Study of physico-chemical characteristics of fly ash, its trace metal contents and their leaching behavior are very important requirement for environmentally safe disposal management strategy (Wasay, 1992).

Study Area

Groundwater sampling site was identified during the site survey done at TPS. Panki thermal power station located between Kalpi road and G.T. road. Anpara TPS is located at the Anpara distt. Sonebhadra(U.P.), situated at longitude of 82° 48′E and the latitude of 24° 12′. The height above sea level is 275 meters. Fly ash sample were collected from both the TPS and analyzed for parameters. At Panki TPS three samples were collected at the nearby area of ash pond. At Anpara TPS two samples were collected from the existing sources at nearby area of ash pond where one sample was collected about 3km. away from the ash pond.

Experimental Procedure

Fly ash sample were collected from both TPS and were analyzed for the parameters particle size, pH, Conductivity, Magnesium, Sodium, Potassium, Total chromium, Copper, Iron, Lead, Nickel, Silica, Zinc, Manganese, Aluminium, Mercury, Arsenic, Phosphorus.

Leachability Studies

Evaluation of groundwater pollution potential require the condition of testing protocols to determine constituent concentrations in produced leachates. The leachability test was performed at different pH and contact time using standard TCLP (Toxicity characteristic leaching procedure) test of EPA. Alternately, batch test and column test were also conducted in order to evaluate the extent of dissolution of fly ash contents in the leachates.

These methods provide useful information as to the degree of leachability of many major and trace elements (APHA, 1998)

TCLP Test

The TCLP test is used in a generic manner for the prediction of leaching trends of wastes. TCLP test is appropriate for the screening of fly ash for potential environmental impact (USEPA Method 1311) (Table 21.1).

Batch Test

Batch test was conducted using 100 g of fly ash with one liter of D.W for both the TPS maintaining liquid solid ratio of 1:10 for making fly ash slurry *i.e.*, being disposed to the ash pond for 1 to10 days at 150 RPM in order to stimulate the actual condition. After the desired contact time, the sample was filtered and analyzed in accordance with the *Standard Method for Water and Wastewater Analysis* (APHA, 1998).

Table 21.1: TCLP Test for Ash Samples of Anpara and Panki TPS

Parameters	*Unit*	*Conditions*	
		Panki	*Anpara*
pH	–	5.1	5.2
Conductivity	M mhos/cm	3.46	3.45
Chloride	mg/l	29.9	54.9
Hardness	mg/l	244	152
Magnesium	mg/l	45.7	28.4
Sulphate	mg/l	40	10
Fluoride	mg/l	0.843	0.299
Copper	mg/l	0.098	0.124
Chromium hexavalent	mg/l	0.06	0.07
Silica	mg/l	7.528	3.386
Manganese	mg/l	0.344	0.111
Zinc	mg/l	0.352	0.32
Iron	mg/l	1.05	2.93
Lead	mg/l	0.75	0.09
Sodium	mg/l	197	209
Potassium	mg/l	12.7	3.8

Column Test

Glass columns of 90 cm height and 6 cm diameter was used to study the influence of flow rate on water quality obtained through leaching of fly ash. The column was filled with fly ash and D.W. and having pH-6.5 and 7.5 was allowed to pass through the fly ash column continuously. Keeping the head constant leachates of sample were collected at regular intervals of 24 hrs.

Each column is leached with approximately 230ml/day using ash of both and analysed for metals sulphate, pH, Alkalinity, Hardness, Calcium, Magnesium, Chloride and Conductivity (APHA, 1998).

Results and Discussion

The analysis of fly ash and Panki TPS and Anpara TPS was conducted in the laboratory. When compared with each other. It was observed that all values of the parameters analysed are high in Panki TPS ash compared to the fly ash of Anpara TPS (Table 21.2).

Characteristics of Groundwater of Panki TPS

The groundwater table encountered at the depth 7.5-9.2m depth from the GL during the period of the field exploration in Nov-2002 adjacent to Panki TPS.

Table 21.2: Comparison of Fly Ash of Panki and Anpara

Parameters	*Unit*	*Values Obtained*	
		Panki TPS	*Anpara TPS*
pH		7.9	7.6
Conductivity	M mhos/cm	0.112	0.062
Magnesium	mg/l	1200	425
Sodium	mg/l	1400	680
Potassium	mg/l	5300	4200
Total Chromium	mg/l	51.0	28.0
Copper	mg/l	37.0	30.0
Iron	mg/l	15000	14800
Lead	mg/l	52.0	47.0
Nickel	mg/l	35.0	30.0
Silica	mg/l	70.0	62.0
Zinc	mg/l	84.0	82.0
Fluoride	mg/l	412	500
Manganese	mg/l	139	110.0
Phosphorus	mg/l	712	680
WHC	Per cent	58.1	27.5
Porosity	Per cent	53.0	29.2

Characteristics of Groundwater of Anpara TPS

The groundwater table encountered at the depth 7.5-9.2m depth from the GL during the period of the field exploration in Nov-2002 adjacent to Anpara TPS.

Observation of Batch Test

If compared with each other it is seen that in case of water of pH 6.5 the total quantity of Conductivity, Alkalinity,Chloride, Hardness, Ca, SO_4, SiO_2, Mn, Zn, Pb released are more compared to the water of 7.5 where as the total quantity of Fluoride, Iron, Cu, Na, Al released more when pH of 7.5 water was used. In terms of toxicity of heavy metals use in the fly ash of Panki TPS it is inferred that 6.5 pH water released comparatively more toxic metals compared to the water of 7.5 pH.

Observation of Column Test

The concentration of Conductivity, Fe, SiO_2, Zn, Cu, Al, and SO_4 showed maximum values in the pH of 6.5. Whereas at pH 7.5 concentration of Alkalinity, Cl, Hardness, Mg, Ca, F, Cr^{+6}, Mn, Pb and Na were found to be more.The variation of release of various substances may be due to the resultant pH available at contact time. Variation of concentration of metal ions exhibit maximum absorbency at a specific pH.

It is possible that some of the heavy metals stripped from fly ash would be reprecipitate within the pile along leachate flow path, particularly if the leachate encountered higher pH material prior to just collection of leachates.

Table 21.3: Physico-Chemical Characteristics of Groundwater Quality of TPS during the Post Monsoon

Parameters	*Obtained Values*			*Desirable Limit*	*Permissible Limit*
	Near Township	*Near Ash Pond*	*100m Away from the Pond*		
pH	7.4	8.0	6.9	6.5–8.5	NR
Conductivity	0.70	1.20	0.97	–	–
Alkalinity	350	380	400	200	600
Hardness	135	190	200	300	600
Calcium	45	69	44	75	200
Magnesium	18	42	40	–	–
Sulphate	10.0	60.0	50.0	200	400
Chloride	5.8	25.0	21.0	250	1000
Copper	ND	0.03	0.01	0.05	1.5
$Chromium^{+6}$	ND	0.02	0.01	0.05	NR
Manganese	0.05	0.10	0.08	0.1	0.3
Zinc	0.04	0.12	0.10	5.0	15.0
Iron	0.12	0.20	0.15	0.3	1.0
Nickel	ND	ND	ND	–	–
Lead	0.02	0.03	0.04	0.05	NR
Arsenic	ND	ND	ND	0.05	NR
Cadmium	ND	ND	ND	0.01	–
Fluoride	0.44	0.43	0.583	1.0	1.5
Total Dissolved Solids	600	689	733	500	2000

Note: All values are in mg/l except pH and conductivity in µmhos/cm.

Comparison of Fly Ash of Panki TPS and Anpara TPS

The experimental results of leachates of both the TPS as indicated in Tables 21.7 and 21.8 reveals that with reference to total release potential of both the fly ash varies with the different metals nonmetals which could be due to the solubility behavior which is a function of particle size characteristic of fly ash.

As depicted from the Tables 21.5–21.8 at pH 7.5 concentration F, Cu, Al and Na indicated elevated concentration in the leachate of both the samples irrespective of variation in the fly ash characteristic. Whereas at the pH 6.5 the increased values of Cl, Ca, SiO_4, Zn, and Pb was recorded.

It is also possible that some of the heavy metals stripped from fly ash would reprecipitate with in the pile along leachate flowpaths, particularly if the leachate encountered higher pH materials prior to just collection. The data indicated that the potential for heavy striping from fly ash if an alkaline excess within these disposal zones is not maintained over the long term.

Table 21.4: Physico-chemical Characteristics of Groundwater Quality of TPS during the Post Monsoon

Parameters	*Obtained Values*			*Desirable Limit*	*Permissible Limit*
	HP Near Ash Pond	*Near Village Anpara*	*Junction on Way of Anpara*		
pH	7.9	7.8	8.3	6.5–8.5	NR
Conductivity	0.133	0.092	0.450	–	–
Alkalinity	76	52	272	200	600
Hardness	88	68	176	300	600
Calcium	30.1	40	62	75	200
Magnesium	10	38.0	42	–	–
Sulphate	86	78	80	200	400
Chloride	4.5	2.5	10.5	250	1000
Copper	30	0.05	0.22	0.05	1.5
Chromium^{+6}	0.08	0.03	0.24	0.05	NR
Manganese	0.07	0.01	ND	0.1	0.3
Zinc	0.22	0.1	0.035	5.0	15
Iron	0.115	0.40	0.01	0.3	1.0
Nickel	0.01	0.01	0.01	–	–
Lead	0.02	0.01	0.01	0.05	NR
Arsenic	ND	ND	ND	0.05	NR
Cadmium	ND	ND	ND	0.01	NR
Fluoride	0.710	0.613	0.643	1.0	1.5
TDS	152	88	471	500	2000

Note: All values are in mg/l except pH and conductivity in μmhos/cm.

Table 21.5: Analytical Results of Leachates Test of Panki TPS

Duration	*pH*	*Conductivity*	*Alkalinity*	*Cl*	*Hardness*	*Mg*	*Ca*	*SO_4*	*F*
1st	7.3	0.28	32	13.9	92	17.3	20.4	55	1.00
2nd	7.5	0.30	36	15.9	94	18.6	22.8	40	1.06
3rd	7.6	0.23	44	16.9	108	21.6	19.2	60	1.00
4th	7.3	0.23	44	12.9	104	20.2	20.8	50	0.96
5th	7.3	0.27	48	6.99	100	18.4	24.1	60	1.52
6th	7.3	0.28	48	4.99	104	18.2	28.9	64	1.63
7th	7.4	0.27	52	3.99	100	20.8	14.4	60	1.50
8th	7.6	0.27	56	3.99	100	20.8	14.4	56	1.88
9th	7.5	0.28	56	3.99	100	20.0	17.6	45	1.49
10th	7.5	0.27	54	3.0	92	20.0	17.0	44	1.34

Contd...

Table 21.5–Contd...

Duration	Fe	Cr^{+6}	SiO_2	Cu	Zn	Pb	Na	Mn	Al
1st	0.77	0.08	4.029	0.04	0.37	ND	17.0	0.16	0.54
2nd	0.68	0.08	4.729	0.02	0.23	ND	20.8	0.076	0.54
3rd	0.66	0.08	5.221	0.04	0.23	ND	21.8	0.062	0.56
4th	0.65	0.06	5.864	0.03	0.21	0.01	20.4	0.033	0.57
5th	0.57	0.05	8.474	0.02	0.09	0.01	20.3	0.005	0.59
6th	0.45	0.08	9.288	0.004	0.006	0.02	20.2	0.005	0.68
7th	0.34	0.03	9.401	0.003	0.003	0.03	2.25	0.004	0.71
8th	0.32	0.04	10.29	0.002	0.002	0.04	2.3	0.001	0.71
9th	0.30	0.04	10.44	0.001	ND	0.04	2.0	ND	0.72
10th	0.29	0.04	10.40	ND	ND	0.05	2.0	ND	0.72

Note: All values are in mg/gm of fly ash except pH and conductivity in μmhos/cm.

Table 21.6: Analytical Results of Leachates Test of Anpara TPS

Duration	pH	Conductivity	Alkalinity	Cl	Hardness	Mg	Ca	SO_4	F
1st	7.6	0.11	36	19.9	80	17.1	9.62	20	0.09
2nd	7.3	0.14	36	18.9	72	15.5	8.02	35	0.17
3rd	7.3	0.18	38	17.9	68	14.5	8.02	40	0.26
4th	7.2	0.19	40	16.9	68	13.4	12.8	45	0.28
5th	7.5	0.19	44	12.9	64	12.4	12.8	45	0.45
6th	7.6	0.20	48	8.9	64	12.1	14.4	46	0.48
7th	7.8	0.16	52	7.9	60	11.5	12.8	25	0.74
8th	7.7	0.14	52	7.7	60	12.2	9.62	15	1.34
9th	7.5	0.11	51	6.9	56	11.3	9.62	13	1.38
10th	7.5	0.12	51	6.6	54	11.0	9.5	13	1.37

Duration	Fe	Cr^{+6}	SiO_2	Cu	Zn	Pb	Na	Mn	Al
1st	0.46	0.06	0.26	1.31	0.321	1.2	40.1	0.132	3.43
2nd	0.49	0.08	0.45	0.02	0.152	1.0	27.8	0.041	1.25
3rd	0.37	0.06	0.47	0.03	0.132	1.3	4.3	0.088	0.35
4th	0.31	0.07	1.00	0.02	0.120	1.2	4.4	0.070	0.49
5th	0.34	0.04	5.62	0.01	0.089	0.080	3.6	0.005	0.59
6th	0.36	0.06	6.60	0.008	0.049	0.049	1.9	0.006	0.77
7th	0.47	0.02	6.52	0.004	0.032	0.039	1.7	0.005	0.98
8th	0.74	0.05	6.69	0.001	0.027	0.041	1.7	0.004	1.19
9th	0.78	0.01	7.07	ND	0.020	0.040	1.6	0.004	1.0
10th	0.77	0.01	7.0	ND	0.010	0.039	1.5	0.003	0.79

Note: All values are in mg/gm of fly ash except pH and conductivity in μmhos/cm.

Table 21.7: Quantification of Substances Released from Fly Ash of Panki TPS at Different pH During 20 Days Contact Period

Condition	*Parameters*								
pH of Deionised Water	*Conductivity*	*Alkalinity*	*Cl*	*Hardness*	*Mg*	*Ca*	*SO_4*	*F*	
6.5	3.034	914	256.42	1432	257.42	372.48	313	22.49	
7.5	3.81	738	221.92	1130	230.55	214.54	235.38	33.69	
Condition	*Parameters*								
pH of Deionised Water	*Fe*	*Cr^{+6}*	*SiO_4*	*Mn*	*Zn*	*Cu*	*Pb*	*Na*	*Al*
6.5	12.95	1.32	91.981	0.183	1.955	2.98	6.02	26.25	18.74
7.5	15.21	1.24	71.578	0.003	0.499	5.34	0.182	370.58	30.83

Table 21.8: Quantification of Substances Released from Fly Ash of Anpara TPS at Different pH During 20 Days Contact Period

Condition	*Parameters*								
pH of Deionised Water	*Conductivity*	*Alkalinity*	*Cl*	*Hardness*	*Mg*	*Ca*	*SO_4*	*F*	
6.5	1.926	371	204.6	456	89.47	87.04	79	7.11	
7.5	2.205	580	201.9	650	139.44	73.86	57.5	14.31	
Condition	*Parameters*								
pH of Deionised Water	*Fe*	*Cr^{+6}*	*SiO_4*	*Mn*	*Zn*	*Cu*	*Pb*	*Na*	*Al*
6.5	15.7	0.42	46.08	0.11	0.764	0.070	0.298	12.84	11.67
7.5	4.73	0.64	31.34	2.12	0.22	2.01	0.054	74.89	26.88

Findings and Recommendations

The variation in the concentration of various substances in the leachate may be due to differences in the particle size distribution of the fly ash. Smaller particles have a greater surface area related to their volume thus tending to become enriched in absorbed volatile elements. The relative distribution of trace element on the surface and in the internal matrix of fly ash particle has important environmental implication. Surface deposited metals may be easily mobilized in leaching waters, while metals in the silica metals are released only after periods extended weathering.

Both the power plants burning coals from these states produced alkaline flyash. Preliminary investigations have shown that the physical and chemical properties of fly ash and its alkaline nature. Different species of metal ion exhibit maximum absorbance at a specific pH. At some point of time it was observed that there is a change in resultant pH and increase in metals content in the water column due to the unloading of fly ash particles. pH is one of the main control of the solubility of metal is also closely related to the surface crystalinity. For some trace metals, the relationship between pH and solubility in an intricate process. The observations obtained suggests that some of the metals are released by dissimilar forms at the beginning and at the end of the experiment.

Thus, despite the presence of the only "trace" amount of heavy metals in coal and coal ash, these material have the potential to accumulate in soil and increase in concentration by food chain biomagnification it is important to use bioassays to evaluate toxicity.

The result indicates that if pH of water 7.5 is used for making ash slurry the ash in contact with eluting media increases the resultant pH of leachate after about 15 days if the sub soil pH is alkaline the leachate percolating down to meet groundwater may increase the pH of the groundwater (WHO, 1993).

References

Arnold, G.K., Dawson, A.R. and Muller, M., 2002. Determining the extent of ground and surface water contamination adjacent to embankments comprising pulverized fuel ash (PFA), 2, Project report by the university of Nottingham, School of Civil Engineering, Nottingham Centre for Pavement Engineering.

MacDonald, W.G., Rozendaal, A. and De Meijer, R.J., 1997. Radiometric characteristic of heavy mineral deposits along the west coast of South Africa. *Miner. Deposita*, 32: 371–381.

Mishra, U., 2004. Environmental impact of coal industry and thermal power plants in India. *J. Environ. Radiact.*, 72: 35–40.

Mishra, U.C. and Ramachandran, T.V., 1991. Environmental impact of coal utilization for electricity generation. In: *Environmental Impact of Coal Utilization from Raw Materials to Waste Resources*, (Ed.) K.C. Sahoo. Proc. Int. Conf., IIT, Bombay, p. 117–125.

Patil, A., Gopal, R., Dube, S.K. and Mandal, P.K., 1991. (Res Dev Div, Natl Thermal Power Corpn Ltd. New Delhi 110048). Characterization and utilization of coal ash in the context of Indian super thermal power station. In: *Proc. Int. Conf. Environ. Impact Coal Utilization*, IIT, Bombay, p. 153–164 [11 Ref].

Ray, Rajan, 1991. (Cheml Engang Sec, Coll Ceramic techno, 79 Ac Banerjee Lane, Calcutta 700010). Utilization of flyash: A review. *Cheml. Ind. New.*, 37(4): 335–339.

Sengupta, J., 1991. (Natl Bldg Orgn, New Delhi). Characteristics of Indian coal ash and its utilization as building material. In: *Proc. Int. Conf. Environ. Impact Coal Utilization*, IIT, Bombay, p. 165–183.

Wasay, S.A., 1992. (Min Env Forests, CGO Complex, Lodi Road, New Delhi 110003). Leaching behaviour of trace toxic metals from flyash. *J. Environ. Sci. Hlth.*, A27(1): 25–39.

WHO, 1993. *Water, Sanitation and Health: Guidelines for Drinking Water Quality*.

Water and Wastewater Management in 2 Vols (2011) *Pages* **131–136**
Editors: **D.R. Khanna, R. Bhutiani & Gagan Matta**
Published by: **DAYA PUBLISHING HOUSE, NEW DELHI**

Chapter 22

A Study of Groundwater Quality of Hand Pumps Situated in the Vicinity of Slaughter Houses at Aligarh, U.P.

D.R. Khanna[1], J.P. Singh[2] and Shakun Singh[3]
[1]Department of Zoology and Environmental Sciences, Gurukula Kangri Vishwavidyalaya, Haridwar – 249 404
[2]U.P. Pollution Control Board, Aligarh
[3]Environmental Protection Society, Aligarh

ABSTRACT

A study has been made to assess the water quality of groundwater of hand pumps situated around the slaughter houses in respect of physico-chemical characteristics. Samples were collected within the range of 500 meter from the Makhdoom Nagar slaughter house situated at Mathura bye pass road, Aligarh and 1 km from slaughter house named as M/s. Hind Agro Industries Ltd., situate towards Anupshahar Road, Aligarh. There are two main slaughter houses out of which one is functioning in open area near Makhdoom Nagar and other M/s Hind agro industries. Samples were collected from the hand pumps situated around slaughter houses and results of these collected samples revealed that water quality of groundwater has been contaminated in most of the hand pumps situated near the slaughter houses in respect of physico-chemical characterisitics like colour, pH, conductivity, dissolved solids, hardness, chloride, DO and BOD. Hence, in the vicinity of these slaguther houses lot of pungent, untolerable, bad odour affects the area up to approximately 5km.

Keywords: *Conductivity, TDS, DO, BOD and Hardness.*

Introduction

Aligarh city is famous for the lock making industries, door fittings and Aligarh Muslim University. A Mathura bye-pass road from Sarsol links to Aligrah Mathura road and a road from Aligarh to Anupshahar goes to Moradabad *via* Dibai and Narora Atomic Power Plant (NAPP). Aligarh city has mainly two types of communities *i.e.* Hindus and Muslims having population more than 8 lacs. There is no water supply from surface water sources. So, majority of people use groundwater through hand pumps and submersible. In the new developed housing colonies public water distribution system is yet to be laid and hence the residents solely depend upon groundwater. The city is yet unsewered. No scientific study of groundwater quality has been carried out so far.

A study was conducted to assess the water quality of groundwater of hand pumps situated in vicinity of stored effluents of slaughter houses in low lying areas. There are two main slaughter houses out of which one is functioning in open area near Makhdoom Nagar situated at Mathura bye-pass road and other named as M/s. Hind Agro Industries situated near Central Dairy Farm (CDF) complex, about 1 km away from Aligarh–Anup Shahar road. In these slaughter houses approximately 2-5 thousand animals are slauhtered daily and their effluents and wastes reach in low lying areas/ fields. Although M/s Hind agro industries has well mechanized closed system and established ETP to treat effluent of their industries but due to malfunctioning of ETP and bye-pass drainage of effluents generates a bad smell. On the other hand, slaughter house situated near Makhdoom nagar has no mechanized closed system and ETP which creates a lot of pungent, untolerable, bad smell and affects inhabitants badly. Three samples were taken from the hand pumps situated within the range of approximately 500 metre from Makhdoom Nagar and four sample from the hand pump situated within the range of about 1 km from M/s. Hind Agro Industries. Number of workers have conducted their studies on water quality of surface and groundwater in terms of physico-chemical characteristics (Singh *et al.*, 1988-89,1991,1993 and 1994); Soren and Julian, 1977; Pande *et al.*, 1979). As far as it is reviewed that no such type of study has been made to assess the water quality of groundwater of hand pumps situated near the slaughter houses. Therefore it is proposed to carry out investigations on water quality of hand pumps in respect of physico- chemical characteristics.

Materials and Method

The groundwater samples were collected in a neat and clean two liter capacity white plastic Jericanes for general parameters and samples for DO were taken in 300 ml capacity borosil glass bottles and DO was fixed by using $MnSO_4$ and alkaline azide reagents.Methods of analysis, sampling and preservation of samples were adopted as per standard methods of APHA (1985), Trivedi and Goel (1984) and Kotiah and Kumaraswamy (1994).

All the chemicals, reagents, and glasswares were used of A.R. grade and glasswares of borosil make. Instruments related parameters were used of best company make. Parameters studied were colour, odour, pH, conductivity, dissolved solids, DO, BOD, total hardness, Ca/Mg hardness and chloride.

Sampling sites were selected as per following points and map.

1. Hand pump situated just adjacent to Makhdoom Nagar slaughter house, Mathura bye pass road, Aligarh. This sampling point is denoted as–(A).
2. Hand pump situated at about 100 metre away from Makhdoom Nagar slaughter house, Mathura bye pass road, Aligarh.This sampling point is denoted as–(B).

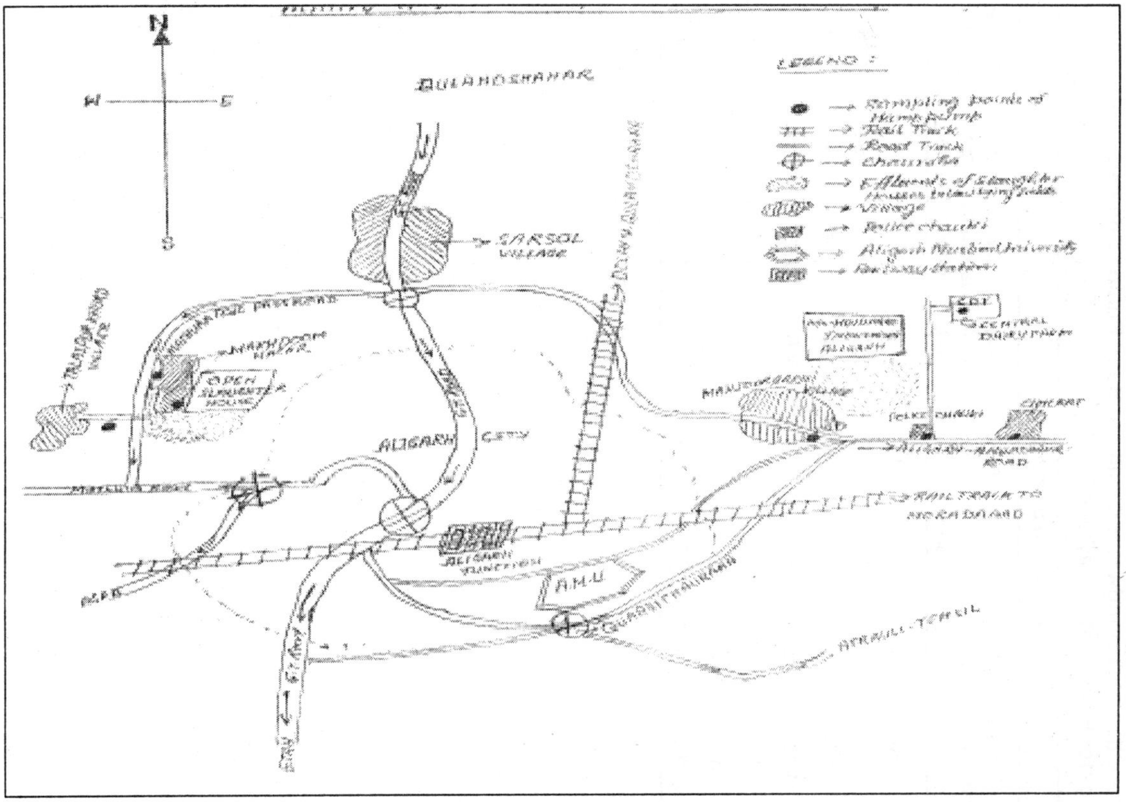

Figure 22.1: Map Showing Situation of Sampling Sites in the Vicinity of Slaugter Houses at Aligarh

3. Hand pump situated at about 250 metre away towards Talaspur Khurd village from Makhdoom Nagar slaughter house, Mathura bye pass road, Aligarh. This sampling point is denoted as–(C).
4. Hand pump situated in the premises of Central Dairy Farm (CDF) at approximately 200 metre away from M/s Hind agro industries, Anupshahar-road, Aligrah. This sampling point is denoted as–(D).
5. Hand pump situated near police check post at approximately 500 metre away from M/s. Hind agro industries, Anupshahar-road, Aligarh. This sampling point is denoted as-(E).
6. Hand pump situated on road at Chherat approximately 1 km away from M/s Hind agro industries, Anupshahar road, Aligarh. This sampling point is denoted as–(F).
7. Hand pump situated on road of Jamalpur bye pass tiraha near Manjoorgarhi approximately 1km away from M/s. Hind Agro Industries, Anupshahar Road, Aligarh. This sampling point is denoted as–(G).

All these hand pumps have been installed approximately 30-40 feet deep.

Table 22.1: Results of Physico-chemical Characteristics of Groundwater Quality of Hand Pumps Situated in the Vicinity of Slaughter Houses at Aligarh (U.P.)

Sl.No.	*Sampling Point*	*Date of Sample Collection*	*Analysed Parameters*										
			Colour	*Odour*	*pH*	*Conductivity (μmhos/cm)*	*TDS (mg/l)*	*DO (mg/l)*	*BOD (mg/l)*	*Total Hardness (mg/l)*	*Ca-Hardness (mg/l)*	*Mg-Hardness (mg/l)*	*Chloride (mg/l)*
1.	A	5/06/05	Straw	Unpleasant	9.3	1.96	2209	2.1	3.8	674	387	287	312
2.	B	5/06/05	Light Straw	Unpleasant	7.7	1.33	1874	3.2	1.8	480	290	190	228
3.	C	5/06/05	Colourless	No specific	7.4	0.819	1640	4.3	1.5	329	215	114	276
4.	D	5/06/05	Redish	Unpleasant	8.7	1.93	2166	4.1	1.4	380	258	122	109
5.	E	5/06/05	Light Straw	No specific	8.6	1.88	1832	4.8	1.6	366	184	182	256
6.	F	5/06/05	Colourless	No specific	7.7	0.812	1422	5.9	1.2	295	177	118	86
7.	G	5/06/05	Colourless	No specific	7.3	0.803	1107	6.7	0.9	336	197	139	94

Results and Discussion

Results of groundwater quality of hand pumps are given in Table 22.1 with respect to colour, odour, pH, conductivity, Total Dissolved Solids (TDS), DO, BOD, Total hardness, Ca/Mg hardness and chloride. To compare these characteristics in respect of groundwater quality for drinking purposes, prescribed standard limits of BIS-1991and CPCB-1997 are given in Table 22.2.As it is evident from the Table 22.1 that colour noted was straw, light straw, reddish, at sampling points A,B,D,E respectively and unpleasant odour at sampling point A,B,D while at other sampling points colour and odour found normal. The values of the parameters pH, conductivity, TDS, DO, BOD, Hardness and chloride ranged from 7.3-9.3, 0.803-1.96 mmhos/cm, 1107-2209 mg/l, 2.1-6.7mg/l, 0.9-3.8 mg/l, 295-674 mg/l, 177-387 mg/l,114-287mg/l and 86-312 mg/l respectively (Table 22.2) within the range of study areas. The values of total dissolved solids were found beyond prescribed standards 500 mg/l (Table 22.2) at all sampling points while the values of pH, Conductivity, BOD, chloride and hardness were observed higher than prescribed standard (BIS,1991and CPCB,1997) at sampling points (A,D,and E); (A,B,D and E); (A); (A,C and E) and (A,B,C,D,E and G,) respectively where as level of Dissolved Oxygen was observed below the standard limit of drinking purposes at all sampling point except sampling point-G. Minimum DO (2.1)mg/l and maximum values of Biochemical Oxygen Demand (3.8 mg/l); pH(9.3); conductivity (1.96 μmhos/cm); TDS (2209mg/l); Total hardness (674mg/l), calcium hardness (387 mg/l),magnesium hardness (287 mg/l) and chloride (312 mg/l), at sampling–A indicates that groundwater quality is most polluted which may be due to percolation of stored effluents of slaughter house in low lying areas/fields. The hand pump (sampling–A) is situated adjacent to slaughter house. The values of these characteristics decreases as distances of hand pumps increases from the main sources of effluents of slaughter house. Groundwater quality of hand pump (sampling–D) installed within the premises of M/s. Central Dairy Farm complex is not fit for drinking purposes in respect of pH, conductivity,TDS,DO,Total hardness calcium hardness, magnesium hardness while groundwater quality of sampling point-C, F and Cr is some what better as compared to other points on respect of pH, colour, odour, conductivity, DO and BOD. The values of pH of groundwater samples are found alkaline in nature at all sampling points.

Table 22.2: Drinking Water Quality Standard-BIS (Bureau of Indian Standard)-IS: 10500-1991 Manual for Water Testing Kit: 1997-CPCB

Sl.No.	*Characteristics*	*Standard*
1.	Colour	Colour
2.	Odour	Unobjectionable
3.	pH	6.5–8.5
4.	Conductivity (μmhos/cm)	I
5.	TDS (mg/l)	500
6.	DO (mg/l)	>6
7.	BOD (mg/l)	2
8.	Total hardness (mg/l)	300
9.	Ca-hardness (mg/l)	200
10.	Mg-hardness (mg/l)	100
11.	Chloride (mg/l)	250

Presences of BOD almost at all sampling points reveals that the groundwater quality upto some extent is being contaminated due to precolation of most polluted stored effluents of slaughter houses which may caues different kinds of diseases to human beings on drinking such contaminated water. Therefore, it is the duty of the concerned authorities to take necessary measures to control further contamination of groundwater by such type of water polluting industries.

References

APHA, AWWA, WPCF, 1985. *Standard Methods for the Examination for Water and Wastewater*, 16th edn. American Public Health Association, Washington DC.

BIS, 1991 and CPCB, 1997. *Drinking Water Quality Standards*. Bureau of Indian Standards, IS: 10500.

Kotaiah, B. and Kumaraswami, N., 1994. *Environmental Engineering: Laboratory Manual*. Charotar Publishing House, Anand, India, P. 115.

Pande, S.P., Narayanaswamy, V.S. and Hasan, M.Z., 1979. Quality of well waters of Nagpur with regard to nitrates, nitrites (Phase 1). *Ind. J. Environ. Hlth.*, 21(1): 35.

Singh, J.P., Yadava, P.K. and Singh, L., 1988. Pollution status on Sangam and its adjoning rivers before the Kumbh Mela at Allahabad. *Indian J. Environmental Protection*, 8(11): 839–842.

Singh, J.P., Yadava, P.K. and Singh, L., 1989. The assessment of water quality of Sangam and its adjoining rivers Ganga and Yamuna after Maha Kumbh Mela at Allahabad. *I.J.E.P.*, 1(5): 372–375.

Singh, J.P., Yadava, P.K. and Singh, L., 1989. Mass bathing effect on water quality of Sangam during Maha Kumbh Mela at Allahabad. *Indian J. Environmental Protection.*, 9(3): 189–193.

Singh, J.P., Yadava, P.K., Singh, Shakun and Prasad, S.C., 1991. BOD contamination in Kali river at Sadhu at Ashram in Aligarh. *I.J.E.P.*, 1(5): 325–326.

Singh, J.P., Upadhyay, R. and Singh, Shakun, 1994. A study of water quality of River Yamuna at Agra with special reference to coliform bacteria. *Res. J.P.I.Envi.*, 10: 53–57.

Singh, J.P., Yadava, P.K., Singh, Shakun and Madhu, Joshi, 1993. Assesement of BOD load in River Yamuna in respect of potable water at Agra. *I.J.E.P.*, 13(2): 119–123.

Soren and Julian, 1977. Groundwater quality near the water table in Suffolk Contury, Long Island, NewYork. Suffolk Country Department of Environemntal Control, *Long Island Water Resour. Bull.*, 8: 33.

Trivedi, R.K. and Goel, P.K., 1984. *Chemical and Biological Methods for Water Pollution Studies*. Karad (India).

Water and Wastewater Management in 2 Vols (2011) *Pages* **137–145**
Editors: **D.R. Khanna, R. Bhutiani & Gagan Matta**
Published by: **DAYA PUBLISHING HOUSE, NEW DELHI**

Chapter 23

Numerical Identification of BOD_5 and DO Profiles Using Beck Modified Khanna Bhutiani (BMKB) Model

D.R. Khanna, R. Bhutiani and Kumar S. Chandra
Department of Zoology and Environmental Science, Gurukula Kangri University, Haridwar

ABSTRACT

Seeing the mounting public pressure at water bodies the need to protect it from pollution is essential where mathematical modeling is the best alternative as accepted by the decision makers. The use of system analysis and mathematical modeling for formulating and solving river pollution is of relatively recent vintage and has been used widely during last three decades. The present study was aimed to verify the BMKB model for River Ganges. The results of BOD_5 shows the correlation $DF_{(1,2)}$ =2, n=3, F=. 588, r = 0.603 between two values *i.e.* obtained by Winkler's method and by model, which is significant. The regression model with the two value for BOD_5 is found as r^2 = 0.307 and BOD_5 (t)= 9.6–0.03 BOD_5 (m), while the result of DO shows the correlation $DF_{(1,2)}$ =2, n=3, t= 0.407, r = 0.120 between the two value which is significant. The regression model with the two value is found as r^2 = 0.300 and DO (t) = 8.4 + 0.30 DO (m).

Keywords: DO, Mathematical models, BOD_5-DO interaction, BMKB model.

Introduction

Water quality can be described in terms of physical, chemical and biological characteristics. Water bodies offers a convenient option for the disposal of domestic, agricultural and industrial effluents and wastewaters, all of which can significantly affect the natural physical, chemical and biological characteristics of receiving waters. Although biologists have been studying the effects of human activities on aquatic systems and organisms for decades, their findings have only relatively recently been translated into methods suitable for monitoring the quality of water bodies. A very fundamental concern with the existing approach is the fact that dissolved oxygen is one of the most important variables in water quality analysis. With most other substances, the less there is in the water, the better is the quality. But the situation is reversed for DO, higher the concentration of dissolved oxygen; the better is the water quality.

Bacteria and other microorganisms use organic substances for growth and these microorganisms consume oxygen dissolved in water to metabolize organic material.The organics are broken down into simpler compounds, such as CO_2 and H_2O, and microbes use the released energy for their growth and reproduction. If oxygen is not continually replaced in the water by artificial or natural means, then the DO level will decrease as the organics are decomposed by the microbes. This need for oxygen is called the Biochemical oxygen demand. Hence it is a sensitive indicator of the health of the aquatic system. Indeed the interaction between BOD_5 and DO is a complex process, which depends upon the several important factor for any given river systems. Complexity is an issue that is gaining much attention in the field of research.

Knowing the significance of BOD_5-DO interaction it is very important to think about BOD_5–DO models based upon mathematical equation and some physical law. Beck (1974) has given a detailed discussion of processes governing BOD_5–DO interaction. Before Beck it seems that Streeter and Phelps (1925) used mathematical models, for BOD_5 and DO for the first time. There after a number of models were suggested/developed and applied by Masch *et al.* (1970), Moley (1979), Orlob (1983), Vogler and Scherfig (2000).

Our study uses the recently developed BMKB model (Bhutiani, 2006) based upon Beck model (1974), regarding DO (Dissolve Oxygen) and BOD_5 (biochemical oxygen demand) to assess the extent of pollution load in River Suswa.

Materials and Methods

Study Sites

Total five Sampling sites were selected to verify BMKB model for the first time in Ganga river from Rishikesh to Haridwar. The sites were selected because at Rishikesh the turbulent velocity of river Ganga is high in comparison to Haridwar. This turbulent velocity plays an important role in BOD_5-DO interaction therefore to know the change in reaeration coefficient it is important to know the complexity of interactions occurring between Rishikesh to Haridwar.

Following five sampling Sites were selected for the study, which are situated at a stretch of approximately 35 kms.

C1: Triveni Ghat, Rishikesh

C2: Lakkar Ghat, Rishikesh

C3: Gohri Ghat, Haridwar

C4: Bhimgoda Barrage, Haridwar

C5: Chandighat, Haridwar

Analysis for State Variables

Two state variables *i.e.* BOD_5 and DO were analyzed in laboratory for twelve months (Jan. to Dec. 2005). BOD_5 and DO were measured by Winkler's method following the standard methods of APHA (1998), Trivedi and Goel (1984) and Khanna and Bhutiani (2005).

Model Application

The BMKB model is the modified form of an earlier developed BOD_5–DO model given by Beck (1974). The two multiple input/single output models given by Beck (1974) is as follows:

$$\text{DO: } y_1(k) = 0.715\, y_1(k-1) + 0.174\, u_1(k-1) + 0.057\, u_3(k) + 0.044\, u_3(k-1) + 0.554\, n_2(k)$$

$$\text{BOD: } y_2(k) = 0.751\, y_2(k-1) + 0.102\, u_2(k-1) + 0.048\, u_3(k-2) + 0.060\, u_3(k-4) + 0.618\, n_2(k) - 0.313\, n_2(k-1)$$

where,

y_1, y_2: Downstream observation of DO, BOD_5 respectively, *i.e.* $y_1(k, z_1), y_2(k, z_1)$

u_1, u_2: Upstream observation of DO, BOD_5 respectively, *i.e.* $u_1(k, z_0), u_2(k, z_0)$

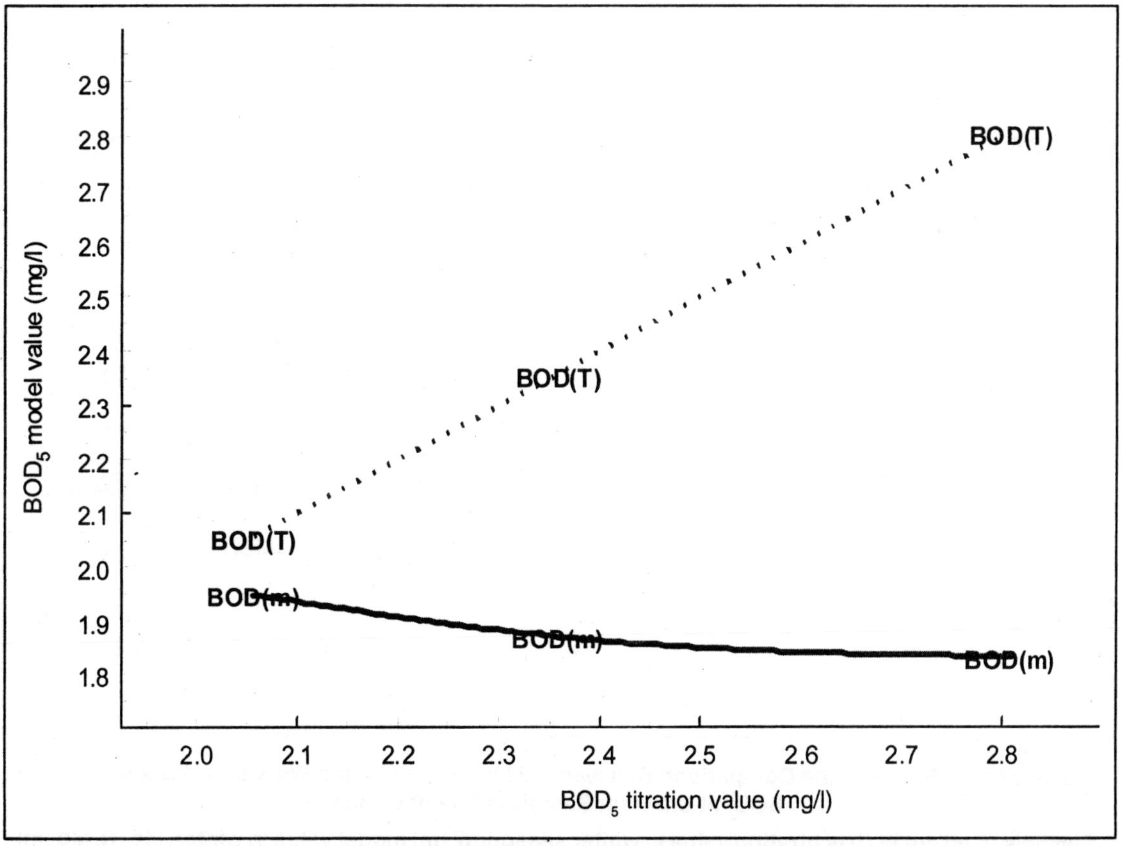

Figure 23.1: Showing the Comparison Between BOD_5 Value Obtained by Winkler's Method in BOD (t) Symbol with Dotted Line; while the BOD (m), Shows the BMKB Model Value with Continues Line

n_1, n_2: Stochastic noise sequences

u_3 (k): An observation of the sunlight incident on the system during the kith day (hr/day).

Using these governing processes, Bhutiani and Khanna (2006) developed BMKB model (Beck modified Khanna Bhutiani Model), which is used to obtain input/output, relationships of BOD_5 and DO between upstream and downstream of river system. BMKB model largely motivated by a case study of river CAM (Beck, 1974) encompasses the parameters λ_1, λ_2, U_1 and e_1.

$$\text{DO: } \lambda_1(S) = C_1\lambda_1(S-1) + C_2U_1(S-1) + C_3e_1(S)$$

$$\text{BOD: } \lambda_2(S) = C_4\lambda_1(S-1) + C_5U_1(S-1) + C_6e_1(S)$$

where,

λ_1 (S): Value of DO (mg/l) to be obtained in particular season.

λ_2 (S): Value of BOD (mg/l) to be obtained in particular season

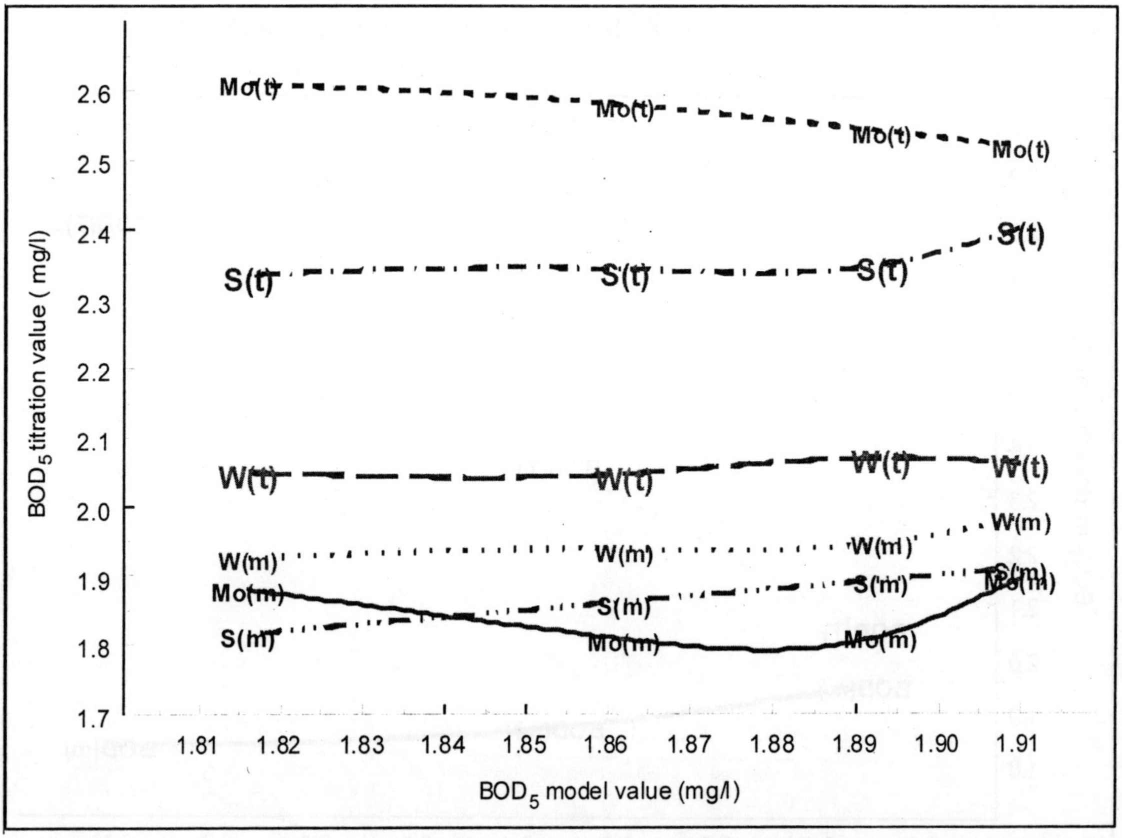

Figure 23.1(a): Showing the Comparison Between the Model Output and Winkler's Method Value for BOD_5 at Different Sampling Site of Ganga

Symbols used are W (t) is titration value in winter season, W (m) model value in winter, Mo (t), Mo (m) titration value and model value in monsoon season respectively, S (t) and S (m) are value of output by titration method and model respectively in summer.

λ_1 (S-1): DO/BOD observation of same place in previous season
U_1 (S-1): DO/BOD observation of upstream in previous season
e_1 (S): Errors of possibility.
C_1, C_2, C_3, C_4, C_5 and C_6 are constants

We use the BMKB model to verify the concentration of BOD_5 and DO of Ganga River from Rishikesh to Haridwar by comparing the model values with manually obtained values using Winkler's method.

Results

Model verification of any system depends upon the experimental result, which shows the model output and its interrelationship with other variables present in that particular system. The analyzed result of BOD_5 and DO were putted in the model.The value of BOD_5 obtained manually by Winkler's method and as obtained by model are shown in Figure 23.1. The result showed the correlation DF $_{(1,2)}$ =2, n=3, F= 0.588, r = 0.603 between these two values, which is significant. The regression model with the two value is found as $r^2 = 0.307$ and BOD_5 (t)= 9.6–0.03 BOD_5 (m).

Similarly, the values of DO obtained by Winkler's method and the model outputs are shown in Figure 23.2. The results showed the correlation DF $_{(1,2)}$ =2, n=3, t= 0.407, r =0.120 between these two value which is significant. The regression model with the two value is found as $r^2 = 0.300$ and DO (t) = 8.4 + 0.30 DO (m).

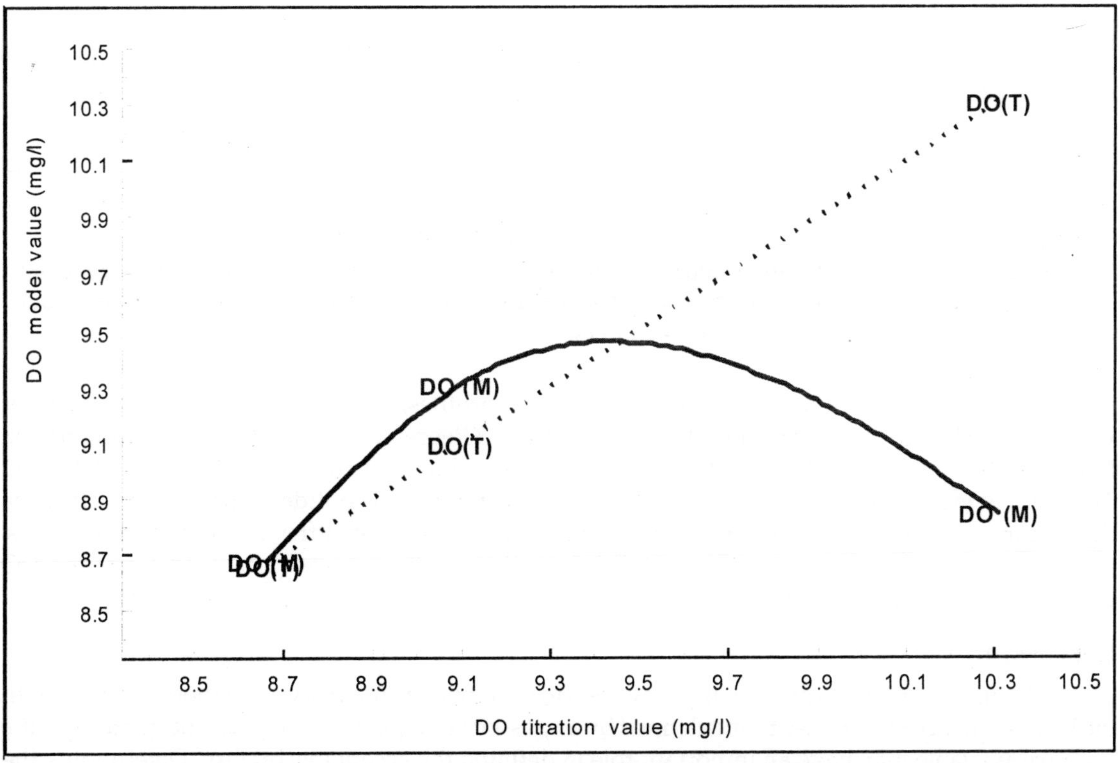

Figure 23.2: Comparison Between DO Values of Winkler's Titration Method in DO (t) Symbol with Dotted Line; While the DO (m), Shows the BMKB Model Value with Continues Line

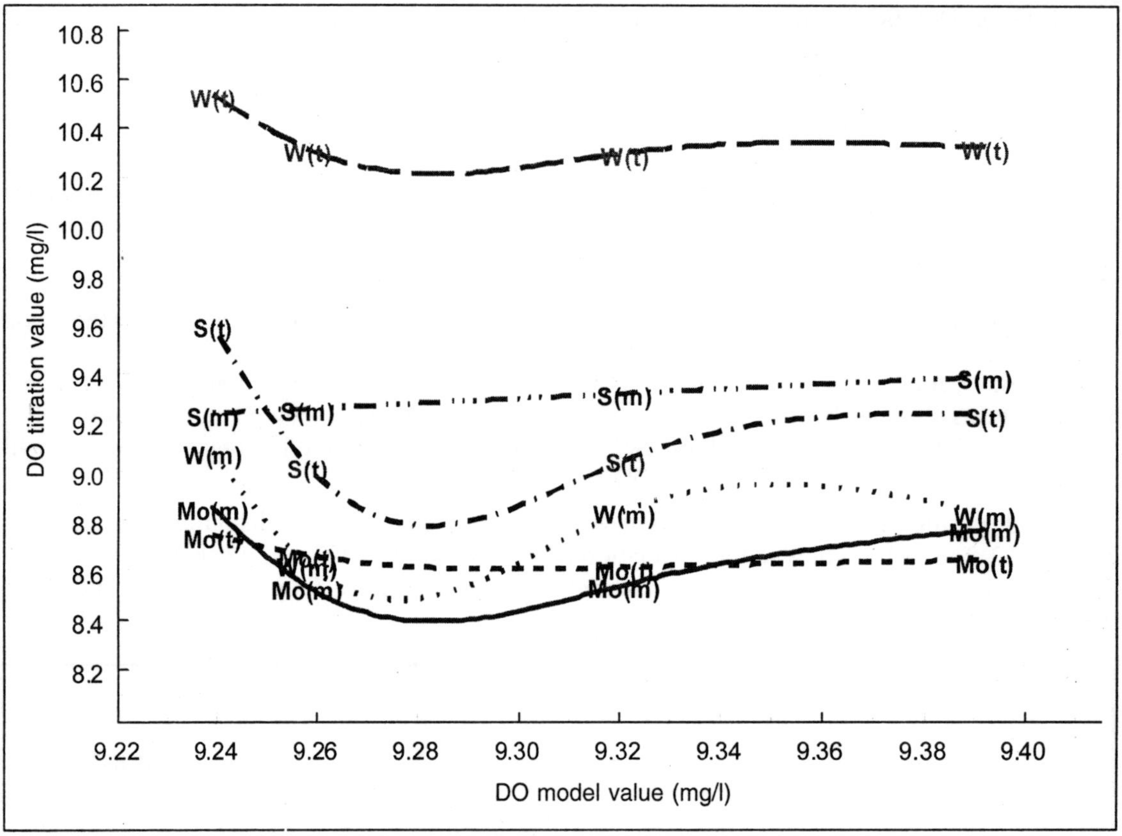

Figure 23.2(a): Comparison Between the Model Output and Winkler's Method Value for DO at Different Sampling Sites of Ganga

Symbols used are W (t) is titration value in winter season, W (m) model value in winter, Mo (t), Mo (m) titration value and model value in monsoon season respectively, S (t) and S (m) are value of output by titration method and model respectively in summer.

Figures 23.3 and 23.4 showed, the BOD_5 and DO profiles, both predicted and measured, for verification of BMKB model for Ganga River. In Figure 23.3 Regression plot of BOD_5-DO concentration has been shown in 95 per cent confidence level. Both the values were derived form the model input. The result in Figure 23.3 showed that all the value are in 95 per cent confidence band that verifies the data accuracy. Figure 23.4 showed the residual plot and the data Fit in different comparative Figures. Nevertheless, correlation between observed and predicted values seems quite good. Ofcourse, DO data shows better agreement in comparison with BOD_5 data, which is expected to be a better one.

Discussion

Although acceptable water quality may need to be defined principally in relation to the use with the highest quality requirements and defined by any associated quality standards, the quantity of the resource available may have an important role in defining the acceptable quality. Determining the degree and structure of the interaction between variables and model components is difficult to quantify.

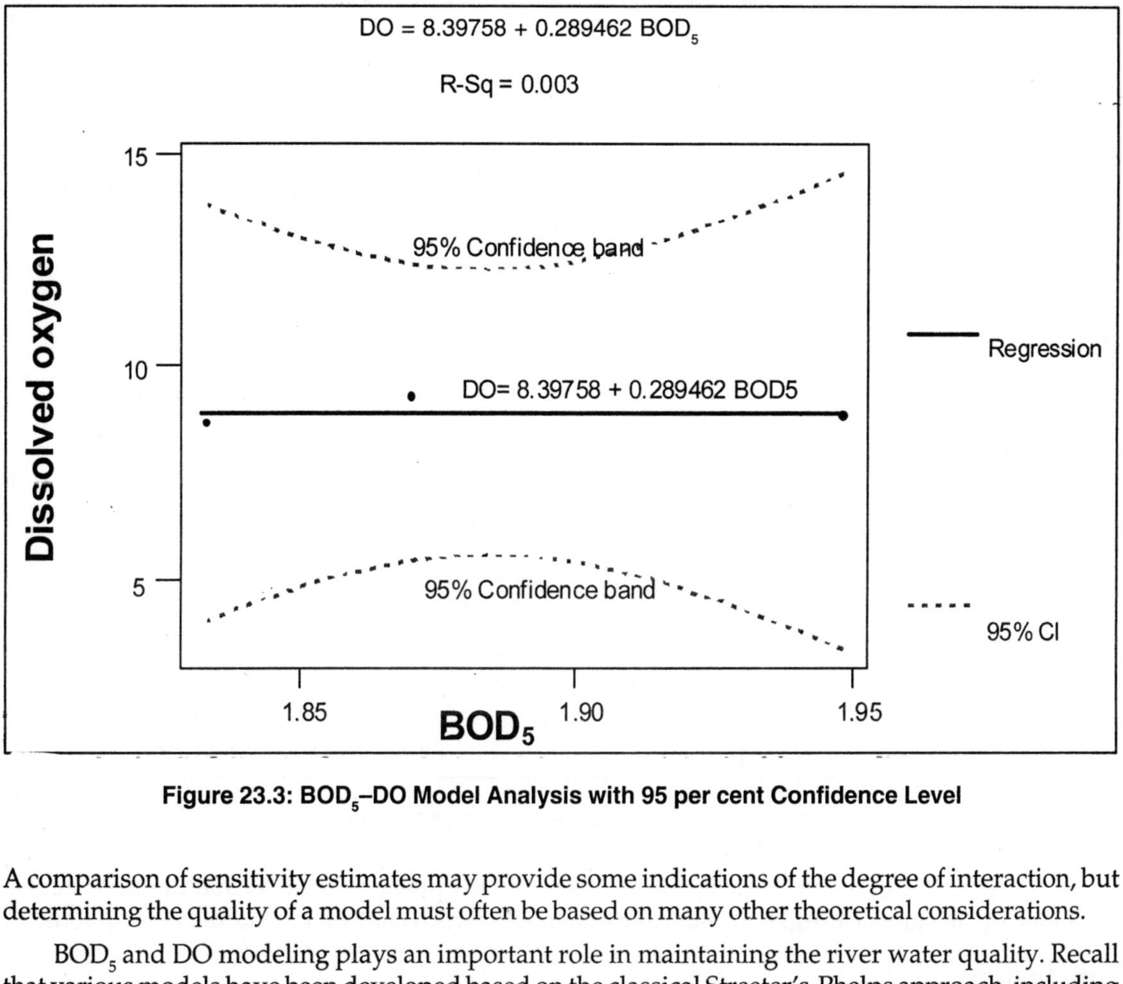

Figure 23.3: BOD_5–DO Model Analysis with 95 per cent Confidence Level

A comparison of sensitivity estimates may provide some indications of the degree of interaction, but determining the quality of a model must often be based on many other theoretical considerations.

BOD_5 and DO modeling plays an important role in maintaining the river water quality. Recall that various models have been developed based on the classical Streeter's-Phelps approach, including models, which incorporate oxygen demands by sediments and oxygen supply by photosynthesis. In addition there are lake and reservoir models as well as models of surface water quality are in Biswas (1981), Grimsrud *et al.* (1976), and a state of the art book on water quality modeling is James (1993). In the present study we used BMKB models for BOD_5 and DO with multiple inputs and single output. These models do not assume explicit knowledge of pre-biochemical reaction, such as the explicit pre-reaction concentration. In this paper firstly, there is a designs of input/output model (Beck model); secondly, it represents each process of material input, interaction and output by an BMKB model respectively, where the additional equations representing processes exploit the information of BOD_5 and DO relation reactions; and thirdly, it shows that the three-steps model has the advantage of prediction and control, using the numerical solutions model. Other parameters *i.e.*, error analysis has little effect on the river DO over the range of variation were the flow rate is tributary. BOD_5 has also effected very little by the error analysis in out study.

The model is considerably in agreement with our findings at all the sampling sites of river Ganga similar results have been obtained by Bhutiani and Khanna (2006) in river Suswa. An internally

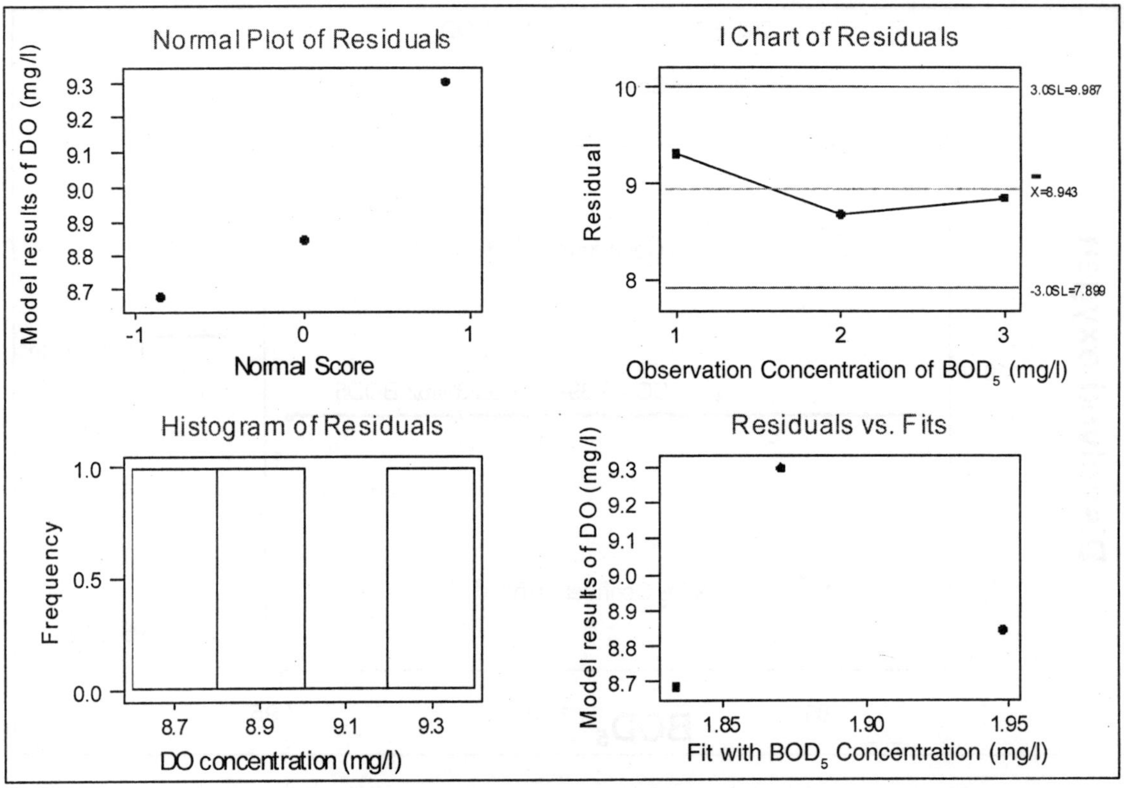

Figure 23.4: Error Analysis of BMKB Model

descriptive model exploits the available information on the phenomena determining the system's behaviors, *e.g.* the physical and biochemical mechanisms which control the internal descriptions.

Conclusion

In this study, we have investigated two water pollution models based on BMKB model (the input/output model (Beck model). The above findings show that BMKB model may be characterized as 'White Box model' and this model is the simplest model used for any river system. The model uses multiple inputs with single output and both BOD_5 and DO models have good potential for control applications. This chapter present the true verification and comparison of BMKB model with the experimental data. It seems that this modified model is the simplified once with competition to the Beck model (1974). Input and output results are quite accurate and equation derivation with data is easier.

References

APHA, AWWA, WPCF, 1998. *Standard Methods for the Examination of Water and Wastewater*, 20th edn. Washington D.C., New York.

Beck, M.B., 1974. *Maximum Likelihood Identification Applied to DO–BOD–Algae Models for a Freshwater Stream*. Report 7431(C), Lund Institute of Technology, Division of Automatic Control, Sweden.

Bhutiani, R. and Khanna, D.R., 2006. Ecological study of River Suswa: Modeling DO and BOD. *Environmental Monitoring and Assessment*, Netherlands (In press).

Biswas, A.K., 1981. *Models for Water Quality Management*. McGraw Hill Company, New York.

Grimsrud, G.P., Finnemore, E.J. and Owen, H.J., 1976. *Evaluation of Water Quality Models: A Management Guide for Planners*. EPA 600/5–76–004, U.S. Environmental Protection Agency, Washington, D.C. pp. 21–35.

James, A., 1993. *An Introduction to Water Quality Modeling*. John Wiley and Sons, West Sussex, England.

Khanna, D.R. and Bhutiani, R., 2005. *Water Analysis at a Glance*. ASEA Publication, pp. 1–116.

Masch, F.D. and Associates, 1970. QUALI, *Simulation of Water Quality in Streams and Canals, Program Documentation and User's Manual*. Report to Texas water development Board, Austin, Texas.

Morley, Donald A., 1979. *Mathematical Modeling in Water and Wastewater Treatment*. Applied Science Publishers Ltd. London, pp. 1–366.

Orlob, G.T., 1983. *Mathematical Modeling of Water Quality: Streams, Lakes and Reservoirs*. John Wiley and Sons, pp. 1–518.

Streeter, H.W. and Phelps, E.B., 1925. *A Study of the Pollution and Natural Purification of the Ohio*. Bulletin No. 146. U.S. Public Health Service.

Trivedy, R.K. and Goel, P.K., 1984. *Chemical and Biological Methods for Water Pollution Studies*. Environmental Publication, Karad, pp. 1–251.

Vogler, E.T. and Scherfig, J.S., 2000. Modeling BOD removal in constructed wetlands with mixing cell method. *Journal of Environmental Engineering*, 125: 782–785.

Water and Wastewater Management in 2 Vols (2011) *Pages* **146–150**
Editors: **D.R. Khanna, R. Bhutiani & Gagan Matta**
Published by: **DAYA PUBLISHING HOUSE, NEW DELHI**

Chapter 24

Correlation of Physico-chemical Parameters in Virla Reservoir, M.P.

S.K. Pathak[1] and L.K. Mudgal[2]
[2]Department of Zoology, Govt. Autonomous Holkar Science College, Indore
[2]Govt. Girls P. G. College, Moti Tabela, Indore

ABSTRACT

Water samples were collected from Virla reservoir during July 2001 to June 2003. In the present study an inter-relations has been established by statistical methods. Statistical analysis of the present data indicated the existence of a positive correlationship among various physico-chemical parameters, like water temperature, transparency, pH, free carbondioxide, alkalinity, dissolved oxygen (DO), biochemical oxygen demand (BOD), hardness and phosphate.

Keywords: *Reservoir, Water, Correlation, Statistical.*

Introduction

Now a days limnology has become a well established interdisciplinary branch of science in ecology (Tiwari and Patel, 1991). The statistical technique to interpret data of various aquatic habitat were also used by Nair *et al.* (1980), Tiwari and Mazoor (1988), Somashehara Rao (1990), Kannan and Rajasekaran *et al.* (1991), Reddy *et al.* (1992), Nagarajan *et al.*(1993), Singanan *et al.* (1995), Prakash (1996), Dnembare and Ponde (1997), Nadoni *et al.* (2001). The correlation analysis method is a very useful tool for assessing the water quality parameters.

Materials and Method

Virla reservoir in the vicinity of Virla village with an expansion of 525 hectares is located at 21°-50'-30° latitude and 75°-23'-30° longitude about 309 M above the MSL. The reservoir is primarily rainfed and has been stagnant since state time.

Table 24.1: Site-wise Range and Average Values of Physico-chemical Characteristics at Virla Reservoir During July 2001 to June 2003

Parameters	Site I		Site II		Site III		Site IV	
	Range	Average	Range	Average	Range	Average	Range	Average
Water temp. (°C)	19.20-32.50	24.93	19.40-33.00	26.27	19.70-33.40	25.75	20.40-34.20	26.65
Transparency (cm)	71-170	116.16	62.165	110.50	60-161	103.62	57-148	97.29
pH	7.38-8.76	8.10	7.45-8.85	8.17	7.52-9.00	8.23	7.58-9.26	8.32
Free Carbondioxide (mg/l)	3.86-4.38	4.10	3.94-4.42	4.24	3.98-4.48	4.23	4.20-4.55	4.37
Alkalinity (mg/l)	133-238	183.95	140-245	190.50	144-252	195.70	147-256	199.54
Dissolved Oxygen (mg/l)	6.39-9.84	8.37	6.23-9.78	8.27	6.10-9.62	8.11	5.80-9.51	8.01
Biological Oxygen Demand (mg/l)	3.26-5.02	3.87	3.28-5.15	3.97	3.33-5.26	4.02	3.37-5.53	4.09
Hardness (mg/l)	150-208	173.33	160-214	181.29	162-222	185.79	165-233	191.79
Phosphate (mg/l)	0.20-0.36	0.27	0.21-0.39	0.29	0.24-0.45	0.34	0.27-0.50	0.37

Water samples were drawn from the surface of sampling site in 5 liters capacity plastic containers,in the first week of every month for 24 months *i.e.,* from July 2001 to June 2003. Analysis of various physico-chemical parameters was done according to the procedure given in the standard method of APHA (1989) and Trivedi and Goel (1986).The results of monthly analysis for the parameters were taken as an average of 24 months study (Table 24.1).The correlation coefficient 'r' was calculated directly by using computer program SPSS (Tables 24.2–24.4).

Results and Discussion

The seasonal data obtained for the various physico-chemical features are presented in Table 24.1. The correlation coefficient among the different parameters are presented in the Table 24.2. All the high degree positive and the high degree negative correlation coefficients 'r' are presented in Tables 24.3 and 24.4 respectively. The analysis shows the high degree positive correlation between alkalinity and phosphate (0.893). There is also a high degree positive correlation between transparency and DO (0.885), pH and alkalinity (0.873), BOD and alkalinity (0.859), pH and phosphate (0.826) and BOD and phosphate (0.764). Similarly a moderate positive correlation exists between transparency and BOD (0.747), and Hardness (0.726), water temperature and BOD (0.652), Hardness and phosphate (0.579) and free carbon-dioxide and DO (0.518).

Table 24.2: Correlation Matrix of Physico-chemical Factors at Virla Reservoir During July 2001 to June 2003

Parameters	*Water Temp.*	*Trans-parency*	*pH*	*Carbon-dioxide*	*Alkalinity*	*Dissolved Oxygen*	*BOD*	*Hardness*	*Phosphate*
Water Temp.	1	–0.651	0.879	–0.605	0.885	–0.706	0.652	0.762	0.821
Transparency		1	–0.718	0.651	–0.827	0.885	–0.839	–0.259	–0.784
pH			1	–0.562	0.873	–0.786	0.747	0.726	0.826
Carbondioxide				1	–0.592	0.518	–0.450	–0.332	–0.511
Alkalinity					1	–0.895	0.859	0.497	0.893
Dissolved Oxygen						1	–0.988	–0.238	–0.794
BOD							1	0.190	0.764
Hardness								1	0.579
Phosphate									1

Table 24.3: High Degree Positive Correlation Values of Physico-chemical Factors at Virla Reservoir During July 2001 to June 2003

Parameters	*Water Temp.*	*Trans-parency*	*pH*	*Carbon-dioxide*	*Alkalinity*	*Dissolved Oxygen*	*BOD*	*Hardness*	*Phosphate*
Water Temp.	1		0.879		0.885			0.762	0.821
Transparency		1				0.885			
pH			1						0.826
Carbondioxide				1					
Alkalinity					1		0.859		0.893
Dissolved Oxygen						1			
BOD							1		0.764
Hardness								1	
Phosphate									1

Table 24.4: High Degree Negative Correlation Values of Physico-chemical Factors at Virla Reservoir During July 2001 to June 2003

Parameters	*Water Temp.*	*Trans-parency*	*pH*	*Carbon-dioxide*	*Alkalinity*	*Dissolved Oxygen*	*BOD*	*Hardness*	*Phosphate*
Water Temp.	1								
Transparency		1			–0.827		–0.839		–0.784
pH			1			–0.789			
Carbondioxide				1					
Alkalinity					1	–0.895			
Dissolved Oxygen						1	–0.988		–0.794
BOD							1		
Hardness								1	
Phosphate									1

The high degree negative correlation coefficient 'r' between DO and BOD (-0.988), DO and alkalinity (- 0.895), transparency and BOD (-0.839) and alkalinity (-0.827), DO and Phosphate (-0.794), DO and pH (-0.786) and transparency with phosphate (-0.784). The moderate negative correlation is also shown between transparency and pH (-0.718), water temperature with DO (-0.706) and transparency (-0.651),free carbon-dioxide with water temperature (-0.605), alkalinity (-0.592), pH (-0.562) and phosphate (-0.511).

Conclusion

The correlation coefficient 'r' among the parameters was negative 18 times and positive 19 times. This showed that positive relationship dominated in the present reservoir. The study helps in predicting compositional structure and their dependence to each other in reservoir ecosystem. It also gives an idea about quick monitoring of water and pollution status.

Acknowledgement

The authors are thankful to the Principal, Govt. P.G. College, Khargone for providing laboratory facilities.

References

APHA, 1995. *Standard Methods for the Examination of Water and Wastewater*, 19th edn. American Public Health Association.

Dnembare, A.J. and Ponde, G.M., 1997. Correlations of groundwater quality parameters of Sonai area (Maharashtra). *Poll. Res.*, 16: 188–190.

Kannan, N. and Rajasekaran, N., 1991. Correlations of water quality parameters of printing industry effluents in Sivakashi (South India). *Ind. J. Environ. Hlth.*, 33: 330–335.

Nagarajan, S., Swaminathan, M. and Sabarathinam, 1993. Correlation study on physico-chemical characteristics of groundwater in and around Chidambaram town, Tamil Nadu. *Poll. Res.*, 12: 245–250.

Nair, B.M.S., Seth, O. N. and Mishra, K.D., 1980. Limnological studies on a village pond, India (Vidisha). I. Physico-chemical characteristics. In: *Proc. Nat. Symp. Past. Present and Feature of Bhopal Lakes*, p. 195–201.

Prakash, M.M., 1996. *Correlation among the Physico-chemical Characteristics of Pond Water Pollution*, (Ed.) S.R. Mishra. APH Publishing Corporation, New Delhi, India, pp. 441–451.

Reddy, P.J., Krishna, D. and Gajghate, D.D., 1992. Correlates of COD and BOD sewage water of some major Indian cities. *Poll. Res.*, 11: 69–72.

Singanan, M.K., Rao, Soma Shekarappa and Ramababy, C., 1995. A correlation study on physico-chemical characteristics of groundwater in Rameswaram island. *Ind. J. Environ. Pro.*, 15: 213–217.

Somashekara Rao, K., 1990. Correlations among water quality parameters of groundwaters of Nazvid town and Nazvid Mandalam. *Ind. J. Environ. Prot.*, 13: 261–266.

Tiwari, T.N. and Ali, Mazoor, 1988. Correlation among water quality parameters of industrial waters. *Ind. J. Environ. Prot.*, 8: 44.

Tiwari, T.N. and Patel, M.K., 1991. Limnological application of some well known Mathematical Techniques. In: *Current Trends in Limnology*, (Ed.) N.K. Shastree. Narendra Publishing House, New Delhi, pp. 301–318.

Trivedy, R.K. and Goel, P.K., 1986. *Chemical and Biological Methods for Water Pollution Studies*. Environmental Publications, Karad, India.

Vadhirajan, R., Pandey, G. and Swami, M.S.R., 1993. Correlation among some physico-chemical parameters of surface water in Manali area, Madras. *IJEP.*, 13(6): 420–430.

Water and Wastewater Management in 2 Vols (2011) *Pages* **151–156**
Editors: **D.R. Khanna, R. Bhutiani & Gagan Matta**
Published by: **DAYA PUBLISHING HOUSE, NEW DELHI**

Chapter 25

Water Quality Status of River Ganges in Respect of Physico-chemical and Microbial Characteristics at Anupshahar, District Bulandshahar, U.P.

J.P. Singh[1], Shakun Singh[2] and D.R. Khanna[2]
[1]*U.P. Pollution Control Board, Aligarh, U.P.*
[2]*Department of Zoology and Environmental Sciences, Gurukula Kangri University, Haridwar*

ABSTRACT

A study was carried out to assess the water quality status of river Ganga with respect to physico-chemical and bacteriological characteristics at Anupshahar town (Tehsil of distt. Bulandshahar), to collect the samples. Before sampling point-B, several sewage drains discharge their effluent into river Ganga and bathing ghat activities are also held after sampling point- A. Parameters studied ranged from pH 7.0-8.97, Total Solids 112-902 mg/l,Total Dissolved Solids 90-262 gm/l, Fixed Dissolved Solids 30-76 mg/l, Conductivity 0.16-0781 mhos/cm, Akalinity 59-110 mg/l, Hardness 61-119 mg/l, DO 9.8-12.6 mg/l, BOD 1.2-5.5 mg/l, COD 12-35 mg/l and Total Coliform 700-24000 MPN/100ml. But out of these parameters the average values of BOD (3.483 mg/l) and total coliform bacteria (5408 MPN/100ML) at sampling point-B and BOD (2.25 mg/l) and total coliform bacteria (3445 MPN/100 ml) at sampling point-A were obtained beyond the tolerance limits of drinking purpose. However, average values of these parameters were found higher at sampling point-B than sampling point-A which may be due to contaminations of sewage water of the town as well as bathing activities held before sampling point-B. Nature of water quality of river Ganga with respect to pH is alkaline. Present paper indicates that water quality of river Ganga is not suitable for drinking purpose with respect to BOD and bacteria within the range of study area.

Keywords: *Conductivity, TDS, FDS, Hardness, DO, BOD and MPN.*

Introduction

Anupshahar is Tehsil of district Bulandshahar (UP) which is situated at the bank of river Ganga where number of people take bath daily. Anupshahar is connected through the roads of Aligarh, Bulandshahar and Narora Atomic Power Plant. Narora Atomic Power Plant is situated 20 km away from Anupshahar in down stream of Ganga. National capital of India (Delhi) is about 130 km away in west and Aligarh city 55 km in south-west from Anupshahar. Ganga river passes through the cities/ towns such as Rishikesh, Haridwar, Bijnor, Garhmukteshwar, Anupshahar and Narora *etc*. There are several bathing ghats and domestic sewage drains from where wastewater reaches the river. Main Hindu festivals are also organized at bathing ghats of the town. Number of workers have carried out their investigations on water quality of River Ganga with respect to different physico-chemical and microbiological characteristics from its origin to merging point (Singh *et al.*, 1988, 89, 91, 93, 94; Khanna *et al.*, 2003; Gautam and Sati, 1987). But no study has been conducted to assess the water quality of river Ganga at Anupshahar. Therefore it was proposed to conduct a study on water quality of river Ganga at this town (Anupshahar). This town is having population approximately 40 thousand. The present investigation includes one year study (April-2001-March-2002).

Materials and Method

Water samples of River Ganga were collected in neat and two litre white plastic Jericanes for general parameters and samples for coliform bacteria were collected in 125 ml sterilized Borosil glass stoppered bottles. Samples for Dissolved Oxygen (DO) were collected in a neat and clean 300 ml capacity Borosil glass stoppered DO bottles. DO was fixed by using 1 ml of each reagents $MnSO_4$ and alkaline azide on the spot of sampling points. Grab sample collection methods was adopted through out the study. Sample preservation and analytical methods were adopted as per APHA, AWWA, WPCF-(1992), Trivedi and Goel (1984), Kotiah and Kumaraswami (1994).

All chemicals, reagents and solvents were used for the analysis of analytical grade procured from Qualigens and SD fine chemicals and glass wares used were of Borosil make. Instruments used were pH meter, Conductivity meter, BOD incubator, Hot air oven, Bacteriological incubator, Water bath and Muffle furnace were of best company make.

Two sample points were selected as U/s-Anupshahar denoted as sampling point-A andD/s Anupshahar denoted as sampling point-B. Sampling point-A is approximately 500 metre in upstream of Anupshahar and sampling point-B is situated 1km in D/s of Anupshahar. There is a bathing ghat at river bank of Ganga just after sampling point-A and several small or medium sewage drains are merging into Ganga river before sampling point-B.

Results and Discussion

Samples analysed for physico-chemical and bacteriological parameters are given in Tables 25.1 and 25.2 and their minimum, maximum and average values are given in Table 25.3. Prescribed standards of tolerance limits for drinking purpose are mentioned in Table 25.4 to compare the results of studied characteristics to evaluate the water quality of River Ganga (BIS-1991 and CPCB-1997).

The values of studied parameters pH, TS, TDS, FDS, Conductivity, DO, BOD, COD, alkalinity, hardness, Ca-Hardness, Mg-Hardness, Chloride and Total coliform varied between 7.42-8.81; 112-883 mg/l; 90-183 mg/l; 30-56mg/l; 0.16-0.65 mhos/cm;10.7-12.6mg/l,1.2-3.6mg/l;12-32mg/l;59-100mg/l;61-119mg/l;49-70mg/l;12-49mg/l;7.2-29.4 mg/l; and 700- 16000 MPN/100 ml respecticvely at sampling point-A where as at sampling point-B the values of these parameters ranged from 7.0-8.97; 122-902 mg/l; 96-262 mg/l; 31-76 mg/l; 0.172-0.781 mhos/cm; 9.8-12.0 mg/l; 2.0-5.6 mg/l; 13-35

Table 25.1: Results of Physico-chemical and Microbial Characteristics at Sampling Point-A (Year 2001–2002)

Parameters	April	May	June	July	Aug	Sep	Oct	Nov	Dec	Jan	Feb	Mar
Physical Parameters												
Ambient Temp. (°C)	28±0.20	38±0.40	41±0.80	32±0.60	29±0.40	30±0.30	31±0.20	30±0.16	25±0.40	18.2±0.20	22.6±0.21	29.4±0.30
Water Temp. (°C)	21±0.40	22.3±0.30	23±0.40	21.1±0.20	20.8±0.20	20.9±0.10	20.6±0.20	20±0.12	19.5±0.07	14.6±0.52	15±0.63	20.4±0.50
pH	8.03±0.06	8.08±0.05	8.19±0.30	8.81±0.30	8.7±0.30	8.64±0.211	8.3±0.02	7.96±0.12	7.61±0.0	7.42±0.08	7.68±0.14	7.7±0.08
T.S (mg/l)	138±6.10	257±7.40	424±10.0	660±18	883±0.22	719±0.211	306±0.12	201±8.0	170±5.6	116±2.4	112±3.0	132±2.6
T.D.S (mg/l	116±4.2	130±3.0	156±4.8	165±3.6	183±3.4	179±3.2	182±3.4	178±3.6	150±2.0	96±1.6	90±1.2	110±2.1
F.D.S (mg/l)	36±0.8	41.6±1.0	46±1.2	48.9±1.4	56±1.3	45±1.6	42.2±1.4	40±1.8	36±1.1	30±0.8	32±1.0	40±0.6
Cond. (mMhos/cm)	0.26±0.02	0.27±0.03	0.40±0.04	0.65±0.03	0.58±0.50	0.521±0.20	0.43±0.01	0.34±0.01	0.241±0.0	0.16±0.0	0.176±0.0	0.209±0.0
Chemical Parameters												
D.O (mg/l)	11.5±0.8	11.2±0.6	10.7±0.5	11±0.8	12.1±0.7	11±0.2	11.2±0.04	11±0.01	12±0.08	12.6±0.2	11.8±0.12	11±0.1
C.O.D (mg/l)	14±0.9	18±1.0	22±1.4	29±1.2	32±1.6	26±1.2	24±0.03	20±0.9	16±0.4	12±0.2	12.4±0.14	13±0.12
B.O.D (mg/l)	1.8±0.1	2±0.3	2.9±0.2	3.6±0.2	3±0.3	2.6±0.2	1.8±0.0	1.6±0.06	1.2±0.01	1.8±0.01	2±0.01	2.4±0.01
Alkalinity (mg/l)	69±0.4	74±1.2	76±1.2	79±0.8	81±1.0	74±0.9	100±3.2	90±0.08	76±1.2	59±1.1	60±0.8	65±1.0
T. Hardness (mg/l)	75±0.6	78±1.4	90±1.4	119±1.8	80±0.8	85±1.0	86±2.4	84±1.3	78±1.0	81±1.3	68±1.0	77±1.3
Ca Hardness (mg/l)	60±0.5	58±0.3	66±0.8	70±0.9	54±0.4	60±0.5	66±1.2	60±0.7	58±0.8	49±0.5	54±0.9	80±1.5
Mg Hardness (mg/l)	15±0.3	20±0.2	24±0.3	49±0.3	26±0.2	25±0.2	20±0.4	24±0.4	20±0.08	12±0.7	14±0.03	17±0.1
Chloride (mg/l)	12.9±0.6	20±0.8	21.5±0.0	29.4±0.3	26±0.3	18.6±0.2	10±0.2	8±0.04	7.4±0.03	7.2±0.2	7.5±0.1	7.9±0.03
NO_2 (mg/l)	0.016±0.0	0.022±0.0	0.0231±0.0	0.039±0.0	0.020±0.0	0.016±0.0	0.014±0.0	0.013±0.0	0.011±0.0	0.01±0.0	0.012±0.0	0.016±0.0
NO_3 (mg/l)	0.085±0.0	0.094±0.01	0.106±0.02	0.412±0.08	0.30±0.01	0.08±0.03	0.09±0.0	0.085±0.0	0.06:±0.0	0.03±0.0	0.04±0.0	0.05±0.0
Metals												
Chromium (mg/l)	0.001±0.0	0.002±0.0	0.002±0.0	0.003±0.0	0.003±0.0	0.002±0.0	0.002±0.0	0.001±0.0	0.001±0.0	0.00±0.0	0.001±0.0	0.001±0.0
Cobalt (mg/l)	0.00±0.0	0.00±0.0	0.00±0.0	0.002±0.0	0.001±0.0	0.00±0.0	0.00±0.0	0.00±0.0	0.00±0.0	0.00±0.0	0.00±0.0	0.00±0.0
Copper (mg/l)	0.020±0.01	0.03±0.0	0.041±0.0	0.048±0.01	0.049±0.0	0.49±0.01	0.036±0.0	0.028±0.0	0.21±0.0	0.02±0.0	0.025±0.0	0.03±0.0
Iron (mg/l)	0.206±0.06	0.312±0.04	0.463±0.06	0.912±0.04	0.865±0.0	0.790±0.02	0.641±0.01	0.460±0.13	0.280±0.01	0.19±0.0	0.189±0.0	0.198±0.0
Lead (mg/l)	0.002±0.0	0.004±0.0	0.005±0.0	0.006±0.0	0.006±0.0	0.004±0.0	0.003±0.0	0.002±0.0	0.00±0.0	0.001±0.0	0.001±0.0	0.0015±0.0
Nickle (mg/l)	0.02±0.0l	0.02±0.0	0.03±0.01	0.06±0.01	0.05±0.0	0.045±0.0	0.036±0.0	0.024±0.0	0.013±0.0	0.01±0.0	0.01±0.0	0.02±0.0
Potassium (mg/l)	2.4±0.09	2.8±0.08	2.8±0.08	5.6±0.10	3.2±0.09	2.9±0.01	2.8±0.01	2.6±0.01	2.2±0.05	2.0±0.01	2.1±0.01	2.2±0.01
Sodium (mg/l)	12.8±1.0	13± 1.4	13.2±0.5	48±2.0	20±0.80	16±0.80	13.1±0.03	13±0.04	12.6±0.4	11.8±0.3	11.6±0.05	12±0.06
Zinc (mg/l)	0.06±0.02	0.065±0.01	0.081±0.04	0.107±0.01	0.098±0.0	0.090±0.01	0.082±0.0	0.074±0.0	0.042t0.0	0.015±0.0	0.02±0.0	0.039±0.0
Microbial Parameters												
Total Coliform (MPN/100 ml)	1100±20	1700±30	1400±50	16000±210	9200±80	3400±60	2500±50	2200±50	1100±51	940±20	700±20	1100±40
S.P.C/ml	34±1.2	56±1.8	78±2.1	161±4.2	130±3.1	95±2.8	41±1.2	36±2.1	25± 1.1	25±0.8	18±0.4	30±1.2

Table 25.2: Results of Physico-chemical and Microbial Characteristics at Sampling Point-B (Year 2001-2002)

Parameters	April	May	June	July	Aug	Sep	Oct	Nov	Dec	Jan	Feb	Mar
Physical Parameters												
Ambient Temp. (°C)	28.2±0.30	38±0.20	42±0.40	32±0.50	29±0.20	30±0.40	31±0.50	30±0.20	25±0.50	18.2±0.30	22.6±0.30	29.4±0.50
Water Temp. (°C)	21.6±0.20	22.4±0.10	22.9±0.30	21.0±0.20	20.6±0.20	20.7±0.10	20.4±0.30	20.1±0.10	19.2±0.10	14.8±0.30	15±0.40	20.3±0.20
pH	8.18±0.20	8.23±0.30	8.39±0.20	8.97±0.50	8.83±0.30	8.64±0.20	8.76± 0.50	8.52±0.30	8.04±0.20	7.61±0.50	7.6±0.30	7±0.50
T.S (mg/l)	198±8.10	282±11.20	590±18.50	902±15	898±21.2	799±2.50	452±3.12	239±4.80	172±3 040	122±2.10	125±1.80	122±2.0
T.D.S (mg/l)	166±2.1	192±3.5	200±4.2	262±5.10	188±7.30	190±4.2	192±2.8	139±4.8	130±5.70	98±3.50	101±2.80	96±1.9
F.D.S (mg/l)	56±1.2	70±1.8	68±1.7	76±1.3	50±2.0	46±2.10	44±2.10	43±1.5	48±1.9	38±1.7	36±1.6	31±1.4
Cond. (mMhos/cm)	0.252±0.0	0.26±0.0	0.484±0.0	0.781±0.02	0.71±0.02	0.702±0.01	0.68±0.01	0.61±0.0	0.424±0.02	0.172±0.1	0.211±0.1	0.21 9±0.1
Chemical Parameters												
D.O. (mg/l)	10.8±0.02	10.1±0.03	9.8±0.2	10.8±0.18	11.6±0.13	11.2±0.12	10.9±0.10	11.2±0.18	12±0.20	12±0.16	11.1±0.17	9.8±0.20
C.O.D (mg/l)	18±0.16	22±0.15	25±0.18	32±0.17	35±0.45	29±0.27	26±0.21	22±0.08	20±0.06	13.8±0.04	13±0.04	13.9±0.05
B.O.D (mg/l)	2.6±0.03	3.9±0.04	4.8±0.03	5.6±0.06	4.5±0.02	4.3±0.03	4.2±0.02	2.1±0.01	2±0.01	2.4±0.01	2.6±0.01	2.8±0.01
Alkalinity (mg/l)	82±1.2	100±1.5	108±1.1	110±1.1	82±0.80	86±0.9	101±0.10	98±0.00	82±2.0	62±1.2	65.1±1.5	66.2±1.4
T. Hardness (mg/l	88±1.4	95± 1.1	100±1.2	110±1.7	81±1.8	98± 1.1	92±1.5	100±1.7	90± 1. 6	65±1.3	69± 1.5	80±1.2
Ca-Hardness (mg/l)	68±204	69±2.0	70±1.8	76± 1.3	55±1.5	76±2.1	72±2.10	72±2.3	64±3.10	51±2.9	54±2.7	66±2.0
Mg-Hardness (mg/l)	20±0.17	26±0.17	30±0.11	34±0.15	26±0.13	22±0.18	20±0.13	28±0.17	26±0.21	14±0.31	15±0.42	14±0.21
Chloride (mg/l)	14±0.21	42±0.23	36±0.18	60.8±0.16	26.6±0.14	22±0.19	18±0.18	16±0.12	14±0.10	8.4±0.11	8.9±0.13	9.2±0.15
NO_2 (mg/l)	0.020±0.0	0.031±0.0	0.042±0.0	0.055±0.0	0.023±0.0	0.025±0.0	0.018±0.0	0.016±0.0	0.013±0.0	0.04±0.0	0.015±0.0	0.017±0.0
NO_3 (mg/l)	0.091±0.0	0.20±0.0	0.14±0.0	0.52±0.0	0.40±0.0	0.29±0.0	0.096±0.0	0.09±0.0	0.08±0.0	0.035±0.0	0.049±0.0	0.052±0.0
Metals												
Chromium (mg/l)	0.002±0.0	0.002±0.0	0.003±0.0	0.003±0.0	0.002±0.0	0.002±0.0	0.001±0.0	0.001±0.0	0.001±0.0	0.00±0.0	0.001±0.0	0.0015±0.0
Cobalt (mg/l)	0.002±0.0	0.003±0.0	0.00l±0.0	0.002±0.0	0.00±0.0	0.00±0.0	0.00±0.0	0.00±0.0	0.00±0.0	0.00±0.0	0.00±0.0	0.01±0.0
Copper (mg/l)	0.03±0.03	0.03±0.0	0.044±0.0	0.072±0.0	0.048±0.0	0.46±0.0	0.049±0.0	0.045±0.0	0.036±0.0	0.015±0.0	0.028±0.0	0.032±0.0
Iron (mg/l)	0.38±0.38	0.406±0.0	0.564±0.01	0.998±0.02	0.89±0.01	0.861±0.01	0.72±0.01	0.67±0.01	0.58±0.01	0.21±0.0	0.206±0.0	0.31±0.0
Lead (mg/l)	0.001±0.0	0.002±0.0	0.009±0.0	0.009±0.0	0.007±0.0	0.007±0.0	0.005±0.0	0.004±0.0	0.003±0.0	0.0012±0	0.0025±0	0.004±0.0
Nickel (mg/l)	0.02±0.0	0.03±0.0	0.035±0.0	0.08±0.0	0.052±0.0	0.051±0.0	0.049±0.0	0.043±0.0	0.037±0.0	0.01±0.0	0.015±0.0	0 021±0.0
Potassium (mg/l)	2.6±0.01	3±0.01	3.9±0.01	6.2±0.20	4.6±0.04	3.7±0.08	3.2±0.01	3.0±0.01	2.8±0.02	1.9±0.01	1.8±0.0	2.4±0.02
Sodium (mg/l)	14±0.08	15.1±0.07	15.9±0.08	57±0.05	30±0.15	18±0.25	13.5±0.88	18±0.09	13.4±0.15	12.6±0.14	12.8±0.15	13.4±0.15
Zinc (mg/l)	0.08±0.0	0.08±0.0	0.096±0.0	0.172±0.0	0.104±0.0	0.096±0.0	0.85±0.0	0.81±0.0	0.072±0.0	0.021±0.0	0.022±0.0	0.039±0.0
Microbial Parameters												
Total Coliform (MPN/100ml)	170000	2200±40	2200±42	24000±140	16000±110	9200±80	2400±46	2200±38	1400±20	1100±16	1100±18	1400±22
S.P.C/ml	49±2.0	76±3.0	110±4.6	220±5.2	173±4.8	131 ±3.0	85±1.8	48±1.6	30±1.1	49±1.8	29±1.2	49±2.15

mg/l; 62-110 mg/l; 65-110 mg/l; 51-76 mg/l; 14-34 mg/l; 8.4 60.8 mg/l; and total coliform bacteria 1100-24000 MPN/100 ml respectively. The average values of pH, TS, TDS, FDS, Conductivity, DO COD, alkalinity, hardness and chloride were obtained within the prescribed tolerance limit of drinking purpose (BIS-1991 and CPCB-1997) where as average values of BOD (2.25 mg/l and 3.48 mg/l) and Total coliform bacteria 3445 MPN/100ml and 5408 MPN/100 observed were beyond the standard limits at both the sampling point-A and B (Tables 25.3 and 25.4).

Table 25.3: Water Quality Status of River Ganga in Respect of Physico-Chemical and Bacteriological Characteristics at Anupshahar (Bulandshahar) during the Study Period 2001–02

Parameters	*2001–02*			*2001–02*		
	Sampling Point-A			*Sampling Point-B*		
	Minimum	*Maximum*	*Average*	*Minimum*	*Maximum*	*Average*
pH	7.42	8.81	8.09	7.0	8.97	8.4
Total solids (mg/l)	112.0	883.0	343.17	122.0	902.0	400.41
TDS (mg/l)	90.0	183.0	144.45	96.0	262.0	162.8
FDS (mg/l)	30.0	56.0	41.14	31.0	76.0	50.5
Conductivity (μmhos/cm)	0.16	0.65	0.350	0.172	0.781	0.45
DO (mg/l)	10.7	12.6	11.5	9.8	12.0	10.94
BOD (mg/l)	1.2	3.6	2.25	2.0	5.6	3.48
COD (mg/l)	12.0	32.0	19.87	13.0	35.0	22.48
Alkalinity (mg/l)	59.0	100.0	75.25	62.0	110.0	86.86
Total hardness (mg/l)	61.0	119.0	81.75	65.0	110.0	89.0
Ca-hardness (mg/l)	49.0	70.0	59.58	51.0	76.0	66.08
Mg-hardness (mg/l)	12.0	49.0	22.16	14.0	34.0	22.917
Chloride (mg/l)	7.2	29.4	14.69	8.4	60.8	29.04
Total Coliform (MPN/100ml)	700.0	16000.0	3445.0	1100.0	24000.0	5408.0

Maximum values of the studied parameters were observed during summer season and shock load of first rain in rainy season except DO and minimum values of these parameters obtained during winter, while higher values of DO were observed during rainy period at both sampling points which may be due to dilution in rainy period and super saturation of oxygen at lower ambient temperature and less human activities like bathing/washing *etc.* in winter. After winter period as temperature of the environment rises bathing washing and other human activities are also increased which gradually deteriorates the water quality of River Ganga. It is very interesting to note that an enhancement in the values of the studied parameters are obtained at sampling point-B more than sampling point-A, It is because

Table 25.4: Standards of Water Quality for Drinking Purpose in Terms of Physico-chemical Characteristics (BIS-1991 and CPCB-1997)

Parameters	*Standard Limits*
Colour	Colourless
Odour	Odourless
pH	6.5–8.5
TDS (mg/l)	500.0
Conductivity (μmhos/cm)	1.0
DO (mg/l)	>6.0
BOD (mg/l)	2.0
Total hardness (mg/l)	300.0
Calcium hardness (mg/l)	200.0
Magnesium hardness (mg/l)	100.0
Chloride (mg/l)	250.0

of several small/big sewage drains discharging their effluents into River Ganga and due to daily conducting bathing washing activities at bathing ghats situated before sampling point-B. Water quality of River Ganga with respect to pH is alkaline in terms of bacteria, may cause diseases that vary in severity from mild gastroenteritis to severe and sometime fatal dysentery, cholera or typhoid (WHO, 1984). Similar trends of water quality of River Ganga were also found by (Singh, 1988, 89 and Khanna, 2003). Therefore, it is the duties of concerned authorities as well as people of the area to protect such contaminations and bathing/washing by soaps because Ganga river is a most sacred river and worshiped by the Hindus. Public awareness is most essential specially for those who are responsible to pollute the natural water quality of River Ganga.

References

APHA, AWWA, WPCF, 1992. *Standard Methods for the Examination of Water and Wastewater*, 18th edn. American Public Health Association, Washington DC.

Abbasi, S.A., Khan, F.I. , Sentilvelan, K. and Shabudden, A., 1999. Modelling of Bunkingham Canal water quality. *Indian J. Environ. Health*, 41(3): 176–183.

BIS–1991 and CPCB–1997. *Drinking Water Quality Standards*. Bureau of Indian Standards, IS:10500.

Khanna, D.R., Singh, Shakun, Gautam, Ashutosh and Singh, J.P., 2003. Assessment of water quality of River Ganga in district Bulandshahar (UP), India. *J.Nat. Con.*, 15(1): 167–175.

Kotaiah, B. and Kumaraswami, N., 1994. *Environment Engineering: Laboratory Manual*. Charotar Publishing House, Anand, India, P. 115.

Singh, J.P., Yadava, P.K. and Singh, L., 1988. Pollution status on Sangam and its adjoning rivers before the Kumbh Mela at Allahabad. *Indian J. Environmental Protection*, 8 11): 839–842.

Singh, J.P., Yadava, P.K. and Singh, L., 1989. The assessment of water quality of Sangam and its adjoining rivers Ganga and Yamuna after Maha Kumbh Mela at Allahabad. *I.J.E.P.*, 1(5): 372–375.

Singh, J.P., Yadava, P.K. and Singh, L., 1989. Mass bathing effect on water quality of Sangam during Maha Kumbh Mela at Allahabad. *Indian J. Environmental Protection.*, 9(3): 189–193.

Singh, J.P., Yadava, P.K., Singh, Shakun and Prasad, S.C., 1991. BOD contamination in Kali river at Sadhu at Ashram in Aligarh. *I.J.E.P.*, 1(5): 325–326.

Singh, J.P., Upadhyay, R. and Singh, Shakun, 1994. A study of water quality of River Yamuna at Agra with special reference to coliform bacteria. *Res. J.P.I.Envi.*, 10: 53–57.

Singh, J.P., Yadava, P.K., Singh, Shakun and Madhu, Joshi, 1993. Assessment of BOD load in River Yamuna in respect of potable water at Agra. *I.J.E.P.*, 13(2): 119–123.

Trivedi, R.K. and Goel, P.K., 1984. *Chemical and Biological Methods for Water Pollution Studies*. Karad (India).

WHO, 1984. *Guidelines for Drinking Water Quality*. World Heath Organisation, Geneva, 2: 264.

Water and Wastewater Management in 2 Vols (2011) *Pages* **157–161**
Editors: **D.R. Khanna, R. Bhutiani & Gagan Matta**
Published by: **DAYA PUBLISHING HOUSE, NEW DELHI**

Chapter 26

Impact of River Flow Regulation on the Planktonic Population in Ganga River at Rishikesh, Uttarakhand

K.S. Negi, D.S. Malik and P.K. Bharti
Department of Zoology and Environmental Science,
Gurukula Kangri University, Haridwar

ABSTRACT

Biological monitoring is an essential element to assess the environmental health of aquatic biotopes. Any stress imposed in aquatic ecosystem manifests its impact on the inhabitant communities and result in restructuring the present biotic components. The present study is an effort to assess the impact of damming of river water on the physico-chemical characteristics and their relations to planktonic population of Ganga river. Results indicated remarkable changes in some physico-chemical factors in upstream and downstream of Ganga river at Pashulok barrage. Rich planktonic concentration were observed as 1273.67 ± 598.124 and 1184.25 ± 588.724 units/l in the upstream and downstream section respectively. Phytoplankton always dominated (88.4 to 91.7 per cent) over zooplankton and mainly constituted by Bacillariophyceae (65.07-78.95 per cent) followed by Chlorophyceae (25.34-17.14 per cent) in upstream and downstream of Ganga river.

Keywords: *Plankton, Ganga river, Pashulok barrage, Bacillariophyceae.*

Introduction

Uttaranchal have included 2700 km. of rivers and their tributaries, which receives water from melting snow in higher peaks and rainfall in the lower altitudes. Most rivers of the state are dammed or regulated for power generation, irrigation and domestic uses for the water. One of them Chilla

Hydroelectric Power Project on the river Ganga extends over a stretch of 20 km. between Rishikesh and Haridwar. The Pashulok barrage at Veerbhadra (Rishikesh) dammed the water and a large amount of water have diverted into the Chilla canal and only little amount of water flows in the mainstream of the Ganga river (Negi, 2002). Due to reduction in water discharge, the temperature of water has affected and the current velocity and other abiotic parameters of river have also changed. Such changes affect the population of plankton and ultimately reduce the population of primary consumer and herbivore animals in the riverine ecosystem.The present chapter describes the correlation between certain physico-chemical factors and planktonic population of river Ganga in upstream and downstream section of Pashulok barrage at Rishikesh.

Materials and Method

The Rishikesh is a township of district Dehradun located in the foothills of Shiwalik Himalaya at 365 m asl (Long. 78° 19' E and Lat. 30° 06' N). Two sites covering upstream and downstream of Pashulok barrage were selected for sampling at monthly intervals over a period of 2000-2002. The physical and chemical parameters selected for the study were analysed following the standard methods (Adoni, 1985 and APHA, 1995). For river plankton, 50 litre of water sample was filtered randomly through the plankton net (25 No.) and preserved in 1 ml. Lugol's solution. The planktonic density was studied with the help of Lackey (1938) and Edmondson (1974).

Results and Discussion

The results of physico-chemical parameters of upstream and downstream section of Ganga river are given in Table 26.1 and the average quantitative analysis and percentage of different groups among phytoplankton are shown in Tables 26.2 and 26.3 respectively.The planktonic life of any aquatic ecosystem is directly influenced by the conditions existing in its ambient environment. The physico-chemical parameters of a river have a direct linkage on its planktonic communities. In the present study average annual water temperature ranged between 16.93–17.32°C observed in the upstream and downstream section of Ganga river, which is favorable for planktonic growth in any aquatic ecosystem. Similar thermal regime (12.0-19.8°C) observed by Sharma (1968) in river Bhagirathi around Tehri, while Das and Srivastava (1959) pointed out the role of temperature as limiting factor for phytoplanktonic production in the riverine ecosystem.Velocity and transparency directly affects the planktonic number in the river. Higher velocity and low transparency in rainy season obscure the sun light penetration to deep water and reduce the photosynthetic activity. In the present study planktonic population was inversely proportional to velocity but showed reverse relationship with transparency which have supported by Allen (1920) and Berner (1951). Similarly Khanna (1993) and Khanna and Singh (2000) observed high plankton peak during winter months when water velocity was lowest and transparency was higher in Ganga and Suswa river respectively.

The average value of total solids and total dissolved solids were recorded 282.85-284.05 mg/l and 84.77-85.55 mg/l in both section of Ganga river. According to Welch (1948), TDS vary quantitatively in different waters depending upon the season, locations and other factors. pH of water is an important factor for plankton production, pH range from 7.2 to 8.5 is favorable for growth of plankton (Das, 1961). In the present study pH of Ganga river analysed 7.43-7.85 in both section. Moitra and Bhattacharya (1965) have stated that high pH coincided with phytoplankton peak and low with zooplankton. Rao *et al.* (1993) also reported that presence of myxophyceae is related with high level of pH and water temperature.

Table 26.1: Physico-chemical Characteristics of Ganga River in Upstream and Downstream Section of Pashulok Barrage

Sl..No.	*Parameters*	*Upstream*	*Downstream*
1.	Water Temperature (°C)	16.93±4.175	17.32±4.282
2.	Transparency (per cent)	60.75±29.749	53.63±27.596
3.	Water Current (m/sec)	0.54±0.301	1.12±0.635
4.	Total Solids (mg/l)	282.85±163.159	284.05±167.272
5.	Total Dissolved Solids (mg/l)	84.77±30.756	85.55±39.375
6.	pH	7.43±0.610	7.85±0.241
7.	Free Carbon Dioxide (mg/l)	2.33±0.976	1.58±0.842
8.	Dissolved Oxygen (mg/l)	8.17±1.479	9.52±1.206
9.	Total Hardness (mg/l)	74.03±8.781	75.70±9.206
10.	Total Alkalinity (mg/l)	80.46±13.790	83.87±13.776
11.	Chloride (mg/l)	13.50±2.493	14.29±2.918

Table 226.: Quantitative Analysis of Planktonic Population of the Ganga River

Sl..No.	*Parameters*	*Upstream*	*Downstream*
1.	Phytoplankton (nos./ml)	1168.58±538.581	1046.50±520.522
2.	Zooplankton (nos./ml)	106.75±64.962	137.75±74.291
3.	Total Plankton (nos./ml)	1273.67±598.124	1184.25±588.724

Table 26.3: Number and Percentage of Different Groups Among the Phytoplankton of Ganga River in Upstream and Downstream Section of Pashulok Barrage at Rishikesh

Sl.No.	*Group*	*Upstream*		*Downstream*	
		Density (Units/l)	*Percentage*	*Density (Units/l)*	*Percentage*
1.	Bacillariophyceae	760.42±330.406	65.07	826.25±348.951	78.95
2.	Chlorophyceae	296.08±135.640	25.34	179.42±143.794	17.14
3.	Myxophyceae	111.75±92.153	9.56	40.83±33.017	3.90

Due to photosynthetic and respiratory activity of biota, the dissolved oxygen and free CO_2 were inversely related in both the sites of Ganga river, which was early reported by Badola and Singh (1981) and Bhowmick and Singh (1985) in different lotic ecosystem. The average value of free CO_2 and DO ranged between 1.58-2.33 mg/l and 8.17-9.52 mg/l respectively. The high level of DO shows good sign of planktcnic communities in the system which is supported by Khanna *et al.* (1998) and reported that DO content usually correlated with the abundance of phytoplankton in Ganga river.

The present study reveals that both sites have few common characteristics as the high value of total alkalinity (74.03-75.70 mg/l) which show highly productive nature of water. Moyle (1946) considered that alkalinity above 48 ppm is a good sign of productivity. Hardness is the property of water and the natural source are sedimentary rocks, seepage and runoff water from soils. While the chloride concentration might be attributed to the increased discharge of domestic waste and sewage which is the main source of chloride, rainfall also contributed to the chloride level in water. Most of the water gets chloride, when it flows through the area where salt is deposited (Gondave, 1985).

The present study revealed that the phytoplanktonic population constituted the bulk contribution 88.4 to 91.7 per cent of total plankton assemblages. The total standing crop of phytoplankton in the river 1168.58 ± 538.581 units/l in upstream and 1046.50 ± 520.522 units/l in downstream of barrage. The important genera encountered were *Acnanthes, Nitzchia, Synedra, Diatoma, Naviculla, Tabellaria, Cocconeis* and *Cymbella* among Bacillariophyceae; *Spirogyra, Zygnema, Ulothrix, Cladophora, Schizogonium* and *Chlorella* among Chlorophyceae and *Anabena, Oscillatoria, Spirulina* and *Rivularia* among Myxophyceae. Bacillarophyceae group was the most dominant group (65.07-78.95 per cent) among the total phytoplankton, which is also supported by Khanna *et al.* (1993) and Joshi *et al.* (1996) in the similar water body. In the present observation, Bacillariophyceae, Chlorophyceae and Myxophyceae are the principle constituents of phytoplankton and are found in decreasing trend in the Ganga river.

References

Adoni, A.D., 1985. *Workbook on Limnology*. Pratibha Publishers, Sagar, M.P. pp. 215.

Allen, W.E., 1920. A quantitative and statistical study of the plankton of the San Joaquin river and its tributaries in and near Stockton California 1913. *Univ. Calif. Publ. Zool.*, 22(1): 1–124.

APHA, 1995. *Standard Methods for Examination of Water and Wastewater*. American Public Health Association, New York, Washington D.C., pp. 1193.

Badola, S.P. and Singh, H.R., 1981. Hydrobiology of the river Alaknanda of Garhwal Himalaya. *Indian J. Ecol.*, 8: 269–276.

Barner, L.M., 1951: Limnology of the lower Missouri river. *Ecology*, 32: 1–12.

Bhowmick, B.N. and Singh, A.K., 1985. Phytoplankton population in relation to physico-chemical factors of river Ganga at Patna. *Indian J. Ecol.*, 12(2): 360–364.

Das, S.M., 1961. *Nature*, 20, Porte South, 4–Crinan Street, London.

Das, S.M. and Srivastava, V.K., 1959. *Proc. Nat. Acad. Sci., India*, 29: 174.

Edmondson, W.T., 1974. A simplified method of counting phytoplankton. In: *A Manual on Methods for Measuring Primary Producers in Aquatic Environment*, (Ed.) R.A. Vollen Weeider. Oxford Blackwell Sci. Pub., pp. 14–16.

Gondave, E.S., 1985. Supply of chloride in groundwater. In: *2nd World Congress on Engg. and Environment*, New Delhi, p. 7–9.

Joshi, T.N., Joshi, B.D. and Singh, R., 1996. A report on the diurnal variations of some eco-biological parameters of Kalpanigad (Distt. Pithoragarh, U.P.). *Him. J. Env. Zool.*, 10: 78–82.

Khanna, D.R., 1993. *Ecology and Pollution of Ganga River*. Ashish Publishing House, New Delhi, pp. 1–241.

Khanna, D.R. and Singh, R.K., 2000. Seasonal fluctuation in plankton of Suswa river at Raiwala (Dehradun). *Env. Cons. J.*, 1(2 and 3): 89–92.

Khanna, D.R., Malik, D.S. and Rana, D.S., 1998. Phytoplanktonic communities in relation to certain physico-chemical parameters of Ganga canal at Haridwar. *Him. J. Env. Zool.*, 12: 193–197.

Khanna, D.R., Badola, S.P., Dobriyal, A.K. and Singh, H.R., 1993. Observations on temporal trends in diatomic diversity in river Ganga at Saptsarovar, Haridwar. In: *Recent Trends of Researches in Coldwater Fisheries*, (Ed.) K.L. Sehgal. Today and Tommorow Printers and Publishers, New Delhi, India, pp. 335.

Lackey, J.B., 1938. The manipulation and counting of river plankton and changes in some oganisms due to formalin preservation. *US Public Health Reports*, 50: 2080–2093.

Moitra, S.K. and Bhattacharya, B.K., 1965. *Ichthyologica*, 4: 7.

Moyle, J.B., 1946. Some indices of lake productivity. *Trans Amer. Fish. Soc.*, 76: 322–334.

Negi, Kamal Singh, 2002. Impact of barrage on the biology of mahseer in river Ganga at Rishikesh. *Ph.D. Thesis*, Gurukula Kangri University, Haridwar, pp. 159.

Rao, V.N.R., Mohan, R., Hariprasad, V. and Ramasubramaniam, R., 1993. Seasonal dynamics of physico-chemical factors in a tropical high altitude lake in relation to phytoplankton. *J. Environ. Biol.*, 14(1): 63–75.

Sharma, R.C., 1986. Effect of physico-chemical factors on benthic fauna of Bhagirathi river of Garhwal Himalaya. *Indian J. Ecol.*, 13(1): 133–137.

Welch, E.B., 1948. *Limnology*. McGraw Hill Book Company Inc., London, pp. 588.

Water and Wastewater Management in 2 Vols (2011) *Pages* **162–167**
Editors: **D.R. Khanna, R. Bhutiani & Gagan Matta**
Published by: **DAYA PUBLISHING HOUSE, NEW DELHI**

Chapter 27

Seasonal Variation in Ionic Fluxing in the Freshwater Bodies of Lucknow City of U.P.

Jatin Srivastava[1], Harish Chandra[2] and Nandita Singh[2]
[1]National Botanical Research Institute, Lucknow, U.P.
[2]Department of Microbiology,
Dolphin (PG) Institute of Biomedical and Natural Sciences, Dehradun

ABSTRACT

Surface water of natural wetlands of Lucknow city was monitored as a part of comprehensive study of natural imbalance of nutrient ions such as PO_4, NO_3, Cl and NH_4. The common seasonal changes were recorded and statistically analyzed exhibiting a correlation of parameters and also the dependency of one on another. In summer where there is a loss of water due to evaporation results in more ionic concentrations, however, the rainy and winter seasons receives a different chemistry after a average spell of rain. Overall study presents a seasonal variation in the ionic species in water and its impact on the water chemistry.

Keywords: *Seasonal variation, Nutrients, Correlation, Parameters.*

Introduction

All natural water contains dissolved ionic constituents. The distribution of these species in water vary with geographic location and the residence time. Nitrate is present in small amount in freshwaters having higher values of oxygen. Nitrogen is a complex element that can exist in seven states of oxidation. For the water quality standpoint nitrogen containing compounds such as ammonia (NH_3) Nitrate (NO_3^-) and Nitrite (NO_2^-) are of great importance. Bacterial life within the water largely affects the nitrogen cycle in water (Water quality, 1985). Phosphates enter waterways from several different sources. An average human body excretes about 1 pound *i.e.*, 454 grams per year as Phosphates (Quality Criteria, 1975). Phosphorus is an essential element for the growth of algae and other aquatic organisms, therefore eutrophication of lakes and ponds and even estuaries may occur if the phosphorous content exceeds 0.015 g/m^3. Eutrophication decreases the O_2 content and increased the NH_4^+ availability in freshwaters. These changes may affect carbon and nitrogen transformation processes and the production of CH_4 and N_2O (Liikanen and Martikainen, 2003).

The ionic species and their properties in water depend on the geographical and topographical conditions. The closed systems (Ponds, Lakes and Pools) entirely depend upon annual average precipitation. The ionic strength of water is also determined by the conductivity, which is the capability of electrical conductance (Quality Criteria, 1975). It has been reported that about the change in the ion chemistry of the water considerably with the change of climatic conditions on regional scale. The ion concentration in the lakes has also been reported more due to the excessive evaporation (Yang and Williams, 2003).

In the present work the emphasis has been given to the ionic fluxing of the ions of biological importance such as chloride, nitrate, ammonia and phosphate along with other parameters like bicarbonates, pH and conductivity. The ion fluxing study and the dilution factor is considered here as a major water quality-affecting phenomenon.

Materials and Method

Three water bodies belonging to the municipal corporation of Lucknow city were selected for the present study. Water samples were collected in different seasons of summer *i.e.*, March–June, rainy *i.e.*, July–October and winter *i.e.*, in month of November to February, in iodide washed white plastic containers. To avoid the loss of dissolved oxygen it was fixed after collecting it in a glass bottles using fixing reagent. The pH and electrical conductivity were recorded at the site itself with help of a portable pH and EC meters however, both the parameters were also tested in the lab conditions too. All the test parameters were performed as per the Standard Method of APHA (1991).

Ions were analyzed with the help of an Ion Selective Electrode meter provided by Thermo Orion model 920 A Plus and heavy metals were analyzed on the Atomic Absorption Spectrophotometer Perkin Elmer model 5100.

Water samples were collected from the different sites from ponds. The statistical calculations were made with the help of computer-aided software SPSS 13.0 version (SPSS, 1997).

Results and Discussion

The results are based on the regular analysis of water collected in different seasons from different water bodies. Table 27.1 shows the value of water quality parameters of the waters collected from the shore, showing higher values of ionic concentrations in almost all the ponds *viz.*, D, F and G. pH ranges between <6 in D and reaches up to > 8.76 in G water body Liikanen *et al.* (2003) found similar trend. Parameters of water quality were showing a tendency of dependence on each other. For which correlation matrix was obtained by the help of computer-aided software SPSS (Tables 27.3 and 27.4). Two-tailed significant value was considered showing a negative correlation between the parameter sets such as pH–EC, PO_4, NO_3 and HCO_3–EC, pH. The negative correlation values showing the reciprocal changes in the nature of chemicals *i.e.*, the increase in one will cause the decrease in the other. Bicarbonate ions are negatively correlated with electrical conductivity with a significant value of–0.980 that is also confirmed by the correlational value of EC and pH. NO_3 is significantly correlated

Table 27.1: Mean and Standard Deviation of Quality Parameters of Water in Different Seasons

Pond	Group (Season)	Parameters							
		pH	EC	DO	HCO_3	PO_4	NH_4	NO_3	Cl^-
D1	Summer	6.99 (0.10)	1800.0 (172.71)	2.33 (1.45)	543.0 (35.17)	4.0 (1.08)	115.9 (5.99)	17.80 (3.75)	109.30 (14.35)
	Rainy	7.14 (0.59)	719.0 (66.20)	3.40 (1.31)	255.0 (21.80)	0.17 (0.01)	1.92 (1.20)	21.60 (1.22)	96.50 (4.09)
	Winter	7.92 (0.16)	778.0 (58.28)	13.30 (1.66)	603.0 (60.31)	1.95 (0.19)	ND	21.10 (2.87)	136.90 (33.12)
F2	Summer	8.16 (0.10)	230.0 (25.51)	5.60 (2.08)	568.13 (23.74)	0.99 (0.31)	26.87 (6.12)	17.03 (3.54)	114.80 (66.56)
	Rainy	7.36 (0.16)	688.3 (39.79)	5.90 (1.13)	201.48 (12.3)	0.12 (0.05)	0.70 (0.7)	23.13 (4.96)	37.25 (4.02)
	Winter	8.38 (0.87)	477.0 (9.54)	8.30 (1.48)	433.0 (15.13)	0.06 (0.05)	ND	13.40 (2.72)	21.80 (1.44)
G3	Summer	8.76 (0.32)	674.33 (42.10)	4.00 (0.35)	266.67 (0.26)	2.74 (0.17)	17.46 (0.02)	39.18 (0.20)	99.57 (0.63)
	Rainy	8.87 (0.32)	537.33 (42.10)	11.60 (0.35)	161.33 (1.26)	0.96 (0.17)	0.07 (0.02)	70.80 (0.20)	27.72 (0.63)
	Winter	8.30 (0.17)	720.0 (92.90)	6.03 (0.12)	336.00 (27.07)	1.48 (0.27)	0.11 (0.01)	68.50 (I.56)	32.21 (1.35)

Table 27.2: Mean and Standard Deviation of Quality Parameters of Water at 2 mt Away from the Shore in Different Seasons

Pond	Group (Season)	Parameters							
		pH	EC	DO	HCO_3	PO_4	NH_4	NO_3	Cl^-
D4	Summer	7.02 (0.27)	1720.0 (160.93)	5.40 (1.25)	588.0 (59.43)	3.62 (0.91)	49.50 (12.42)	33.30 (2.86)	131.13 (46.12)
	Rainy	7.91 (0.18)	706.0 (56.67)	3.00 (0.35)	280.0 (69.46)	0.10 (0.04)	4.03 (2.61)	19.0 (2.78)	98.0 (4.36)
	Winter	7.83 (0.22)	765.0 (59.81)	3.96 (1.80)	614.0 (118.7)	2.23 (0.52)	ND	18.30 (1.86)	118.1 (30.64)
F5	Summer	8.20 (0.05)	225.0 (29.51)	5.97 (2.14)	615.0 (195.2)	1.04 (0.13)	9.15 (1.56)	15.60 (4.91)	72.60 (19.63)
	Rainy	7.59 (0.46)	590.0 (134.3)	7.37 (0.90)	319.0 (68.94)	0.191 (0.14)	ND	19.92 (2.18)	23.30 (3.24)
	Winter	8.27 (0.23)	480.0 (2.65)	8.60 (1.40)	423.0 (15.4)	0.06 (0.05)	ND	14.50 (1.65)	20.80 (0.04)
G6	Summer	8.04 (0.09)	537.0 (41.73)	6.23 (0.12)	286.3 (5.69)	2.07 (0.10)	9.56 (0.68)	43.99 (3.40)	109.4 (9.35)
	Rainy	7.68 (0.33)	480.0 (8.0)	9.43 (0.80)	152.6 (8.3)	1.29 (0.3)	0.14 (0.03)	68.83 (2.32)	92.4 (2.7)
	Winter	7.62 (0.29)	587.3 (24.34)	8.17 (0.06)	207.6 (13.28)	11.69 (0.25)	0.13 (0.05)	63.73 (3.86)	88.03 (6.27)

EC: Electrical conductivity in µmhos/cm; DO: Dissolved oxygen in mg/l; HCO_3: Bi-carbonates in mg/l; PO: Phosphate in mg/l, NH_4: Ammonium in mg/l; NO_3: Nitrate in ppm; Cl^-: Chlorides in mg/l. Values in parenthesis are std. dev.

with the dissolved oxygen with a value of 0.999 showing the perfect relation of simultaneous increase of each parameter. Similarly phosphate and ammonium ions have been found to be correlated with the same value of 0.999, Yang and Williams (2003) reported same correlation. Table 27.2 represents the value of quality parameters of water collected from 2 meters away from the shores. Tables 27.5 and 27.6 represents the correlation matrix of each parameters season and water body wise. The results showed almost the same correlational values. However, in the waters collected from 2 meters away from the shores showed variable trends.

Table 27.3: Correlation Coefficient between Water Quality Parameters

	Group 1							
	pH	*EC*	*DO*	*HCO_3*	*PO_4*	*NH_4*	*NO_3*	*Cl^-*
pH		–0.585	0.556	0.509	–0.190	–0.635	0.525	0.893
EC	–0.714		–0.999*	0.398	0.907	0.998*	–0.997*	–0.159
DO	0.577	0.158		–0.431	–0.921	–0.995*	0.999*	0.123
HCO_3	0.836	–0.980*	0.036		0.747	0.340	–0.464	0.841
PO_4	0.256	–0.859	0.640	0.743		0.878	–0.935	0.271
NH_4	0.290	–0.876	–0.876	0.766	0.999*		–0.990*	–0.221
NO_3	–0.985*	0.584	–0.708	–0.730	–0.087	–0.122		0.086
Cl^-	0.160	–0.805	–0.712	0.674	0.995*	0.99	0.010	

* Significant at 0.05 probability level.

Table 27.4: Correlation Coefficient between Water Quality Parameters

	pH	*EC*	*DO*	*HCO_3*	*PO_4*	*NH_4*	*NO_3*	*Cl^-*
pH		–0.812	0.431	–0.896	0.052	0.332	–0.271	0.281
EC			–0.875	0.986*	0.538	0.279	–0.339	0.330
DO				–0.786	–0.878	–0.707	0.751	–0.744
HCO_3					0.395	0.120	–0.182	0.173
PO_4						0.959	–0.975	0.973
HN_4							–0.998*	0.998*
NO_3								–0.999*
Cl^-								

* Significant at 0.05 probability level.

The ionic strength remains normally elevated in the summer and get diluted in rainy season. The microbes however, consume the ions in summer the value never goes down. The pH value of natural water depends on the carbonate system which is present naturally and get disturbed in case of the higher levels of biostimulants such as PO_4, NO_3, NH_4 and Cl that causes the growth of algae and other phytoplankton in water that removes almost all the carbonate ions that elevated the pH of water see Table 27.1. The water collected from the 2 meter inside the water body shows a trend of dilution in the concentrations of ions. More water leads to the reduction in the microbial count, as most of the microbes needs substratum to grow. Least microbial activity reduces the demineralization process and hence the reduction in the ionic contents. The run off from the surroundings of the fields nearby the water body let it to get enriched but in the same time most of the phytoplankton grow and remove most of the ions from the water.

Table 27.5: Correlation Coefficient between Water Quality Parameters (2 mt away from shore line)

	Group 4							
	pH	*EC*	*DO*	*HCO_3*	*PO_4*	*NH_4*	*NO_3*	*Cl^-*
pH		–0.999	–0.946	–0.509	–0.845	–0.988	–0.992	–0.844
EC	–0.665		0.937	0.484	0.829	0.992	0.995	0.827
DO	0.056	0.707		0.759	0.972	0.886	0.900	0.971
HCO_3	0.705	–0.998	–0.667		0.890	0.371	0.400	0.891
PO_4	0.302	–0.912	–0.934	0.888		0.753	0.773	0.999
NH_4	0.416	–0.955	–0.884	0.938	0.992		0.999	0.751
NO_3	–0.995	0.587	–0.155	–0.631	–0.206	–0.324		0.772
Cl^-	0.377	–0.942	–0.903	0.922	0.996	0.999	–0.283	

* Significant at 0.05 probability level.

Table 27.6: Correlation Coefficient between Water Quality Parameters (2 mt away from shore line)

	pH	*EC*	*DO*	*HCO_3*	*PO_4*	*NH_4*	*NO_3*	*Cl^-*
pH		–0.094	–0.860	0.850	0.783	0.991	–0.946	0.998
EC			–0.425	0.443	0.544	0.036	–0.231	–0.156
DO				–0.999	–0.990	–0.920	0.978	–0.827
HCO_3					0.993	0.912	–0.974	0.815
PO_4						0.858	–0.941	0.743
NH_4							–0.980	0.981
NO_3								–0.924
Cl^-								

Conclusion

The water quality of ponds of different nature were analyzed in different seasons and found to have a correlation in the parameters. The ionic fluxing in the water depends up on the geographical, chemical and biological factors. However, there is significant impact of each parameter on the water chemistry. Some correlations are naturally occurring such, as the dissolved oxygen is associated with the nitrate and ammonia content.

References

APHA, 1992. *Standard Methods for the Examination of Water and Wastewater*, 16th edn. American Public Health Association, Washington D.C.

Liikanen Anu and Martikainen Pertti, J., 2003. Effect of ammonium and oxygen on methane and nitrous oxide fluxes across sediment-water interface in a eutrophic lake. *Chemosphere*, 52: 1287–1293.

Quality Criteria for Water, 1979. U.S. Environmental Protection Agency, Washington D.C.

SPSS Inc., 1997. *SPSS Base Statistics*, version 13.0 SPSS INC. Chicago, USA.

Water Quality, 1985. Vol 1, Addison Wesley Publishing Company USA. pp. 43–122.

Yang Xeroxing and Williams Martin, A.J., 2003. The ion chemistry of lakes and late holocene desiccation in the Badain Jaran Desert, Inner Mongolia, China. *Catena*, 51: 45–60.

Water and Wastewater Management in 2 Vols (2011) *Pages* **168–174**
Editors: **D.R. Khanna, R. Bhutiani & Gagan Matta**
Published by: **DAYA PUBLISHING HOUSE, NEW DELHI**

Chapter 28

Benthic Fauna and its Ecology of River Ganga from Rishikesh to Haridwar, Uttarakhand

D.R. Khanna and R. Bhutiani
Department of Zoology and Environmental Sciences,
Gurukul Kangri University, Haridwar

ABSTRACT

Benthic Fauna of river Ganga in relation to physico-chemical characteristics of water has been investigated. A total of 10 groups of benthic fauna were collected. Ephemeroptera was found as dominant during the course of study. Benthos were found maximum in winter months. The physico-chemical characteristics of river Ganga was found almost in limits.

Keywords: Benthos, Ecology, Physico-chemical, Ganga.

Introduction

Due to unproportional growth of population and industries, the quality and purity of Ganga water has deteriorated considerably. The religious importance of the Ganga may exceed than that of any other river in the world. It has been revered from the earliest times, and Hindus regard it today as the holiest of all rivers.

At Rishikesh-Haridwar, the Ganga cuts across the Shivalik hills and for the first time enters the plains. The discharge of domestic and industrial sewage, makes water unpotable and also severely affects the bio-productivity of the aquatic system. These discharge consists of large number of chemicals and heavy metals. These waste materials react with each other as a result, the water pollutes and may become toxic and effects the macrobenthic diversity of river.

The aquatic fauna are the natural indicators of water quality.Today no proper information is provided regarding the macro-invertebrates of river Ganga at Rishikesh-Haridwar. In the present study an attempt has been made to collect information regarding the macrobenthic diversity of river Ganga.

Materials and Method

For limnological study of river Ganga the water samples were collected fortnightly from different sampling stations for a period of one year.

Various physico-chemical parameters were analyzed following the standards methods of APHA (1998), Trivedi and Goel (1986), Mathur (1982) and Khanna and Bhutiani (2004). While the benthos were identified with the help of Day (1878), Hutchinson (1957), Needham and Needham (1972).

Results and Discussion

The results of the present study are given in Tables 28.1–28.3 and Figures 28.1 and 28.2. In the present study the minimum (11.25ºC± 0.35) and maximum (20.00 ºC ± 0.00), temperature was observed in the month of January and September respectively. The water temperature showed an upward trend from January to April followed by a downward trend from May onwards. Similar trend was also reported by Khanna (1993), Khanna and Chugh (2004), Badola and Singh (1981).

Temperature showed an inverse relationship with dissolved oxygen through out the year, this is due to maximum photosynthesis when temperature was minimum. Maximum (11.15 mg/l ± 0.07) DO was observed in January and minimum (7.60 mg/l ± 0.14) in July. From May onwards there is decrease in DO due to increased turbidity which retarded the photosynthesis in aquatic flora. Maximum (485.00

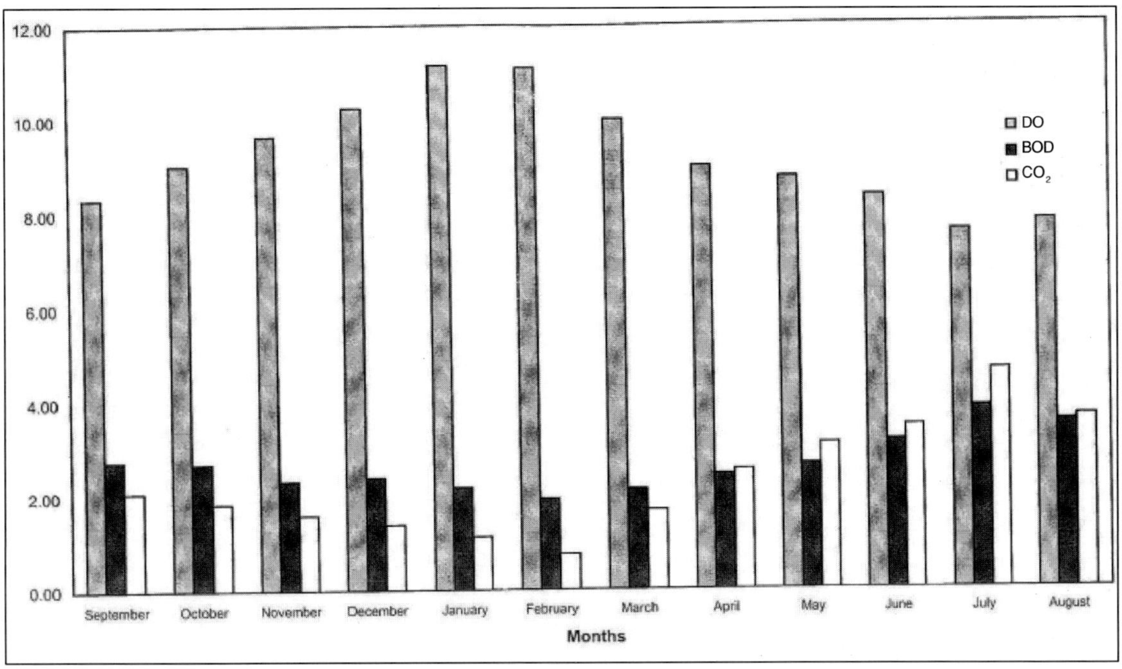

Figure 28.1: Comparative Study of Free CO_2, Dissolved Oxygen and BOD

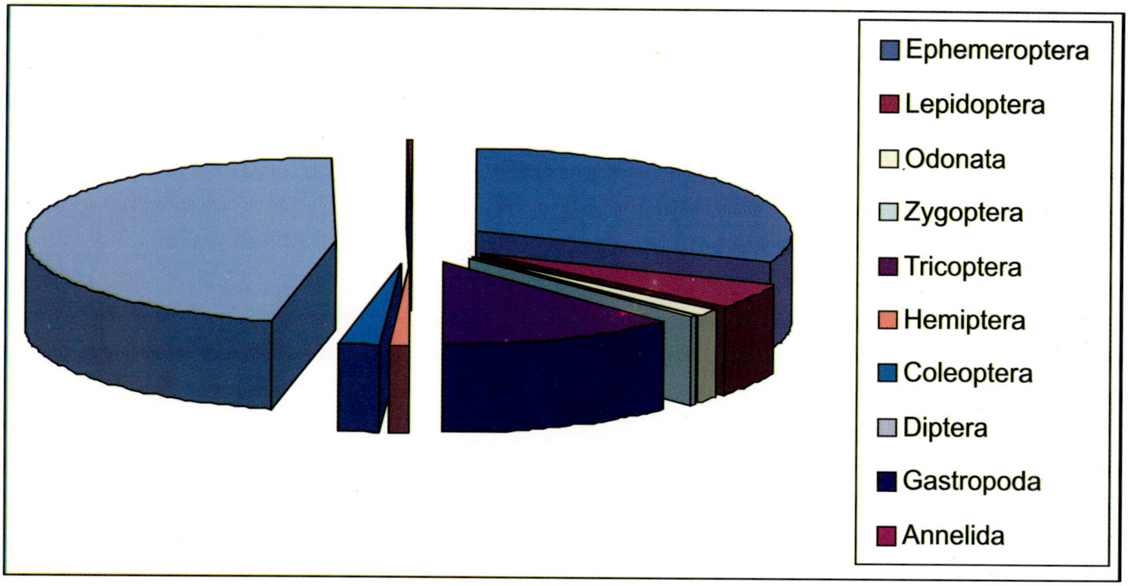

Figure 28.2: Percentage Composition of Benthic Fauna

J.T.U ± 162.63) turbidity was observed in July and minimum (0.00 J.T.U ± 0.00) in the month of January, February and March. Singh *et al.* (1982), Khanna and Chugh (2004) reported similar results.

The value of BOD was observed maximum (3.85 mg/l ± 0.70) in July and minimum (1.95 mg/l ± 0.07) in February. A negative relationship has been observed between BOD and DO contents. Similar pattern was observed by Verma *et al.* (1984).

Table 28.1: Monthly Variation in Physical Parameters of River Ganga

Months	*Temperature (ºC)*	*Turbidity (JTU)*	*Velocity (m/sec)*	*Transparency (cm)*	*Total Residue (mg/l)*
September	20.00±0.00	375.00±21.21	1.78±0.11	6.70±l.56	1019.50±419.31
October	18.75±0.35	75.00±7.07	1.03±0.46	67.50±10.61	193.50±98.29
November	15.00±0.00	57.50±3.54	0.65±0.00	97.50±17.68	172.50± 10.61
December	11.75±0.35	32.50±3.54	0.55±0.14	127.75±3.89	160.00±0.00
January	11.25±0.35	0.00± 0.00	0.40± 0.07	152.62±3.32	198.50±3.54
February	13.25±1.06	0.00±0.00	0.37±0.00	170.10±2.69	129.50± 23.33
March	15.50±0.71	0.00±0.00	0.35±0.00	158.00±3.54	170.00± 14.14
April	19.75±0.35	27.50± 3.54	0.60±0.00	134.30±10.04	179.00± 4.24
May	19.25±0.35	92.50±10.61	1.l0± 0.00	16.00±0.00	204.00±8.49
June	19.00±0.00	200.00± 42.43	1.16±0.08	11.00± 4.24	752.00± 565.69
July	18.00±0.00	485.00±162.63	1.98±0.11	3.00± 0.71	3048.00±862.67
August	18.75±0.35	450.00±0.00	2.10±0.00	2.00±1.27	1740.50± 741.76
Annual Average	16.69±3.20	12.08±151.63	1.00±0.63	78.87±68.37	663.88±898.76

Table 28.2: Monthly Variations in Chemical Parameters of River Ganga

Months	*pH*	*DO (mg/l)*	*BOD (mg/l)*	*Free CO_2 (mg/l)*	*Alkalinity (mg/l)*	*Total Nitrogen (mg/l)*
September	7.78±0.04	8.35±0.07	2.75±0.07	2.01±0.00	57.50±3.54	0.40±0.00
October	7.78±0.04	9.05±0.78	2.70±0.14	1.85±0.07	65.50±4.95	0.38±0.30
November	7.87±0.06	9.65±0.21	2.33±0.00	1.60±0.00	80.50±2.21	0.30±0.01
December	7.83±0.00	10.25±0.35	2.40±0.14	1.40±0.00	75.50±0.71	0.23±0.04
January	8.04±0.04	11.15±0.07	2.20±0.06	1.15±0.07	97.50±6.36	0.10±0.01
February	8.06±0.06	11.10±0.01	1.95±0.07	0.75±0.21	87.50±3.54	0.19±0.01
March	7.93±0.04	10.00±0.71	2.15±0.07	1.70±0.14	89.50±0.71	0.45±0.07
April	7.99±0.20	9.00±0.71	2.45±0.07	2.55±0.07	92.50±3.54	0.50±0.14
May	7.80±0.00	8.75±0.07	2.66±0.14	3.10±0.00	67.00±2.83	0.45±0.07
June	7.73±0.04	8.35±0.21	3.15±0.35	3.45±0.07	52.50±0.71	0.60±0.00
July	7.63±0.01	7.60±0.14	3.85±0.7	4.65±0.07	46.50±2.21	0.35±0.07
August	7.71±0.01	7.80±0.14	3.55±0.07	3.65±0.07	44.00±1.41	0.63±0.07
Annual Average	7.85±0.14	2.69±1.50	2.33±0.56	2.33±1.77	71.33±18.56	0.36±0.14

According to Ray *et al.* (1966) and Khanna (1993) maximum free CO_2 have been observed in rainy season in Ganga river. Maximum (4.65 mg/l ± 0.07)and minimum (1.15 mg/l ± 0.07) value of free CO_2 had been reported in the month of July and January respectively. Quadri and Shah (1984) reported similar observations.

The maximum (2.10 m/s ± 0.00) and minimum (0.35 m/s ± 0.00) value of velocity was observed in the month of August and March respectively. The velocity started increasing from May onwards due to melting of snow at the place of origin of river. Again beyond September velocity starts decreasing. Similar results were given by Khanna (1993). In the present investigation residues were maximum (3048.00 mg/l ±62.67) in July and minimum (129.50 mg/l ± 23.33) in February this was due to maximum velocity in rainy season. Similar conditions were reported by David (1956) and Verma and Shukla(1969).

Transparency was maximum (170.10 cm ± 2.69) and minimum (2.00 cm ±1.27) in the month of February and August. Badola and Singh (1981), Chugh (2000) reported similar pattern.

During the course of study the pH was observed slightly alkaline. Maximum (8.06 ± 0.06) and minimum (7.63 ± 0.01) value of pH was recorded in February and July respectively. Similar observations were reported by Sangu and Sharma (1985) and Bhutiani (2004).The highest concentration (97.50 mg/l ± 6.36) and lowest concentration (44.00 mg/l ± 1.41) of alkalinity was observed in the months of January and August respectively. Holden and Green (1960), Abden (1948 a), Khanna and Chugh (2004) gave similar findings.

The value of total nitrogen was observed maximum (0.63 mg/l ± 0.07) in August and minimum (0.10 mg/l ± 0.01) in January. Swarup and Singh (1979) observed similar results. Khanna (1993) observed minimum value of total nitrogen in rainy season.

The total percentage of Benthic fauna in river Ganga was observed as Ephemeroptera (30.52 per cent), Lepidoptera (25.07 per cent), Odonata (2.03 per cent), Zygoptera (0.08 per cent), Tricoptera (12.64 per cent), Hemiptera (1.05 per cent), Coleoptera (2.24 per cent), Diptera (45.21 per cent), Gastropoda (0.25 per cent) and Annelida (0.17 per cent).

Table 28.3: Benthic Fauna in the River Ganga

Months	*Ephemero-ptera*	*Lepido-ptera*	*Odonata*	*Zygoptera*	*Tricoptera*	*Hemiptera*	*Coleoptera*	*Diptera*	*Gastropoda*	*Annelida*	*Average*
September	30	2	–	–	3	1	–	13	–	1	5.00±9.63
October	57	8	1	–	18	2	1	50	–	–	13.7±21.77
November	70	12	3	–	32	1	3	96	–	–	21.70±34.31
December	91	17	3	–	44	3	6	150	–	–	31.40±50.77
January	120	30	3	1	97	3	11	180	2	–	44.70±64.35
February	145	21	9	1	50	5	15	275	2	–	52.30±89.80
March	85	15	6	–	30	2	9	148	1	–	21.60±49.01
April	50	9	3	–	15	3	5	102	1	–	18.80±32.86
May	37	5	6	–	6	1	2	47	–	–	10.40±16.99
June	29	1	2	–	4	2	1	17	–	–	5.60±9.67
July	5	–	–	–	–	1	–	6	–	1	1.30±2.26
August	3	–	–	–	–	1	–	2	–	2	0.80±1.14
Annual Average	60.16±43.98	10.0±9.40	3.00±2.79	0.16±0.38	24.91±28.49	2.08±1.24	4.41±4.94	90.50±84.73	0.41±0.66	0.33±0.65	19.60±16.95
Total Percentage	30.52	25.07	2.03	0.08	12.64	1.05	2.24	45.21	0.25	0.17	

In the present investigation Ephemeroptera was found as dominating. The benthic fauna of river Ganga were attached by increased turbidity which reduces the benthos as also reported by Das and Pathani (1978).

The index of similarity of Benthic fauna during different months was found to be very close to 1.0 in most of cases except in the month of July and August.

The results indicates that Benthos were maximum in January and February *i.e.*, winter months which was due to low temperature, high dissolve oxygen, low velocity and transparency of water along with other suitable conditions. Similar findings have been reported by Khanna (1993) in river Ganga.

References

Abdin, G., 1948a. Physical and chemical investigation relating to algal growth in the river Nile. *Cairo Bull. Inst.*, Egypt, 29: 20–24.

APHA, 1995. *Standard Methods for the Examination of Water and Wastewater*. American Public Health Association 1015 Fifteen Street NW, Washington, 20: 1–1134.

Badola, S.P. and Singh, H.R., 1981. Hydrobiology of the river Alaknanda of Garhwal Himalaya. *Indian J. Ecol.*, 8(2): 269–276.

Bhutiani, R., 2004. Limnological status of river Suswa with reference to its mathematical modeling. *Ph.D. Thesis*, G.K.V., Haridwar.

Das, S.M. and Pathani, S.S., 1978. A study on the the effect of lake ecology on productivity of Mahaseer (*Tor tor* and *Tor putitora*) in Kumaon lakes, India. *Matsya*, 4: 25–31.

David, A., 1956. Studies on the pollution of the Bhadra river at Bhadravati fisheries effluents. *Proc. Nat. Inst. Set.* India, 93(3):132–160.

Day, F., 1878. *The Fishes of India: Being a Natural History of the Fishes known to Inhabit the Seas and Freshwater of India, Burma and Ceylon*, Vols. I and II. William Dawson and Sons Ltd., London, p. 778.

Holden, J.M. and Green, J., 1960. Hydrology and plankton of the river Sokoto. *J. Anim. Ecol.*, 29(1): 65–84.

Hutchinson, G.C., 1957. *A Treatise on Limnology, Geography, Physical and Chemical*, Vol. 1. John Wiley and Sons Inc., New York, p. 1–1015.

Khanna, D.R., 1993. *Ecology and Pollution of River Ganga*. Ashish Publishing House, New Delhi, p. 1–241

Khanna, D.R. and Bhutiani, R., 2004. *Water Analysis at a Glance*. ASEA Publication, Rishikesh, p. 1–116.

Khanna, D.R. and Chugh, Tarun, 2004. *Microbial Ecology*. Discovery Publishing House, New Delhi.

Mathur, R.P., 1982. *Water and Wastewater Testing*. Neem Chand and Bros. Publishers, Roorkee, p. 1–54.

Needham, J.G. and Needham, P.R., 1972. *A Guide to the Study of Freshwater Biology*. Holden–Day I.N.C. San Francisco. Calif., 94(3): 1–108.

Quadri, M.Y. and Shah, G.M., 1984. Hydrobiological features of Haossar: A typical wetland of Kashmir. I. Biotape. *Indian Journal Ecol.*, 2(2): 203–206.

Ray, P., Singh, S.B. and Sehgal, K.L., 1966. A study of some aspects of the river Ganga and Jamuna at Allahabad (U.P.) in 1958–59. In: *Proc. Nat. Acad. Sci., India*, 36B(3): 235–272.

Sangu, R.P.S. and Sharma, K.D., 1985. Studies on water pollution on Yamuna river at Agra. *Indian J. Environ. Hlth.*, 27(3): 257–261.

Singh, H.R., Badola, S.P. and Dobriyal, A.K., 1987. Geographical distributional list of ichthyo fauna of the Garhwal Himalaya with some new records. *J. Bombay Nat. Hist. Society*, 84(1): 126–132.

Swarup, Krishan and Singh, S.R., 1979. Limnological studies of Suraha lake (Ballia). *J. Inland Fish Soc. India*, 2(1): 22–33.

Trivedi, R.K and Goel, P.K., 1984. *Chemical and Biological Methods for the Water Pollution Studies*. Environmental Publications, Karad, p. 1–251.

Verma, S.R., Sharma, P., Tyagi, A., Rani, S., Gupta, A.K. and Dalela, R.C., 1984. Pollution and saprobic status of eastern Kalinadi. *Limnologica* (Berlin), 15(1): 69–133.

Verma, S.R. and Shukla, G.R., 1969. Pollution in a perennial stream Khala, by the sugar factory effluent near Laksar (Dist. Saharanpur) U.P., India. *Env. Health*, 11: 145–162.

Water and Wastewater Management in 2 Vols (2011) *Pages 175–185*
Editors: D.R. Khanna, R. Bhutiani & Gagan Matta
Published by: DAYA PUBLISHING HOUSE, NEW DELHI

Chapter 29

Rational Design of Non-convective Zone of Salt Gradient Solar Pond Considering Turbidity and Biological Growth in Water

A.K. Saxena[1], S. Sugandhi[2] and M. Husain[3]
[1]*Samrat Ashok Technological Institute, Vidisha, M.P.*
[2]*Jawaharlal Institute of Technology, Borawan, Khargone, M.P.*
[3]*SSBT's College of Engineering and Technology, Bambhori, 94 Jalgaon, M.S.*

ABSTRACT

The optimum size of non-convective zone of a Salt Gradient Solar Pond is determined from the heat balance considerations of the storage zone. The equation governing heat flow in storage zone are simplified considering a linear heat flow through the storage zone to the upper convective zone.The radiation absorption in pond liquid is estimated considering the effect of various environmental parameters in pond like turbidity and biological growth in it. This has made the analysis more confirming to the real work situation for a pond. The ambient parameter like temperature and radiation are accounted to vary hourly. One significant observation is that the influence of turbidity on pond's performance is so drastic that at turbidity value higher than 0.8 NTU, the optimum Non-convective Zone (NCZ) size is not obtained analytically.

Keywords: Non-convective Zone, Optimum size, Salt gradient solar pond, Stability, Thermal performance, Turbidity.

Introduction

Salt gradient solar ponds are known as reliable and economical source for long-term heat collection and storage since over hundred years (Weinberger, 1964). With the dawn of new millennium as the energy and environmental crises have got exaggerated, ponds have appeared as environmental friendly and reliable energy alternative. Israel is working with an ambitious plan of fulfilling its entire energy demand by solar pond in near future (Amnon Einav, 2004). Substantial research work has been done on thermal as well as stability aspects of solar ponds. (Zangrando, 1991), (Singh *et al.*, 1994), (Punyasena *et al.*, 2003), (Angeli and Leonardi, 2004), (Jaefarzadeh, 2004), (Angeli and Leonardi, 2005) have done pioneer work on the stability aspect. For the maintenance of experimental solar ponds with variety of salts and different operating conditions, (Ouni *et al.*, 2003), (Huseyin, 2006) have contributed their research experiences. (Huanmin *et al.*, 2004) has given a glossary of major works in the maintenance of pond. The first analytical solution of pond was obtained by (Weinberger, 1964). (Tybout, 1966) proposed use of iterative methods for analyzing the pond. Estimation of radiation flux is a very important aspect of pond thermal behavior analysis. Many researchers have explored this aspect (Hull, 1982; Husain *et al.*, 2004 and Kooi, 1981). Few investigators observed the effect of turbidity on performance of pond (Husain *et al.*, 2004; Sugandhi *et al.*, 2006; Wang and Yagoobi, 1994,1995).

Kooi (1979) has proposed an analytical approach for determining optimum size of non-convective zone (NCZ) of the pond. With this optimum size (x_M), the pond retrieves maximum heat at steady state. However there is a limitation in the approach of Kooi. The pond takes a very long span of time for its warm up. Kooi's optimum size does not give efficient performance of the pond during warm up phase. Husain *et al.* (2003) have proposed an approach for optimum size of NCZ for rapid warm up of pond. All these researchers considered the pond liquid to be clear water. However in reality, pond is exposed to ambient and receives dust load continuously (Wang and Yagoobi, 1994). Also some biological growth in the pond is unavoidable. These parameters create turbidity in water which affects the radiation transmission so drastically that no realistic analysis of the pond can be done ignoring turbidity (Wang and Yagoobi, 1995; Husain *et al.*, 2004).

In the present work, an expression is developed to estimate optimum size of NCZ for rapid warm up (x_m), considering the turbidity in water. Thermal performance of the pond is analyzed with this Optimum (x_m) size of NCZ. Further, it is proposed that in the beginning (during maturation phase) the NCZ size may be kept as x_m and may be later change to x_M.

Theory

Heat balance in the storage zone of SGSP is given as (Sukhatme, 1994; Duffie *et al.*, 1981)

$$r\,C_p\,l_3\,dT/dt = -KdT/dx - Q_{LOSS} - Q_{LOAD} + I_{STZ} \tag{1}$$

Where r is density of pond liquid, C_p is specific heat, K is Conductivity, T is Temperature, t is time, l_3 is size of STZ, x is thickness of NCZ, Q_{Loss} is losses towards ground, Q_{LOAD} is heat extraction, I_{STZ} is radiation absorbed in STZ. Since pond is not loaded in the warm up phase, $Q_{LOAD} = 0$. Bottom and sides are considered as insulated. Hence $Q_{LOSS} = 0$. Equation (1) is rewritten as

$$r\,C_p l_3 (T_{STZ} - T_{amb})/Dt = -K[(T_{STZ} + T_{amb})/2 - T_{amb}]/(l_1 + x_n) + I_{STZ} \tag{2}$$

I_{STZ} is calculated by considering turbidity in water. x_n is the thickness of NCZ.

Equation (2) is differentiated with respect to x_n to obtain its optimum value. This equation deals the over all heat balance for the period Dt. In which initial temperature of STZ is increased from T_{amb} to T_{STZ}. Where T_{amb} is ambient temperature, x_n is thickness of NCZ. The term $K[(T_{STZ} + T_{amb})/2 - T_{amb}]/(l_1 + x_n)$ account for the conductive heat losses from STZ to surface, through NCZ assuming a linear temperature profile. Rewriting equation (2) as

$$(T_{STZ} - T_{amb})/Dt = (-KDT_m/(l_1 + x_n) + I_{STZ)/} rCpl_3 \tag{3}$$

where,

$$DT_m: (T_{STZ} + T_{amb}/2) - T_{amb}$$

In equation (3), x_n is variable. Equation (3) is differentiated with respect to x_n and equated to zero to find its maxima. Maxima is denoted by x_m.

$$d/dx_n(-K (DT_m/(l_1 + x_n) + I_{STZ})/rCpl_3 = 0 \quad (4)$$

$$\text{Say } f(x) = 0$$

Solving which, the optimum NCZ thickness x_m is obtained. The analysis is given in Appendix-I. Using this value of x_m, the warm up time is calculated. Optimum size of NCZ is also calculated by Kooi's method for comparison. Analysis is presented in Appendix-II.

Input Data

Hourly varying ambient temperature and solar radiation is considered for city of Jalgaon (21°N, 75°W) (Mani and Rangrajan, 1982). UCZ temperature of pond is estimated by Weinberger's, 1964 approach. Bottom and sides are considered to be insulated. Bottom is considered to be black. Turbidity of water is varied between 0.3 NTU to 1.0 NTU. Pond depth is taken as 1.75 m and UCZ size is taken as 0.2m. The target (desirable) temperature of STZ is varied in the range of 70°C to 85°C.

Results and Discussion

1. Figure 29.1 shows the influence of NCZ size on warm up time of pond. It is seen that, with an increase in NCZ size, the warm up time is getting reduce continuously. While in case of no turbidity the warm up time comes out to be minimum wit h about $0.8x¢_m$, where $x¢_m$ is the optimum size of NCZ as proposed by Husain *et al.* (2003).
2. At calculated x_m value and values of NCZ lower than this, the target temperature is not achieved. At NCZ values higher than x_m, the target STZ temperature achieved as shown in Figures 29.1–29.7.

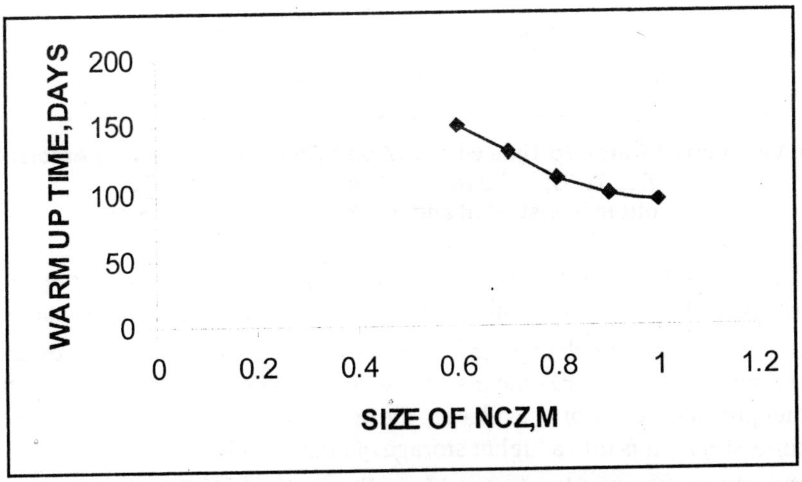

Figure 29.1: Variation of Warm Up Time with NCZ Size Considering Varying Ambient Parameters. T_{STZ}: 70°C; l_1: 0.2 m, L: 1.75 m. Turbidity: 0.3 NTU. Bottom is Insulated and black. x_m: 0.5 *m*; x_M: 0.5 m.

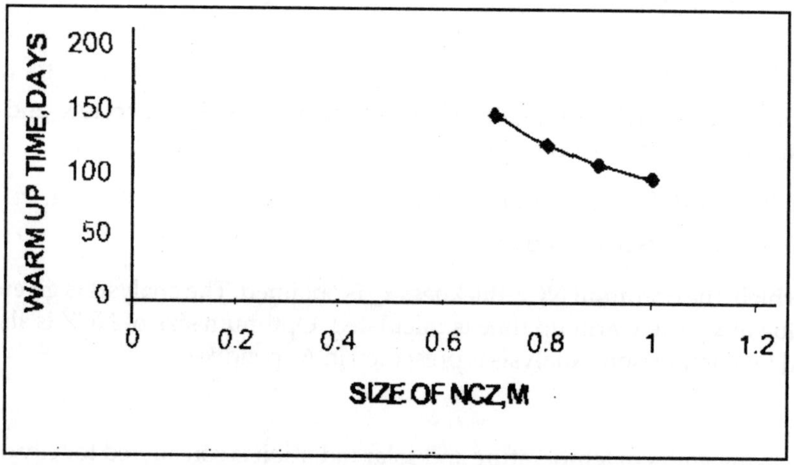

Figure 29.2: Variation of Warm Up Time with NCZ Size Considering Varying Ambient Parameters. T_{STZ}: 75°C; l_1: 0.2 m, L: 1.75 m. Turbidity: 0.3 NTU. Bottom is Insulated and black. x_m: 0.6 m; x_M: 0.5 m.

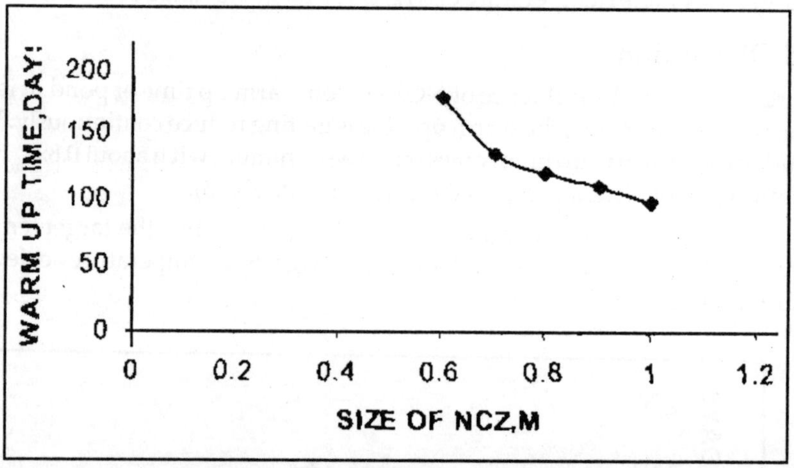

Figure 29.3: Variation of Warm Up Time with NCZ Size Considering Varying Ambient Parameters. T_{STZ}: 70°C; l_1: 0.2 m, L: 1.75 m. Turbidity: 0.5 NTU. Bottom is Insulated and black. x_m: 0.45 m; x_M: 0.5 m.

3. In over all, the use of value close to x_m is advantageous because, too large size of NCZ will be practically difficult and infeasible to be maintained from stability considerations also.
4. Because the overall depth of pond is constant, an increase in NCZ size results in the decrease of STZ size. Lower STZ size means, in over all reduction in storage of quantum of heat. This further justifies the use of x_m as close to the optimum size of NCZ, which gives a lower warm up time, still results into a higher storage of heat in STZ.
5. Similar observation can be obtained from Figures 29.1–29.7, in which other parameters are varied like turbidity, target temperature of STZ *etc.*

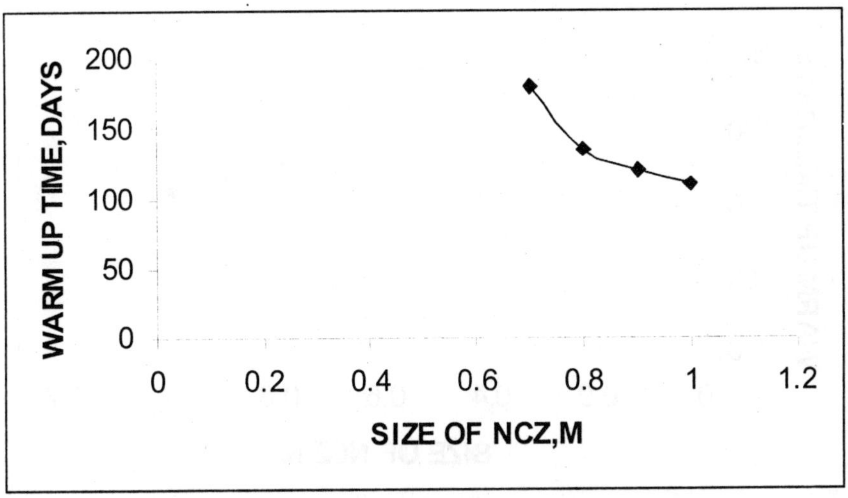

Figure 29.4: Variation of Warm Up Time with NCZ Size Considering Varying Ambient Parameters. **T_{STZ}: 75°C; l_1: 0.2 m, L: 1.75 m. Turbidity: 0.5 NTU.** **Bottom is Insulated and black. x_m: 0.5 m; x_M: 0.5 m.**

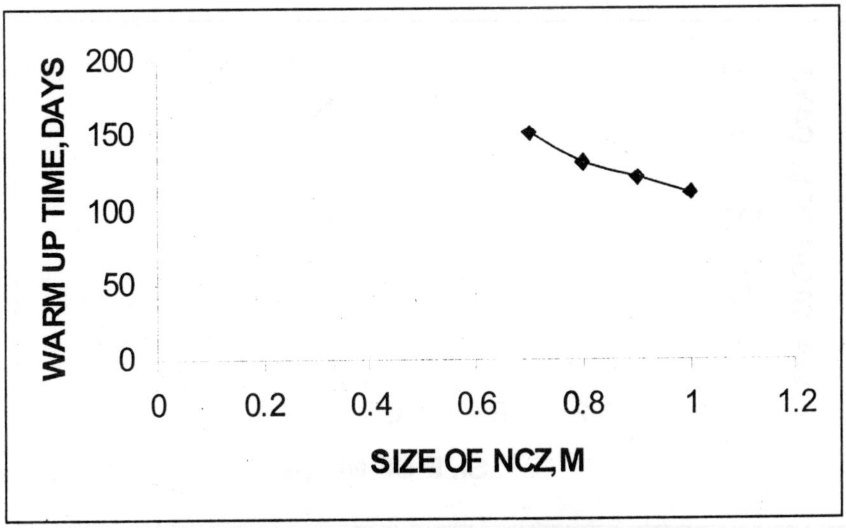

Figure 29.5: Variation of Warm Up Time with NCZ Size Considering Varying Ambient Parameters. **T_{STZ}: 70°C; l_1: 0.2 m, L: 1.75 m. Turbidity: 0.8 NTU.** **Bottom is Insulated and black. x_m: 0.21 m; x_M: 0.5 m.**

6. The optimum size of NCZ, x_M according to the Kooi's approach (Kooi 1979), is always constant for Figures 29.1–29.7, irrespective of target temperature of STZ.
7. The optimum size of NCZ obtained by Kooi's approach does not result into target STZ temperature within reasonable time duration. This is a significant observation with variable ambient parameters while considering constant parameter optimum NCZ size has always obtained target temperature.

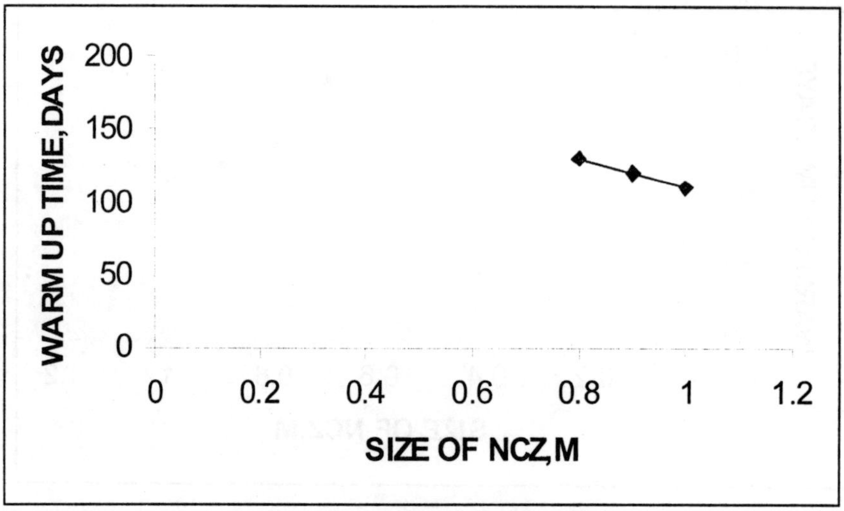

Figure 29.6: Variation of Warm Up Time with NCZ Size Considering Varying Ambient Parameters. T_{STZ}: 75°C; l_1: 0.2 m, L: 1.75 m. Turbidity: 0.8 NTU. Bottom is Insulated and black. x_m: 0.25 m; x_M: 0.5 m.

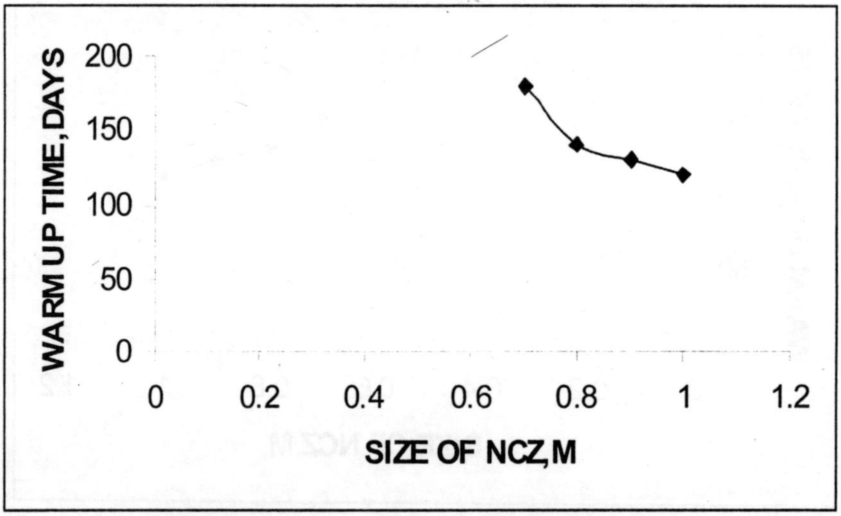

Figure 29.7: Variation of Warm Up Time with NCZ Size Considering Varying Ambient Parameters. T_{STZ}: 70°C; l_1: 0.2 m, L: 1.75 m. Turbidity: 1.0 NTU. Bottom is Insulated and black. x_m: 0.1 m; x_M: 0.5 m.

8. At lower turbidity range, the optimum size of NCZ (x_m) according to present approach is coming out to be always close to the size (x_M) proposed by (Kooi, 1979). It is different to the findings of Husain *et al.* (2003) for turbidity free water in which $x¢_m$ is always less than x_M.
9. For higher STZ temperature and higher turbidity values, optimum value of x_m does not exist. While considering constant ambient conditions, optimum value of x_m is always obtained. [Sugandhi *et al.*, 2006 communicated]. This is a typical observation with variable parameters.

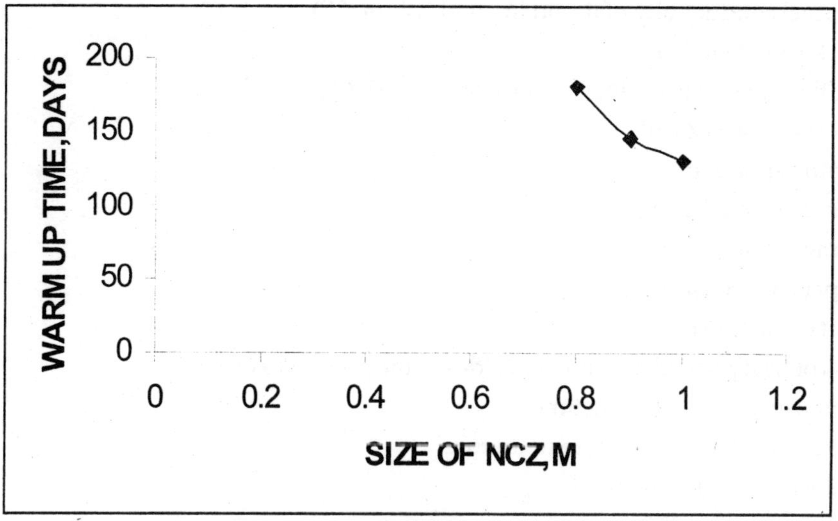

Figure 29.8: Variation of Warm Up Time with NCZ Size Considering Varying Ambient Parameters. T_{STZ}: 75°C; l_1: 0.2 m, L: 1.75 m. Turbidity: 1.0 NTU. Bottom is Insulated and black. x_m: 0.1 m; x_M: 0.5 m.

10. For higher turbidity values like 0.8 and 1.0 NTU, The x_m, values are not found to be in practical range. The values obtained are too small like 10 cm or so. This strongly suggests that the pond turbidity must be maintained below 0.8 NTU. This is an important guideline for pond's operation and maintenance.

Conclusion

Turbidity is an unwanted reality of SGSP. The foregoing analyses explicitly maintain that the turbidity has a very profound influence on optimum NCZ size. In the present case, 0.3 to 1.0 NTU turbidity is considered. The analysis indicates that turbidity greater than 0.8 NTU becomes too high for the pond and the optimum size of NCZ is not obtained. Any real pond exposed to ambient and maintained regularly by coagulant, will always have turbidity; but it must be maintained below 0.8 NTU. For a realistic analysis of SGSP, turbidity must be accounted. The present paper provides an analytical rational approach of determining optimum size of NCZ for faster warm up and over all optimum performance of the pond, considering the turbidity aspect, with hourly varying ambient data. This makes the analysis more realistic as compared to considering constant ambient parameters.

Nomenclature

- C_p: Heat capacity of liquid in pond (kJ/kg/°C)
- h: Transmission function defined by Wang and Yagoobi
- I: Symbol for radiation flux
- Io: Radiation flux incident at surface (W/m²)
- I_{STZ}: Radiation energy absorbed in STZ (W/m²)
- I_{xin}: Radiation flux incident at depth x (W/m²)

K: Thermal conductivity of liquid in pond (W/m/°C)
l_1: Thickness of UCZ (m)
l_2: Depth from surface to the interface of NCZ-STZ (m)
l_3: Thickness of STZ (m)
L: Depth of pond (m)
NCZ: Non-convective zone
STZ: Storage zone
UCZ: Upper convective zone
Q_{LOAD}: Heat extraction rate (W/m²)
Q_{LOSS}: Loss of heat through bottom and sides of the pond (W/m²)
t: Symbol used for denoting time
T: Symbol used for denoting temperature
T_{amb}: Ambient temperature (°C)
T_{STZ}: STZ temperature (°C)
xn: Thickness of NCZ
xm: Optimum Thickness of NCZ for rapid warm up while considering the turbidity
$x¢_m$: Optimum Thickness of NCZ proposed by Husain *et al.* (2003)
x_M: Optimum Thickness of NCZ according to the Kooi's approach
f: Turbidity of water in NTU
r: Density of liquid in pond (kg/m³)

Appendix–I

The equation (1) is solved here. It requires estimation of I_{STZ}. Which is done by (Wang and Yagoobi, 1995) method considering turbidity. Wang and Yagoobi proposed following correlation based upon their experimental investigation to estimate radiation flux.

$$I_{xin}(\mathrm{f}, x) = I_0 h(, x)$$

where, I_0, f and x are the radiation flux incident at the surface, turbidity of water in Nephelo-metric turbidity units (NTU), and depth in meter, respectively. The non-dimensional transmission function h is defined as

$$h(\mathrm{f}, x) = h(0.3, x)\, r(\mathrm{f}, x)$$

where, $h(0.3, x) = 0.58–0.076 \ln (100\, x)$ and

$r(\mathrm{f}, x) = 1.0–0.1975\, x\, (\mathrm{f}–0.3) + 0.0144\, x\, (\mathrm{f}–0.3)^2$

$r(\mathrm{f}, x)$ accounts for turbidity greater than 0.3 NTU.

Solution of equation (4) is done as

$$f(x) = Dx_n\, [-KDT_{m/}(l_1 + x_n) + I_{STZ}]/rCpl_{3=0}$$

$$f(x) = d/dx\, (-KDT_m/(l_1 + x_n) + d/dx\, I_{STZ} = 0$$

Because $rCpl_3$ is a constant.

$$f(x) = KDT_m/(l_1 + x_n)^2 + d/dx\, Io[(a-b\log_e(100(l_1 + x_n))$$
$$(1-0.1975(l_1 + x_n)(f-0.3) + 0.0144(l_1 + x_n)(f-0.3)^{2]}$$

$$f(x) = K\,DT_m/(l_1+ x_n)^2 + Io[(a- b\log_e(100(l_1 + x_n)](0-0.1975(f-0.3) + 0.0144(f-0.3)^2) + (1-0.1975(l_1 + x_n)(f-0.3) + (0.0144(l_1 + x_n)(f-0.3)^2)(0-b\, d/dx(\log_e 100(l_1 + x_n))]$$

$$f'(x) = d/dx\, f(x)$$

$$f'(x) = -2KDT_m/(l_1+ x_n)^3 + Io[b/(l_1 + x_n)[0.1975(f-0.3)$$
$$- 0.0144(f-0.3)^2] + b/(l_1 + x_n)^2$$

The solution is obtained numerically with the help of computer program

Appendix- II

Kooi's expression for steady state optimum thickness of NCZ x_M given as

$$?\,(I_T)dx - x_M\, I_{(l1+xM)} = KDT/I_0 \quad (5)$$

Say $y(x) = 0$

For determination of $x_{M,}$ equation (5) is solved using Wang and Yagoobi correlation.

$$I_{xin}(f, x) = Io\, h(f, x) \text{ as given in appendix (1)}$$

$$y(x) = ?\,(I_T)dx - x_M\, I_{(l1+xM)} - KDT/I_0 = 0$$

$$y(x) = Io\; ?\;(a-b\log_e 100x)(1-0.1975x(f-0.3) + 0.0144x(f-0.3)^2$$
$$-Io\, x_M\, [a- b\log_e 100\,(l_1+ x_M)][1-0.1975(l_1+ x_M)(f-0.3)$$
$$+ 0.0144(l_1+ x_M)(f-0.3)^2] - KDT/Io = 0$$

$$y(x) = Io[a-b\log_e 100[x_M-(x_M{}^2 + 2l_1\, x_M)/2\,(0.1975(f-0.3)$$
$$- 0.0144\,(f-0.3)^2)] - b[\,(l_1+ x_M)\log_e (l_1+ x_M) - l_1\log_e l_1 - x_M]$$
$$+ b\,(0.1975\,(f-0.3) - 0.0144(f-0.3)^2)$$
$$[(l_1+ x_M)^2/2\,(\log_e(l_1+ x_M) - 1/2) - l_1^2/2\,(\log_e l_1 - 1/2)]]$$
$$- Io\, x_M\, [a-b\log_{e100}(l_1+ x_M)]\,[1-0.1975(l_1+ x_M)(f-0.3)$$
$$+ 0.0144(l_1+ x_M)(f-0.3)^2] - KDT/Io = 0$$

$$y'(x) = d/dx\, y(x)$$

$$d/dx\, y(x) = Io\,[\,(a-b\log_e 100)d/dx\, x_M - [(a-b\log_e 100)(0.1975(f-0.3) -$$
$$0.0144f-0.3)^2)]\, d/dx(x_M{}^2 + 2l_1\, x_M)/2 - b\,(d/dx\,(l_1+ x_M)$$
$$loge(l_1+ x_M) - d/dx\, l_1\log_e l_1 - d/dx\, x_{M)} + b\,(0.1975\,(f-0.3) -$$
$$0.0144(f-0.3)^2)d/dx(l_1+ x_M)^2/2\,(\log_e(l_1+ x_M) - 1/2) - d/dx$$
$$l_1^2/2(\log_e l_1 - 1/2)] - Io\, d/dx(x_M\, [a-b\log_{e100}(l_1+ x_M)]\,[1 -$$

$$0.1975(l_1+x_M)(f-0.3)+0.0144(l_1+x_M)(f-0.3)^2])]-d/dx\ KDT/Io$$

$$y'(x) = Io\ [a-b\ \log_e 100\ [1-(0.1975(f-0.3)-0.0144(f-0.3)^2)(l_1+x_M)]$$

$$-b[\log_e(l_1+x_M)] + b[0.1975(f-0.3)-0.0144(f-0.3)^2][(l_1+x_M)$$

$$(\log_e(l_1+x_M)-1/2)+(l_1+x_M)/2]]$$

$$-Io\ [[a-b\ \log_e 100\ (l_1+x_M)[1-0.1975(l_1+x_M)(f-0.3)$$

$$+0.0144(l_1+x_M)(f-0.3)^{2]}$$

$$-x_M\ [b(0.1975(f-0.3)-0.0144(f-0.3)^2)][1+\log_e(100(l_1+x_M)]$$

$$-a\ x_M\ (-0.1975(f-0.3)+0.0144(f-0.3)^2)+b\ x_M/(l_1+x_M)]$$

The solution is obtained numerically with the help of computer program

References

Agha, K.R., Abughres, S.M. and Ramdan, A.M., 2004. Design methodology for a salt gradient solar pond coupled with an evaporation pond. *Solar Energy*, 73(5): 447–454.

Amonon Enav, 2004. Solar energy research and development achievements in Israel and their practical significance. *Journal of Solar Energy Engineering*, 126(3): 921–928.

Angeli, C. and Leonardi, E., 2004. One-dimensional numerical study of the salt diffusion in a salinity-gradient solar pond. *International Journal of Heat and Mass Transfer*, 47(1): 1–10.

Angeli, C. and Leonardi, E., 2005. The effect of thermo diffusion on the stability of a salinity gradient solar pond. *International Journal of Heat and Mass Transfer*, 48: 4633–4639.

Duffie, J.A. and Beckman, W.A., 1981. *Solar Engineering of Thermal Processes*. John Wiley, p. 78–80.

Hull, J.R., 1980. Computer simulation of solar pond thermal behavior. *Solar Energy*, 25: 33–40.

Hull, J.R., 1982. Calculation of solar pond thermal efficiency with a diffusively reflecting bottom. *Solar Energy*, 29(5): 385–389.

Husain, M., Patil, S.R., Patil, P.S. and Samdarshi, S.K., 2003. Optimum size of non-convective zone for improved thermal performance of salt gradient solar pond. *Solar Energy*, 45: 73–81.

Husain, M., Patil, S.R., Patil, P.S. and Samdarshi, S.K., 2004. Combined effect of water turbidity and bottom reflectivity on thermal performance of salt gradient solar pond. *Energy Conversion and Management*, 45: 73–81.

Husain, M., Patil, S.R., Patil, P.S. and Samdarshi, S.K., 2004. Simple methods for estimation of radiation flux in salt gradient solar pond. *Energy Conversion and Management*, 45: 303–314.

Huseyin, Kurt and Ozkaymark, Mehmet, 2006. Performance evaluation of a small-scale sodium carbonate salt gradient solar pond. *International Journal of Energy Research*, 30(11): 905–915.

Huanmin, Lu, Andrew, H.P., Swift, Hobert and Hein, D., 2004. Jr and John Walton. *Journal of Solar Energy Engineering*, 126(2): 759–767.

Jaefarzadeh, M.R., 2004. Thermal behavior of a small salinity-gradient solar pond with wall shading effect. *Solar Energy*, 77(3): 281–290.

Kooi, C.F., 1979. The Steady state salt gradient solar pond. *Solar Energy*, 25: 7–45.

Kooi, C.F., 1981. Salt gradient solar pond with reflective bottom: Application to the 'saturated' pond. *Solar Energy*, 26: 113–120.

Mani, A. and Rangrajan, S., 1982. *Solar Radiation Over India.* Allied Publishers Private Limited, New Delhi, India.

Ouni, M., Guizani, A., Lu, H. and Belghith, A., 2003. Simulation of the control of a salt gradient solar pond in the south of Tunisia. *Solar Energy*, 75(2): 95–101.

Punyasena, M.A., Amarasekara, C.D., Jayakody, J.R.P., Perera, P.A.A. and Ehamparam, P., 2003. An investigation of rain and wind effects on thermal stability of large-area saltpan solar pond. *Solar Energy*, 74(6): 447–451.

Singh, T.P., Singh, A.K. and Kaushika, N.D., 1994. Investigations of thermodynamic instabilities and ground storage in a solar pond by simulation model. *Heat Recovery Systems and CHP*, 14(4): 401–407.

Sugandhi, S., Saxena, A.K. and Husain, M., 2006. Influence of turbidity on optimum size of non-convective zone of salt gradient solar pond. *Energy Conversion and Management*, Communicated.

Sukhatme, S.P., 1994. *Solar Energy: Principles of Thermal Storage and Collection.* Tata McGraw Hill Publishing Co., New Delhi, India.

Tybout, R.A., 1966. A recursive alternate to Weinberger's model of the solar pond. *Solar Energy*, 11: 109–111.

Wang, J. and Yagoobi, S., 1994. Effect of water turbidity and salt concentration levels on penetration of solar radiation under water. *Solar Energy*, 52(5): 429–438.

Wang, J. and Yagoobi, S., 1995. Effect of water turbidity on thermal performance of a salt gradient solar pond. *Solar Energy*, 54(5): 301–308.

Weinberger, H., 1964. The physics of the solar pond. *Solar Energy*, 8: 45–46.

Zangrando, F., 1991. On the hydrodynamics of salt gradient solar ponds. *Solar Energy*, 46(6): 323–341.

Water and Wastewater Management in 2 Vols (2011) *Pages* ***186–189***
Editors: **D.R. Khanna, R. Bhutiani & Gagan Matta**
Published by: **DAYA PUBLISHING HOUSE, NEW DELHI**

Chapter 30

Study on Macrophytes in Ramala Lake, Dist. Chandrapur, M.S.

P.J. Khinchi[1], P.M. Telkhade[2], N.R. Dahegaonkar[2] and S.B. Zade[3]

[1]Janta Mahavidyalaya, Chandrapur, M.S.
[2]Arts, Commerce and Science College, Tukum, Chandrapur, M.S.
[3]M.B. Patel Science College, Sakoli, Dist. Bhandara, M.S.

ABSTRACT

The present piece of investigation deals with aquatic macrophytes of Ramala Lake, Dist. Chandrapur, Maharashtra during the period May-2006 to April-2007. The lake is in the heart of city receiving sewage from maximum part of city through Mucchi Nala. During the study period, total of 18 species representing families of aquatic macrophytes. Among these, species *Eichhornia* and *Pistia* showed their dominance, indicating highly eutrophic condition of lake.

Keywords: Aquatic macrophytes, Ramala lake, Diversity.

Introduction

Aquatic macrophytes are large predominantly angiospermic plants having considerable importance. There are some as good source of food and shelter to periphyton, ichthyofauna and other aquatic invertebrates. They may also serve as good source of fertilizers and some of the aquatic plants are being cultivated for their astonishing diversity of medicinal and aesthetic values (Bardach, 1968). Persual of literature on macrophytes indicates that, many reports were available particularly from India and abroad, but no such reports are available from this area, so present piece of work was undertaken. Several workers reported on macrophytes diversity in different freshwater bodies of India and abroad. Pearsall (1921), Russel Hunter (1970), Zutshi *et al.* (1980), Unni (1991) and Okram *et al.* (1996).

Materials and Method

During the period of investigation, macrophytes were collected by hand picking method from three different sites of lake. They were collected, washed and brought to laboratory in polythene bags and preserved in formaline and identified by Edmondson (1959) and Kodarkar (1992)

Results and Discussion

During the period of investigation, 18 species of macrophytes were observed, among which 5 were of free floating, 8 submerged, 2 rooted with floating leaves and 3 emergent. The site wise distribution of macrophytes showed presence of 9 species at Site-I, 15 species at Site-II and 16 species at Site-III.Out of 18 species recorded *Lemna* sp., *Azolla* sp., *Pistia* sp., *Eichornia* sp., *Trapa* sp., *Wolffia* sp., *Hydrilla* sp. and *Vallisneria* sp. were found at all the sites throughout the different seasons. Species like *Sagittaria* and *Ipomea* showed their presence only at Site-III while *Nymphaea* sp. at Site-I.

The floristic distribution of plant species of present investigation tabulated in Table 30.1and were categorized into 4 groups *i.e.* free floating (5 species), submerged (8 species), rooted with floating leaves (2 species) and species emergent (3 species).

Table 30.1: Name of Different Types of Macrophytes and their Site-wise Presence or Absence

Sl.No.	*Name of Macrophytes*	*Site-I*	*Site-II*	*Site-III*
	Free floating			
1.	*Lemna*-nee floating	+	+	+
2.	*Azolla*-free floating	+	+	+
3.	*Pistia*-free floating	+	+	+
4.	*Eichornia*-free floating	+	+	+
5.	*Wolffia*-free floating	+	+	+
	Submerged			
6.	*Najas*-submerged	–	+	+
7.	*Typha*-submerged	–	+	+
8.	*Potamegeton*-submerged	–	+	+
9.	*Sagittaria aquatica*-submerged	–	–	+
10.	*Hydrilla*-submerged	+	+	+
11.	*Vallisneria*-submerged	+	+	+
12.	*Ceratophyllum*-submerged.	–	+	+
13.	*Marsilea*-submerged	–	+	+
	Rooted with floating leaves			
14.	*Nelumbo*-rooted with floating leaves	–	+	–
15.	*Nymphaea*-rooted with floating leaves	+	–	–
	Emergent			
16	*Cyperus*-emergent	–	+	+
17.	*Trapa*-emergent	+	+	+
18.	*Ipomoea*-emergent	–	–	+

+: Present; –: Absent.

The present findings are in conformity with Purushottama *et al.* (2005) in Kanale tank, Karnataka, who reported 19 species of which 2 were free floating, 5 rooted floating, 6 submerged, 1 emergent and 5 semiaquatic. Mishra and Tripathi (2004) also reported 12 species in the unpolluted site of Ganga river in Varanasi. A comparatively higher number of macrophytes species were reported by Devi (1993) with 86 species in Laktak lake, Manipur. Out of which 73 species were reported from non-phundic floating mat zone of which 6 species were submerged, 4 species free floating and 3 rooted with floating leaves species.

Recently, Devi and Sharma (2007), observed 36 macrophytes species of which 20 emergent, 8 submerged, 4 rooted with floating leaves and 4 free floating species, in Awangsoipat lake Bishnupur, Manipur. It was noticed during present investigation that flourishing growth and frequency of distribution of macrophytes was correlated with an increase in phosphate and nitrogen contents of the water bodies as these nutrients stimulate rapid organic production by aquatic macrophytes. Sarkar *et al.* (2002), similarly the increased dissolved oxygen (DO) and free CO_2 in water bodies gradually favoured the growth and distribution of macrophytes. In present investigation, a unique observation indicates the eutrophic nature of the lake. The lake has got amorphous growth of vegetation with rich biodiversity. The highest growth of submerged species indicates eutrophic nature of lake. A clear cut zone of floating and submerged species cannot be observed and hence all such communities were found in intermixed mats.

However, the *Eichornia* sp. showed its dominance heights may be due to increased load of sewage added by Mucchi nala and decomposition activities. This has resulted serious threat to the ecosystem stabilization and the very existence of lake.

Acknowledgement

The authors are thankful to Dr. M.M. Wankhede, Principal, A.C.S. College, Tukum, and Dr. Subhash, Principal, Janta Mahavidyalaya, Chandrapur. (M.S.) for their kind co-operation during the present piece of work.

References

Bardach, J.E., 1968. Aquaculture. *Science*, 161: 1098–1112.

Devi, N.B., 1993. *Ph.D. Thesis.* Manipur University, Manipur.

Devi, N.P. and Sharma, B.M., 2007. Investigation on the pattern of distribution of microphytes in Hiden compat lake, Bishnupur, Manipur. *Ind. J. Env. and Eco.*, 14(1–2): 109–114.

Edmondson, W.P.,1959. *Freshwater Biology*. Seattle, Washington, pp. 420–495.

Kodarkar, M.S., 1992. *Methodology for Water Analysis, Physico-chemical, Biological and Microbiological Methods*. Indian Association of Aquatic Biologists, Hyderabad. Publ. pp. 2–50.

Mishra, B.P. and Tripathi, B.D., 2004. Distribution of macrophytes and phytosociology of *Hydrilla verticellata* and *Lemna minor*, Linn. in lotic and lentic aquatic ecosystem. *Ecol. Env. and Cons.*, 10: 37–41.

Okram, I.D., Sharma, B.M. and Singh, E.J., 1996. Ecology study of Waithu lake, Manipur: Morphometry and qualitative analysis of macrophytic vegetation. *J. Freshwater Biol.*, 8(4): 177–189.

Pearsall, W.H., 1921. A suggestion to factors influencing distribution of free floating vegetation. *J. Ecol.*, 9: 243–253.

Purushottam, R., Narayana, J., Kiran, B.R. and Harish, K.K., 2005. Nutrient station of Kanale tank at Sagara taluk in Karnataka with reference to diversity of aquatic macrophytes. *J. Curr. Sci.*, 161–164.

Russel Hunter, W.D., 1970. *Aquatic Productivity and Introduction to Some Basic Aspect of Biological Oceanography and Limnology*, Macmillan Publishing Co. Inc., New York, pp. 1–249.

Sarkar, A., Singhal, V. and Arora, M.P., 2002. Physico-chemical characteristic of Hindon and Narmada river. *J. Exp. Zoo., India*, 5(1): 107–112.

Unni, K.S., 1991. An ecology study of the macrohytes vegetation Doodhadhari lake, Raipur, M.P., India J. Distribution and seasonal changes in aquatic plants. *Hydrobiol.*, 37(6): 139–155.

Zutshi, D.P., Subla, B.A., Khan M.A. and Wanganed, A., 1980. Comparative limnology of 9 lakes of Jammu and Kashmir, Himalaya. *Hydrobiol.*, 72: 101–112.

Water and Wastewater Management in 2 Vols (2011) *Pages* **190–200**
Editors: D.R. Khanna, R. Bhutiani & Gagan Matta
Published by: DAYA PUBLISHING HOUSE, NEW DELHI

Chapter 31

A Study of Biotic and Abiotic Factors of Song River at Dehradun, Uttarakhand

D.R. Khanna, Vikas Singh, R. Bhutiani, Kumar Satish Chandra, Gagan Matta and Dheeraj Kumar
Department of Zoology and Environmental Science, Gurukula Kangri Vishwavidyalaya, Haridwar

ABSTRACT

The present study deals with the analysis of physico-chemical parameters of the Song river at Dehradun, Uttarakhand during the pre-autumn, autumn and post-autumn months of 2007. In the present study of the Song river its physico-chemical characteristics *viz.*, temperature, velocity, total solids, total suspended solids, total dissolved solids, turbidity, pH, dissolved oxygen, free CO_2, BOD, COD, alkalinity, chlorides, sulphate and identification of fishes were done. The data revealed that 36 species of fishes are present in the Song river of Dehradun, Uttarakhand.

Keywords: *Physico-chemical, Abiotic, Biotic, Song river, Dehradun.*

Introduction

Through out history water has been considered a natural resource critical to human survival. Human history can infact be written in terms of interactions and interrelations between human and water (Biswas, 1970). Rivers are the most important resources in the world and in India in particular. Unfortunately, the same rivers are being polluted by indiscriminate disposal of sewage and industrial wastes as a plethora of human activities. Even piped water which is available in big cities becomes also mixed with number of impurities causing waterborne diseases like jaundice, cholera, typhoid and gastroenteritis (Kudesia, 1988). Pollution of a river first affects its chemical quality and then systematically destroys the community disrupting the delicate food web. River pollution has several dimensions; and effective monitoring, and control of river pollution requires the expertise from various disciplines (Trivedy, 1990).

Song River is a spring fed river originated from different small rivulets of the mountainous range of Dhanolti, crossing with Sahastradhara streams flow downward towards Doon valley basins and finally it assimilates into river Ganga at Raiwala. The river Song is located at 30°28′ latitude and 78°8′ longitude, with which peoples of Raiwala, Doiwala, Chiddarwala, and Lacchiwala are very much attached because this river is the only ultimate source of water for them travelling a total distance of approximately 42.5 km. It merges into river Ganga at 78°48′27″ longitude and 30°2′ latitude after crossing Satyanarayana area. Therefore, in the present study of the Song river its physico-chemical characteristics *viz.* temperature, velocity, total solids, total suspended solids, total dissolved solids, turbidity, pH, dissolved oxygen, free CO_2, BOD, COD, alkalinity, chlorides, sulphate and identification of fishes were done.

Accurate assessment of water quality, whether in relation to the requirements of intended water uses or in order to determine the impacts of an activity on the water resource depends on the results generated by specific monitoring activities which define the physical, chemical and/or biological condition of the resource. Many workers from India and abroad discussed the different aspects of water quality time to time. Khanna (1993) made a study on ecology and pollution of River Ganga at Haridwar. Pandey *et al.* (1993) studied on the physico-chemical quality of water of river Kosi at Purnia. Khanna *et al.* (1999) studied fishes and their ecology of the river Suswa at Raiwala, Dehradun. Khanna and Singh (2002) published a book on pond fish ecology and economics. Khanna *et al.* (2003) described a study of assessment of water quality of river Ganga in district Bulandshahar of Uttar Pradesh. Kaur and Joshi (2003) described seasonal variation in some physico-chemical parameters of river Ganga. Khanna and Bhutiani (2003) worked on limnological status of Satikund pond at Haridwar. Semwal and Akolkar (2006) had done the water quality assessment of sacred Himalayan Rivers of Uttarakhand. Bhutiani and Khanna (2007) have done ecological study of Suswa River with main reference to BOD and DO modelling.

Methodology

The water samples were collected from three sampling stations "A"- Raiwala, "B"- Chiddarwala and "C"- Lacchiwala in morning hours (From 8:00 A.M. to 10:00 A.M.). The samples were taken in Borosil glass bottles of 300 ml capacity and plastic containers. Fishes were collected in plastic jars and polythenes.The procedure, which was applied to analyse the sample for different physico-chemical parameters, were taken from APHA (1998), Trivedi and Goel (1986), Khanna and Bhutiani (2004), Santra (2004).

Results and Discussion

The physico-chemical parameters and fish species obtained during the study period are tabulated in Tables 31.1–31.15. In the Song River at Dehradun a difference in the fluctuation of water temperature was observed maximum (23.38°C) in post-autumn month and minimum (18.36°C) in pre-autumn month. The water temperature showed an upward trend from winter season to summer season. A more or less similar trend has been observed in the river Yamuna by Chakrabarty *et al.* (1959). Badola and Singh (1981) reported similar trend in river Alakhnanda.

The velocity started increasing after winter season. The maximum (0.95 m/sec) velocity was recorded in post-autumn month and minimum (0.82 m/sec) in pre-autumn month. In the present study, it has been observed that the velocity and the total solids showed positive relationship. Maximum concentration (380 mg/l) of total solid was found during the post-autumn month which may be due to soil erosion or side cutting caused by the main stream and the minimum (300 mg/l) during the pre-autumn month as also observed by Joshi and Bisht (1993) in Ganga canal at Jwalapur, Haridwar.

Table 31.1: Seasonal Variations of Different Parameters of Song River at Dehradun

Parameter	*Pre-Autumn*	*Autumn*	*Post-Autumn*	*Average*
Temperature (°C)	19.50	22.83	23.16	21.16±2.02
Velocity (m/sec)	0.82	0.87	0.95	0.88±0.07
Total Solids (mg/l)	300.00	330.00	380.00	336.66±40.41
TDS (mg/l)	290.00	315.00	335.00	313.30±22.55
TSS (mg/l)	17.33	15.33	35.33	22.66±11.02
Turbidity (JTU)	15.33	20.00	26.66	20.66±5.69
pH	6.90	6.90	8.56	7.45±0.96
DO (mg/l)	9.25	8.90	8.38	8.85±0.44
Free CO_2 (mg/l)	2.56	2.76	3.31	2.88±0.39
BOD (mg/l)	2.30	2.80	3.25	2.78±0.47
COD (mg/l)	2.30	2.68	3.16	2.71±0.43
Sulphate (mg/l)	1.65	2.27	2.76	2.22±0.56
Alkalinity (mg/l)	180.80	158.16	141.50	160.15±19.73
Chlorides (mg/l)	18.36	22.90	23.38	21.54±2.78

Table 31.2: The Value of Temperature (°C) Found at Different Sampling Stations of Song River

Sampling Site	*Pre-Autumn*	*Autumn*	*Post-Autumn*	*Average*
Raiwala	19.0	21.0	23.5	21.16±2.25
Chiddarwala	20.0	22.0	24.0	22.00±2.00
Lacchiwala	19.5	19.5	22.0	20.33±1.44

Table 31.3: The Velocity (m/sec) of Water Found at Different Sampling Stations of Song River

Sampling Site	*Pre-Autumn*	*Autumn*	*Post-Autumn*	*Average*
Raiwala	0.90	0.95	1.00	0.95±0.05
Chiddarwala	0.85	0.85	1.00	0.90±0.087
Lacchiwala	0.73	0.82	0.85	0.80±0.062

Table 31.4: The Value of Total Solids (mg/l) Found at Different Sampling Stations of Song River

Sampling Site	*Pre-Autumn*	*Autumn*	*Post-Autumn*	*Average*
Raiwala	305	345	385	345±0.00
Chiddarwala	310	335	375	340±32.79
Lacchiwala	300	310	380	330±3.59

Table 31.5: The Value of Total Dissolved Solids (mg/l) Found at Different Sampling Stations of Song River

Sampling Site	Pre-Autumn	Autumn	Post-Autumn	Average
Raiwala	285	320	340	315±27.84
Chiddarwala	300	325	335	320±18.03
Lacchiwala	285	300	330	305±22.91

Table 31.6: The Value of Total Suspended Solids (mg/l) Found at Different Sampling Stations of Song River

Sampling Site	Pre-Autumn	Autumn	Post-Autumn	Average
Raiwala	20	25	45	30±3.61
Chiddarwala	10	10	40	20±9.17
Lacchiwala	15	10	50	24±21.79

Table 31.7: The Value of Turbidity (JTU) Found at Different Sampling Stations of Song River

Sampling Site	Pre-Autumn	Autumn	Post-Autumn	Average
Raiwala	16	20	30	23±7.21
Chiddarwala	20	25	30	25±5.00
Lacchiwala	10	15	20	15±5.00

Table 31.8: The Value of Chlorides (mg/l) Found at Different Sampling Stations of Song River

Sampling Site	Pre-Autumn	Autumn	Post-Autumn	Average
Raiwala	17.60	24.23	27.65	23.16±5.11
Chiddarwala	19.50	24.50	23.50	22.50±2.65
Lacchiwala	18.00	20.00	19.00	19.00±1.00

Table 31.9: The Value of pH Found at Different Sampling Stations of Song River

Sampling Site	Pre-Autumn	Autumn	Post-Autumn	Average
Raiwala	7.25	6.75	8.20	7.40±0.74
Chiddarwala	7.00	6.50	9.00	7.50±1.32
Lacchiwala	6.50	7.50	8.50	7.50±1.30

Table 31.10: The Value of Dissolved Oxygen (mg/l) Found at Different Sampling Stations of Song River

Sampling Site	Pre-Autumn	Autumn	Post-Autumn	Average
Raiwala	9.25	8.75	8.40	8.80±0.43
Chiddarwala	9.50	9.00	8.50	9.00±0.50
Lacchiwala	9.00	9.00	8.50	8.83±0.29

Table 31.11: The Value of Free Carbon Dioxide (mg/l) Found at Different Sampling Stations of Song River

Sampling Site	*Pre-Autumn*	*Autumn*	*Post-Autumn*	*Average*
Raiwala	2.95	3.05	3.45	3.15±0.265
Chiddarwala	2.75	2.75	3.50	3.00±0.433
Lacchiwala	2.00	2.50	3.00	2.50±0.50

Table 31.12: The Value of Biochemical Oxygen Demand (mg/l) Found at Different Sampling Stations of Song River

Sampling Site	*Pre-Autumn*	*Autumn*	*Post-Autumn*	*Average*
Raiwala	2.65	2.95	3.55	3.05±0.46
Chiddarwala	2.25	2.95	3.50	2.90±0.63
Lacchiwala	2.00	2.50	2.70	2.50±0.50

Table 31.13: The Value of Chemical Oxygen Demand (mg/l) Found at Different Sampling Stations of Song River

Sampling Site	*Pre-Autumn*	*Autumn*	*Post-Autumn*	*Average*
Raiwala	2.20	2.90	3.15	2.75±0.49
Chiddarwala	2.45	2.65	2.85	2.65±0.20
Lacchiwala	2.25	2.50	3.50	2.75±0.66

Table 31.14: The Value of Sulphate (mg/l) Found at Different Sampling Stations of Song River

Sampling Site	*Pre-Autumn*	*Autumn*	*Post-Autumn*	*Average*
Raiwala	1.85	2.42	2.73	2.32±0.45
Chiddarwala	1.60	2.15	2.85	2.20±0.63
Lacchiwala	1.50	2.25	2.70	2.15±0.61

Table 31.15: The Value of Alkalinity (mg/l) Found at Different Sampling Stations of Song River

Sampling Site	*Pre-Autumn*	*Autumn*	*Post-Autumn*	*Average*
Raiwala	192.50	164.50	142.26	166.42±25.18
Chiddarwala	185.00	165.00	140.75	163.58±22.16
Lacchiwala	165.00	145.00	141.50	150.50±12.68

During the study T.D.S. was found minimum (290 mg/l) in pre-autumn month and maximum (335 mg/l) in post-autumn month and increased value were probably due to heavy rain causing soil erosion and adding more suspended or dissolved solids from the nearby area as also observed by Khanna (1993) in the river Ganga at Haridwar, Logankumar *et al.* (1989) in the river Bhavani at Sirumugai, Tamil Nadu. During the study total suspended solid was found maximum (45 mg/l) in

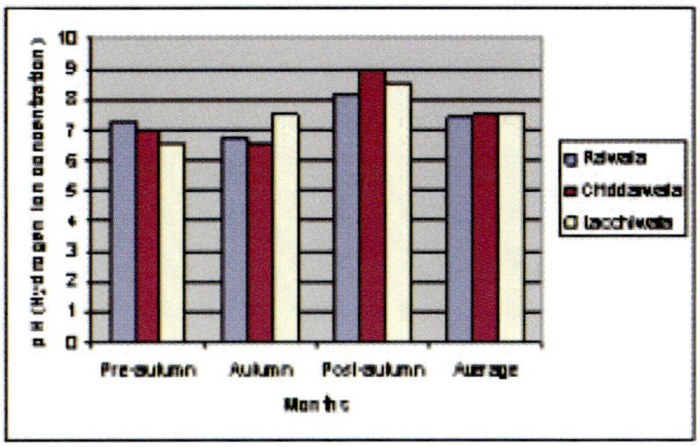
p H (Hydrogen ion concentration)
Pre-autumn
Autumn
Post-autumn
Average
Months
Raiwala
Chiddarwala
Lacchiwala

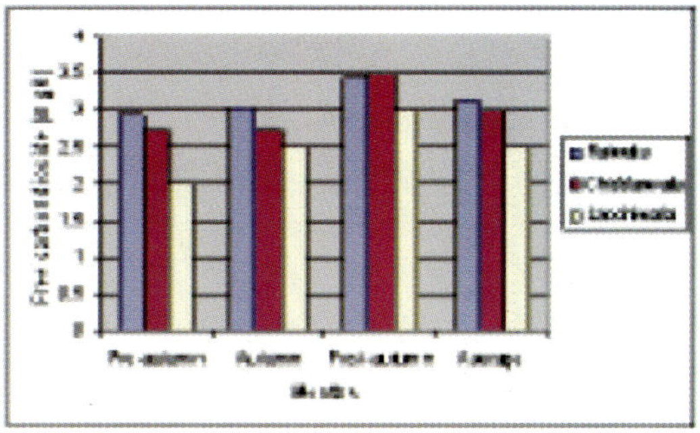

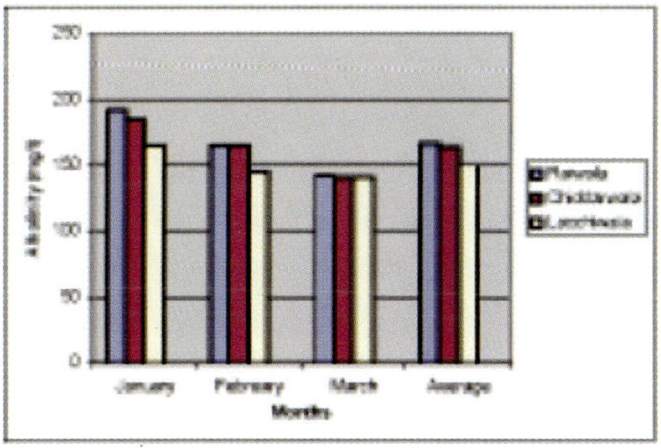
January
February
March
Average
Months

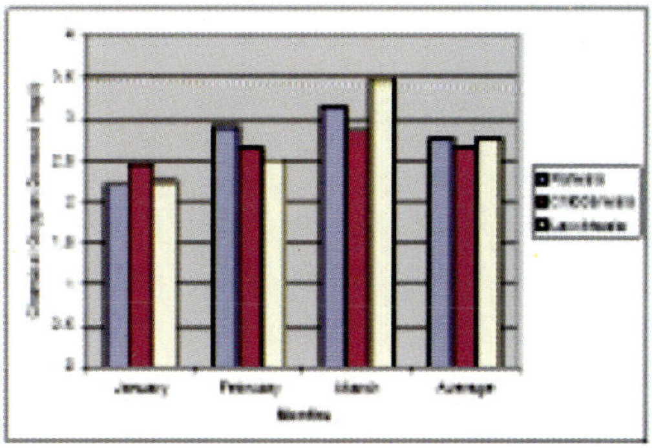

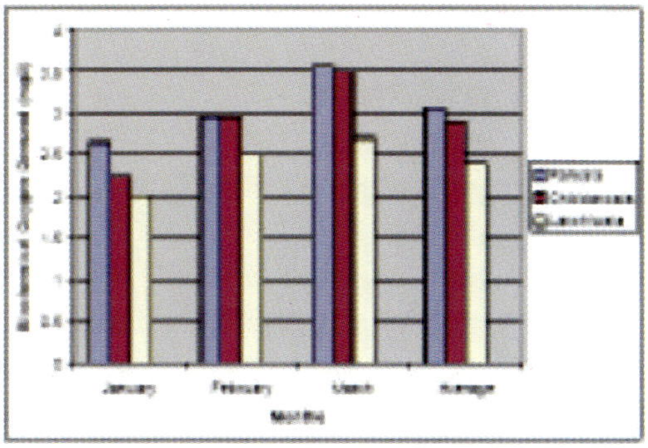

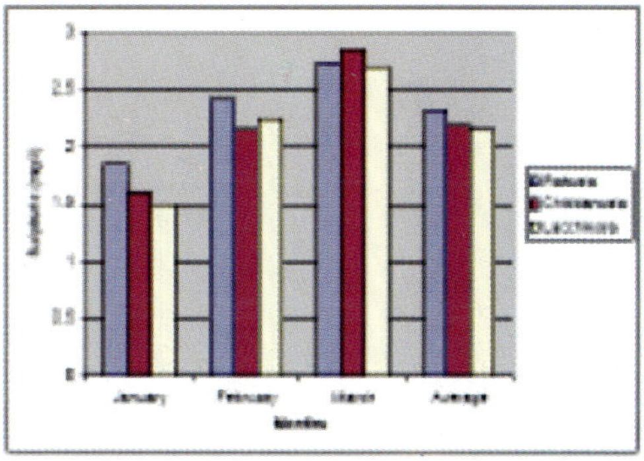

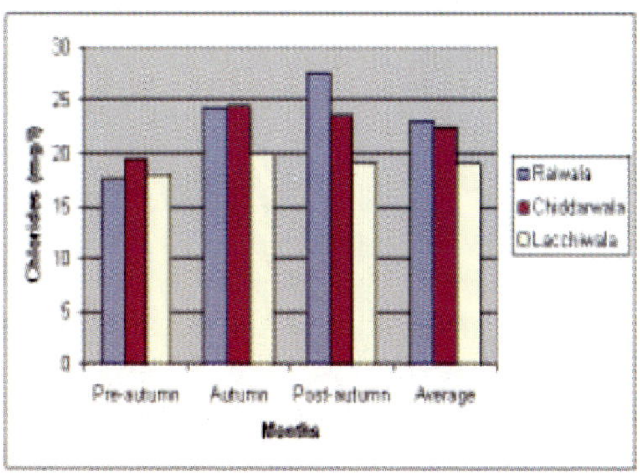
Chlorides (mg/l)
Pre-autumn
Autumn
Post-autumn
Average
Months
Raiwala
Chiddarwala
Lacchiwala

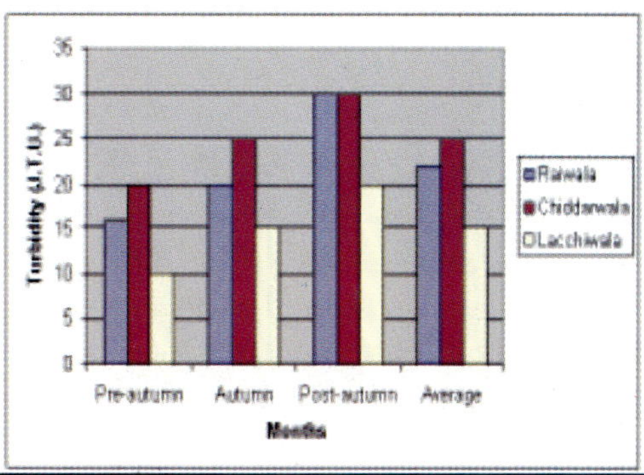
Turbidity (J.T.U.)
Pre-autumn
Autumn
Post-autumn
Average
Months
Raiwala
Chiddarwala
Lacchiwala

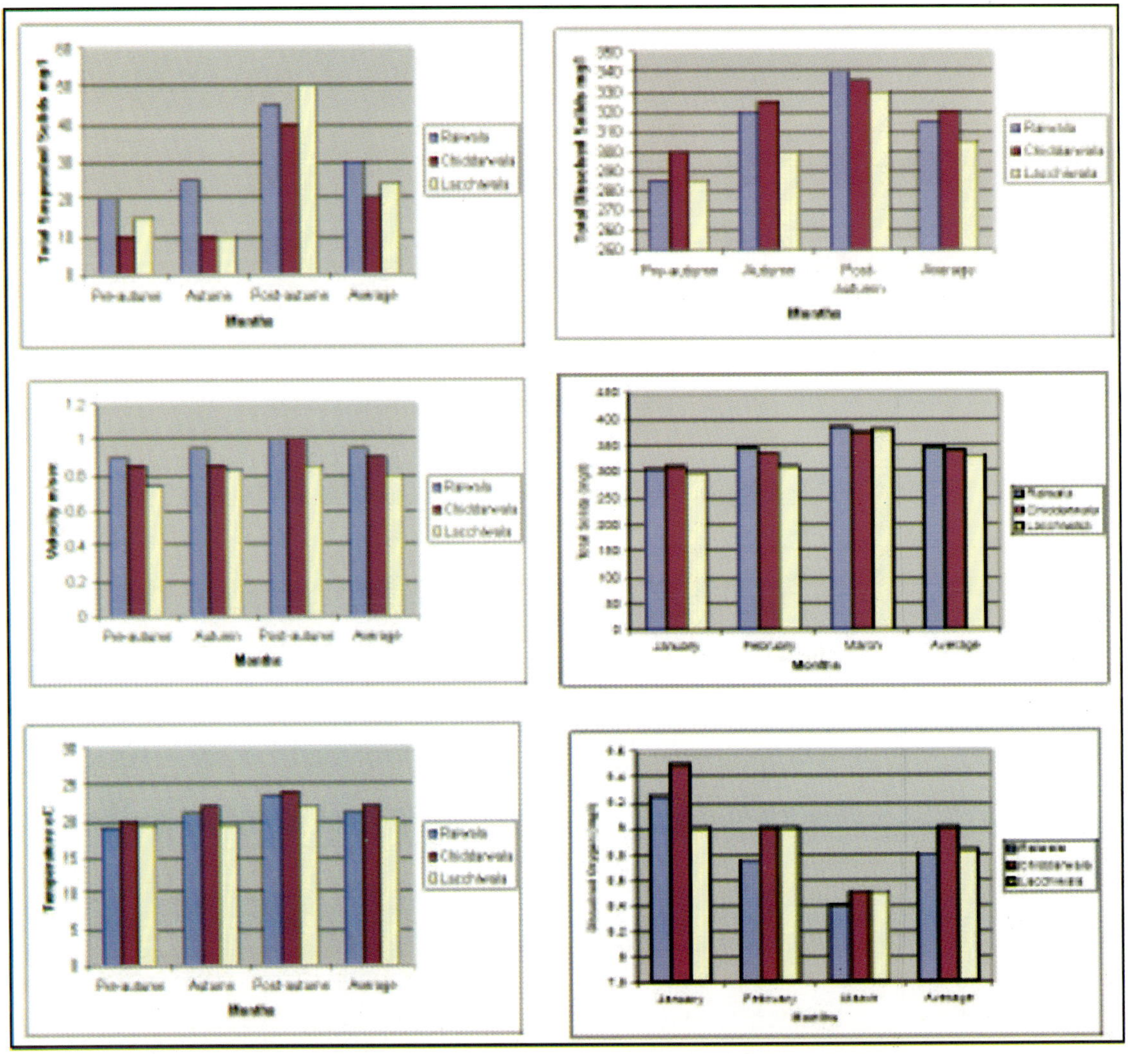

Graphs Showing Fluctuation in Parameters in Different Seasons

post-autumn month and minimum (15 mg/l) in pre-autumn and autumn month and increased value was probably due to heavy rain causing soil erosion and adding more suspended or dissolved solids from the nearby area as also observed by Khanna (1993) in the river Ganga at Haridwar.

During this study it was found that turbidity was higher (26.66 J.T.U.) during post-autumn month which may be due to heavy soil erosion and lower (15.33 J.T.U.) during pre-autumn month as also reported by Khanna (1993) in the river Ganga, Joshi and Bisht (1993) in Ganga canal at Jwalapur, Haridwar. The turbidity and total solids were closely interrelated with one another and cause common effect upon the river and aquatic life as also stated by Verma *et al.* (1984). The maximum (8.5) pH was recorded in post-autumn month, which might be due to increased chemical load in the river or due to increased photosynthetic activities and low dilution capacity of water. The minimum pH was recorded

(6.9) in pre-autumn month which was due to decreased photosynthetic activities which was due to increase in turbidity affecting the penetration of light as also reported by Badola and Singh (1981) reported similar trend in river Alakhnanda river of Garhwal Himalaya. Similar observations were reported by Sangu and Sharma (1985) in river Yamuna. Khanna (1993) in the river Ganga, Joshi and Bisht (1993) in Ganga canal at Jwalapur, Haridwar. Dissolved oxygen is usually related to ambient temperature, current velocity, rainfall and turbulence besides some biological components:-*viz.* planktonic blooms. Higher value (9.25 mg/l) of D.O. was observed during pre-autumn month, which permit retention of higher amount of air at lower temperature. Thus low temperature has high turbulence usually cause rise in oxygen level. Lower value (8.38 mg/l) of D.O. was observed during post-autumn month due to low O_2 retention capacity of water with the rise in temperature and higher TDS status as also observed by Badola and Singh (1981) reported similar trend in river Alakhnanda river of Garhwal Himalaya. Similar observations were reported by Sangu and Sharma (1985) in river Yamuna. Khanna (1993) in the river Ganga and Kataria *et al.* (1995) in Kubza River at Hoshangabad. The BOD was observed maximum (3.25 mg/l) in post-autumn month and minimum (2.3 mg/l) in pre-autumn month. A similar pattern has been reported by Khanna(1993) in the river Ganga, Chugh (2000) in the river Ganga at Haridwar and Kataria *et al.*(1995) in Kubza River at Hoshangabad.

The maximum value (3.16 mg/l) of COD observed during post-autumn month and minimum (2.3 mg/l) in pre-autumn month. A positive relationship has been observed between COD and water temperature. Free Carbon Dioxide was observed maximum (3.31 mg/l) in post-autumn month due to high temperature, and minimum (2.56 mg/l) in pre-autumn month when temperature was low and turbidity was found lowest. Pahwa and Mehrotra (1966) have reported that the Ganga River contains maximum free carbon dioxide in rainy season at Allahabad. Charakbarty *et al.* (1959) also observed the maximum free CO_2 in Yamuna during Monsoon at Allahabad.

Sulphate concentration was found to be minimum (1.65 mg/l) in pre-autumn month and maximum (2.76 mg/l) during post-autumn, also observed by Kataria (1966) in borewells of Bhopal city. Maximum value (23.38 mg/l) of chlorides was observed in Post-autumn month and minimum (18.36 mg/l) in Pre-autumn month.The term alkalinity is defined as the quality if ions in water, which react to neutralize hydrogen ions. The maximum value of alkalinity was found in pre-autumn month and minimum in post-autumn month. The cause of increasing alkalinity was the decomposition of the organic water.

A total number of 36 fish species were observed in river Song while Badola and Singh (1981) observed 31 species of fish in river Alakhnanda of Garhwal Himalaya. Bhatt *et al.* (1984) reported 26 fish species in river Kosi of Kumaon Himalayas. Khanna and Badola (1991) observed 30 species of fish fauna in the river Ganga at Haridwar.

Systematic List of Fishes

Present Scientific Names and Classification

Sub- phylum	Craniata
Super-class	Gnathostomata
Phylum	Vertibrate
Series	Pisces
Class	Teleostomi
Order	Cypriniformes
Family	Cyprinidae
Division	Cyprini

Name of Species	Local Name
Labeo rohita	Rohu
Labeo dero	Aragi
Labeo calbasu	Rohu
Labeo gonious	Rohu
Punitus sarena	Darahi
Punitus ticto	Shidhari
Punitus chola	Shidhari
Punitus sophore	Shidhari
Punitus conchonius	Darahi
Rasbora daniconius	Dandwa
Barilius bola	Bola
Barilius bendelisis	Bola
Barilius vagra	Bola
Danio devario	Fatukari
Brachydanio reria	-
Crossocheilus	-
Garra gotyla gotyla	-
Schizothorax sinatus	-
Schizothorax richardsoni	-
Tor putitora	Mahsheer
Tor tor	Mahsheer
Family : Cobitidae	
Botia dario	
Lepidocephalus guntea	Hakati
Neomacheilus botia	Natwa
Neomacheilus rupicala	Natwa
Family : Bagridae	
Mystus seenghala	Tengar
Mystus teengara	Tengar
Mystus vittatus	Tengar
Family : Saccobranchidae	
Heteropneustes fossilis	Singhi

Family : Clariidae

Clarius batarachus	Mangur

Order : Beloniformes

Family : Belonidae

Xenentodon cancila	Kauwa

Family: Channidae

Channa striatus	Saur
Channa gachuo	Girae

Family: Anabantidae

Colisa fasciatus	Khosti

Family: Mastacembelidae

Mastacembelus armatus	Baam
Mastacembelus pancalus	Patya

References

APHA, AWWA, WPCF, 1998. *Standard methods for examination of water and wastewater*, 20th Eds. American Public Health Association; 1015, Fifteen Street, New Washington., 15: 1–1134.

Badola, S.P. and Singh, H.R., 1981. Hydrobiology of the river Alakhnanda of Garhwal Himalaya. *Ind. J. Eco.*, 8(2): 269–276.

Bhatt, S.D., Bisht, Y. and Negi, U., 1984. Ecology of Limnoflora in river Koshi of the Kumaon Himalaya (U.P.). *Proc. Ind. Natu.*, S.C., 50(4).

Bhutiani, R. and Khanna, D.R., 2007. Ecological study of Suswa River: modeling DO and BOD. *Environ. Monit. and Assesment*, 125: 1–3.

Biswas, J., 1970. Interactions and interrelations between human and water. *Indian Journal Environ. Health*, 14: 297–309.

Chakrabarty, R.D., Ray, P. and Singh, S.B., 1959. A quantitative survey of plankton and physiological conditions of the river Yamuna at Allahabad. *Indian. J. Fish.*, 6(1): 186–203.

Chugh, Tarun, 2000. Seasonal variation in the microbial ecology of river Ganga at Haridwar. *Ph.D Thesis*, Gurukula Kangri University.

Joshi, B.D. and Bisht, R.C.S., 1993. Some aspects of physico-chemical characteristics of western Ganga canal near Jwalapur at Haridwar. *Him. J. Env. Zool.*, 7(1): 76–82.

Kataria, H.C., 1966. Sulphate content in bore well water of Bhopal city. *Ultra. Scientist. Phyl. Sci.*, 8(1): 97–100.

Kataria, H.C., Jain, O.P., Gupta, S.S., Srivastava, R.M. and Shandilya, A.K., 1995. Physico-chemical analysis of water of Kubza river of Hoshangabad. *Oriental. J. Chem.*, 11(2):157–159.

Kaur, Surendra and Joshi, B.D., 2003. Describes seasonal variation in some physico-chemical parameters of Ganga. *Him. J. Env and Zoo.*, 17(1): 45–55.

Khanna, D.R., 1993. *Ecology and Pollution of Ganga River*. Ashish Publishing House, Delhi, p. 1–241.

Khanna, D.R. and Badola, S.P., 1991. Fish fauna of the river Ganga at Haridwar. In: *Aquatic Environment*. Ashish Publishing House, New Delhi, p. 90–93.

Khanna, D.R. and Bhutiani, R., 2003. Worked on limnological status of Satikund pond at Haridwar. *Indian J. Env. Sci.*, 7(2): 131–136.

Khanna, D.R. and Bhutiani, R., 2004. *Water Analysis at a Glance*. ASEA Publications, p. 1– 115.

Khanna, D.R. and Singh, S., 2002. *Pond Fish Ecology and Economics*. Daya Publishing House, Delhi, p. 1–81.

Khanna, D.R., Seth, T.R., Malik, D.S. and Singh, R.K.,1999. Fishes and their ecology of river Suswa at Raiwala (Dehradun). In: *Sw. Eco. Sys. and Env.*, (Eds.) D.R. Khanna and A. Gautam, p. 105–110.

Khanna, D.R., Singh, S., Gautam, A. and Singh, J.P., 2003. Study of assessment of water quality of river Ganga in Distt. Bulandshahar (U.P.). *J. Nat. Con.*, 15(1): 167–175.

Kudesia, P.R., 1988. Studied the diseases caused because of polluted water in New Delhi. *Indian J. Ecol.*, 12(4): 236–254.

Logankumar, K., Sivakumar, A.A., Logasamy, S. and Aruchami, M., 1989. Seasonal variation in the nutrients of the river Bhavani at Sirumugai, Coimbatore district, Tamil Nadu. *Ind. J. Ecol.*, 16(1): 60–63.

Pahwa, D.V. and Mehrotra, S.M., 1966. Observation on fluctuations in abundance of plankton in relation to certain hydrological conditions of river Ganga. *Proc. Nat. Acad. Sci.*, 36B(2): 157–159.

Pandey, B.N., Kumar, K., Lala, A. and Das, P.K.L., 1993. A preliminary study on the physico-chemical quality of water of the river Kosi. *J. Ecobiol.*, 5: 237–339.

Sangu, R.R.S. and Sharma, K.D., 1985. Studies on water pollution on Yamuna river at Agra. *Indian J. Env. Hlth.*, 27(3): 257–261.

Santra, S.C., 2004. *Environmental Science*. New Central Book Agency (P) Ltd., p. 1–944.

Semwal, N. and Akolkar, P., 2006. Water quality assessment of sacred Himalayan Rivers of Uttarakhand. *Current Science*, 91(4).

Trivedi, R.K. and Goel, P.K., 1986. *Chemical and Biological Methods for Water Pollution Studies*. Environmental Publications, Karad, p. 1–251

Trivedy, P.K., 1990. Studied the monitoring and control of river pollution. *Water Pollution and Management*, p. 245–265.

Verma, S.R., Sharma, P., Rani, A., Gupta, A.K. and Dalela, R.G., 1984. Pollution and saprobic status of eastern Kalinadi. *Limnologica (Berlin)*, 15(1): 69–133.

Water and Wastewater Management in 2 Vols (2011) *Pages* **201–207**
Editors: D.R. Khanna, R. Bhutiani & Gagan Matta
Published by: DAYA PUBLISHING HOUSE, NEW DELHI

Chapter 32

Study of Physico-chemical Parameter for a Reservoir at Khandwa District, M.P.

Vivek Kesre[1], L.K. Mudgal[2], D.R. Khanna[3], Gagan Matta[3] and Dheeraj Kumar[3]

[1]Department of Zoology, Govt. M.L.C. Girls P.G. College, Khandwa, M.P.
[2]Department of Zoology, Govt. Girls P.G. College, Indore, M.P.
[3]Department of Zoology and Environmental Science, Gurukul Kangri University, Haridwar, Uttarakhand

ABSTRACT

Importance of water for the living being could not be neglected, with this view and the impact of the reservoir from the ecological and local point, this study was carried out. Reservoir studied is located at Khandwa, 3 km away from city, Khandwa on a local nallha named Barood nallah. It is a rainwater reservoir, the primary source of water to the reservoir is a 6.4 km long canal called Ajanti canal. Analysis of the reservoir water is carried out to observe the effect of decreasing catchment's area, due to siltation and decreasing water quantity on the physico-chemical condition of the reservoir water and its seasonal variation. Physico-chemical parameter analyzed were; temperature, pH, turbidity, transparency, TDS, silicate, free CO_2, DO, BOD, COD, total hardness, Ca hardness, Mg hardness, sulphate, fluoride, chloride and potassium from August 2006 to July 2007. Statistical analysis of the obtained data is carried out to draw conclusion.

Keywords: Reservoir, Ecological, Eutrophication, Siltation.

Introduction

Water plays a significant role in different vital and structural activities. Water is inevitable for all living organisms as it has a great social and economical value ultimately affecting men health. It is essentially required for irrigation, industrial development, hydro-electrical generations, fisheries, human life survival and domesticated animals. As the water is very important reservoirs have multidimensional approach including flood control, hydropower generation, water supply, navigation, restoration, *etc* (Ngo, 2006). Moghat reservoir is a tropical rainwater reservoir located in Khandwa district, near Moghat village, 3 km away from the Khandwa city. It was constructed in 1897 by the British engineers, initially to full fill the water requirement of the Khandwa city. Catchment area of the reservoir was 23.30 square kilometer and to this water is added by local nalah named Ajanti canal and seasonal river, but now it has been reduced greatly due to natural and anthropogenic activity. With the time and due to improper management water holding capacity of the reservoir decreased and ultimately failed to complete the motive for which this reservoir came into existence. Till now no such limnological study has been carried out on the Moghat reservoir therefore no review on this reservoir was found. Limnological study on pond and other reservoir is carried out like Adefemi *et al.* (2007) worked on the physico-chemical condition of a dam in Nigeria and showed the level of significance between different physico-chemical state-variable. Limnological features of man made lakes were analysed and observed by Ayoade *et al.* (2006). Polak studied the nitrifying activities in Wloclawek dam reservoir in 2004. Van Den Bos (2003) released his work on the water quality of the Occoquan reservoir and its tributary watershed. Daily physico-chemical dynamics of Siemianowka reservoir lies in northeastern Poland was studied by Jekatierynczuk-Rudczyk *et al.* (2002). Khanna *et al.* (2000) analyzed the abiotic factors of a pond at Haridwar.

Purpose of this Study

1. The present water reservoir is not studied so far.
2. Conduct and update water quality assessment of the Moghat reservoir.
3. Assess the present physical and physico-chemical status of reservoir.
4. Draw attention towards the problem faced by the reservoir.

Materials and Method

Study Site

Moghat reservoir is located 3 kilometer away from the Khandwa city in northwest. The geographic position of this reservoir is 21–49′ N and 76–20′ E. It is situated 1071 ft. above from mean sea level. The maximum depth 5.2 m and mean depth 1.2 m it has 5.36 kilometer long shoreline. Source of water to the reservoir is a canal named Ajanti canal which collect rainwater from various sources and drain it to the reservoir. Since the time human activity nearby to the reservoir increased problem of eutrophication and siltation to the reservoir increased.

Sampling

Monthly sampling was done from July 2006 to August 2007. Samples were taken between 7 A.M. to 9 A.M. but transparency was measured between 10 A.M. to 2 P.M. Water samples were taken on one liter plastic bottles. Temperature, pH, fixing of DO and transparency were measured at the spot.

Analysis

Samples were analyzed according to "*Standard Methods for the Examination of Water and Wastewater (APHA)" 20th edition* (1998). Published by American Public Health Association.

Figure 32.1: Moghat Reservoir

Figure 32.2: Ajanti Canal

Figure 32.3: Siltation Effect on Moghat Reservoir

Figure 32.4: Eutrophicaton in Moghat Reservoir

Results and Discussion

In the investigated physical trait of the reservoir water it is observed that water temperature was maximum during July (29.63 ºC ±0.29) and minimum during December (17.93 ºC ±0.19), transparency was maximum in December (8.98 cm ±3.10) and minimum in September (1.52 cm ±0.99), turbidity in August was maximum (138 JTU ±26.95) and minimum during the month of January (22.75 JTU ±3.86) and TDS maximum in the month of July (218 mg/l±9.42) and minimum during February (105.80 mg/l±2.20). Average observed for Temperature, Transparency, Turbidity and TDS for the year 2006-07 were 25.69 ºC ±4.27, 5.15 cm ±3.63, 68.04 JTU ±39.39, 166.81±33.60 mg/l respectively. Observations of physical parameter are given in Table 32.2. Chemical parameters analyzed for reservoir water were pH, Free CO_2, Total hardness, Ca hardness, Mg hardness, DO, BOD, COD, Sulphate, Fluoride, Silicate, Potassium and Chloride. Observed values are given in Table 32.1. The maximum and minimum value for pH was (8.82±0.02) in January and (7.72±0.17) in August respectively. Free CO_2 was maximum (3.63 mg/l ±0.83) in May and minimum (1.22 mg/l ±0.08) in October, Ca and Mg hardness found in the reservoir water maximum was (37.25 mg/l ±1.76) in June and (26.05 mg/l ±1.69) in June respectively and the minimum value of them was (26.90 mg/l ±0.82) in February and (11.08 mg/l ±1.29) in November respectively. Total hardness found was maximum (154.33 mg/l ±1.85) in May (92.83 mg/l ±0.70) in

Table 32.1: Seasonal Variation in Physico-Chemical Parameters of Moghat Reservoir Water for 2006–2007

Months	pH	Free CO_2 (mg/l)	Hardness (mg/l)			DO (mg/l)	BOD (mg/l)	COD (mg/l)	Sulphate (mg/l)	Fluoride (mg/l)	Silicate (mg/l)	Potassium (mg/l)	Chloride (mg/l)
			Total	Ca	Mg								
July	7.84±0.23	2.74±0.54	104.45±1.72	29.35±1.80	12.30±143	5.50±0.29	4.50±0.56	11.48±0.68	33.65±1.65	0.30±00	9.88±0.50	1.00±0.02	28.95±4.14
August	7.72±0.17	2.92±0.95	96.38±1.50	27.43±1.30	11.93±2.17	5.70±0.32	4.18±0.71	10.65±0.79	30.10±028	0.30±00	9.10.0.56	097.001	23.33±4.66
September	8.12±0.08	2.59±0.98	92.83±0.70	27.88±1.84	12.07±2.32	6.43±0.29	3.85±0.66	10.18±0.71	25.45±1.77	0.30±00	8.33±0.34	1.20±0.03	22.05±3.64
October	8.27±0.02	2.22±1.02	122.45±1.07	28.78±1.68	12.73±1.34	6.48±0.43	3.28±0.28	9.48±0.50	22.65±1.31	0.30±00	7.38±0.26	1.23±0.04	18.85±1.53
November	8.24±0.02	2.81±0.76	118.28±0.98	28.33±0.92	11.08±1.29	6.43±0.17	313±0.21	8.73±0.65	22.60±1.97	0.30±00	6.75±0.17	123±0.05	17.15±0.53
December	8.23±0.03	1.22±0.08	120.48±0.93	27.50±0.75	14.25±1.44	8.90±0.22	2.93±0.15	8.98±1.10	16.03±1.07	0.30±00	7.23±0.24	1.15±0.03	17.13±2.16
January	8.82±0.02	1.37±0.20	121.50±2.27	27.43±1.03	14.53±1.31	8.25±0.31	3.18±0.15	9.83±0.40	16.78±0.87	0.30±00	7.83±0.13	1.10±0.04	10.05±3.51
February	8.22±003	1.38±0.19	122.38±1.11	26.90±0.82	16.28±1.47	7.70±0.71	3.45±0.13	10.43±0.25	17.68±0.67	0.30±00	9.58±0.28	1.17±0.05	2.33±3.45
March	8.32±0.06	1.92±0.69	127.10±1.78	29.55±0.70	16.79±1.25	6.63±0.61	3.73±0.15	10.90±0.27	19.28±0.46	0.30±00	10.30±0.23	1.21±0.05	23.58±4.30
April	8.37±0.16	2.81±0.83	134.85±1.61	32.98±1.40	18.05±1.26	5.55±037	4.05±0.26	10.90±0.29	20.78±0.82	0.30±00	11.20±0.28	1.23±0.03	25.64±3.50
May	8.34±0.04	3.63±0.83	154.33±1.85	36.10±1.56	24.98±1.48	4.98±0.61	4.28±0.22	11.40±0.28	27.33±0.98	0.30±00	10.83±0.15	1.25±0.02	25.64±1.70
June	8.28±0.03	2.79±0.60	153.65±1.70	37.25±1.76	26.05±1.69	5.13±0.43	`4.73±0.35	11.73±0.46	31.63±1.42	0.30±00	10.60±0.52	1.28±0.03	32.30±2017
Average ±SD	8.18±0.20	2.37±0.75	121.64±18.01	29.95±3.53	15.91±4.98	6.64±1.37	3.77±0.59	10.41±1.00	23.66±5.97	0.30±58.17	9.08±1.54	1.17±0.09	23.13±4.88

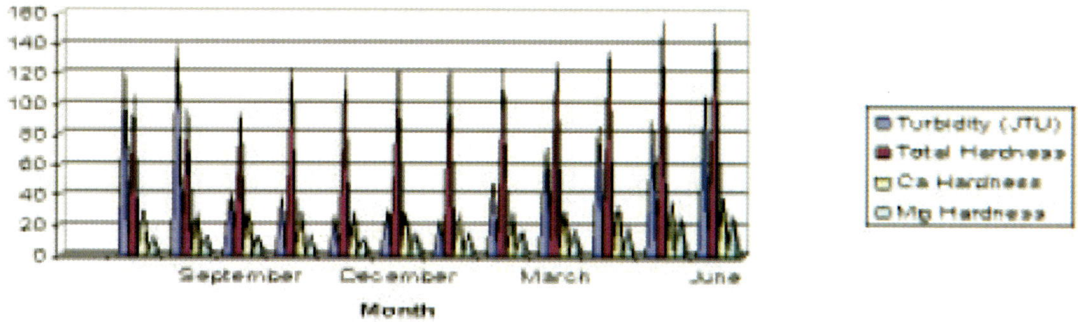

Figure 32.5: Variation Between Turbidity, Total Hardness, Ca Hardness and Mg Hardness

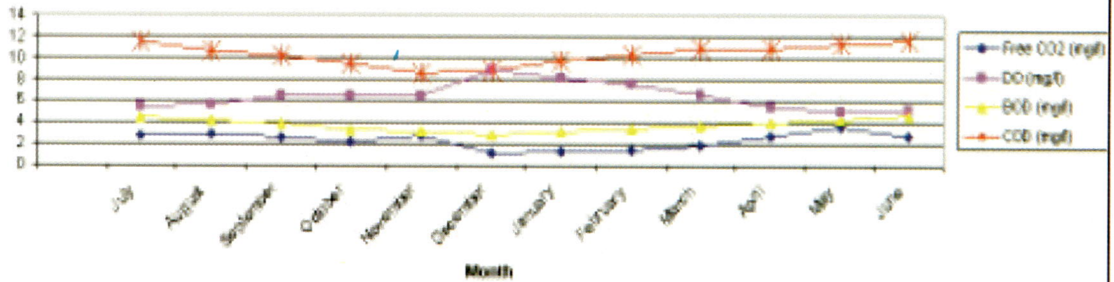

Figure 32.6: Comparison Between Free CO_2, DO, BOD and COD

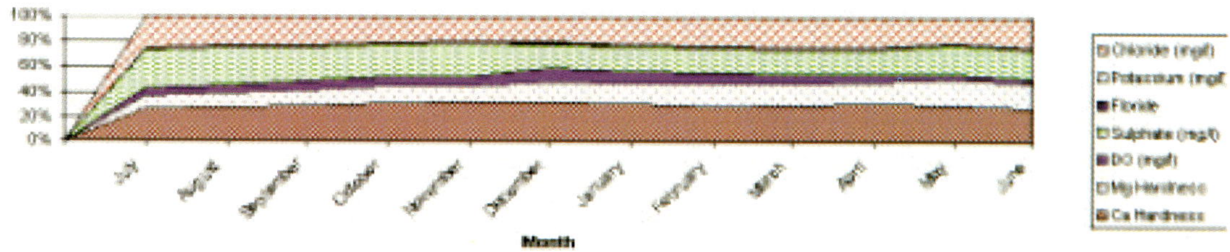

Figure 32.7: Comparative Study Between Ca, Mg, DO, Sulphate, Fluoride, Potassium and Chloride

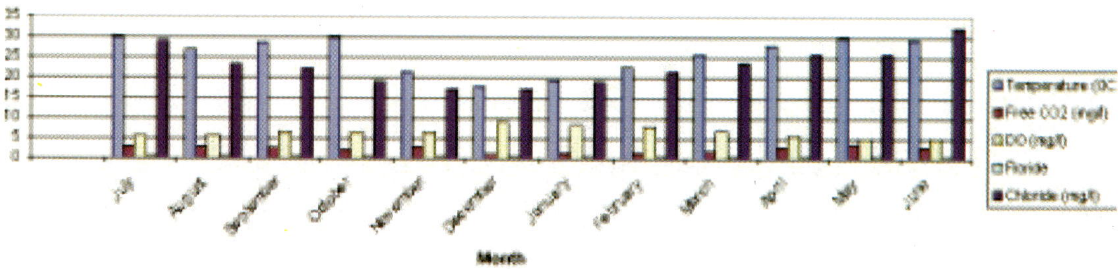

Figure 32.8: Comparison Study Between Temperature, Free CO_2, DO, Fluoride and Chloride

September. Minimum and maximum value of DO found was (4.98 mg/l ±0.61) in April and (8.90 mg/l ±0.22) in December respectively. The minimum value of BOD found for the reservoir water was 2.93 mg/l ±0.15) in December and maximum value was (4.73 mg/l ±0.35) in June. COD observed was (8.73 mg/l ±0.65) in November and minimum was (11.73 mg/l ±0.46) in June. The minimum value of Sulphate, Silicate Potassium and Chloride was (16.03 mg/l ±1.07) in December, (6.75 mg/l ±0.17) in November, (0.97 mg/l ±0.01) in August, (17.13 mg/l ±2.16) in December respectively. The maximum value observed for the reservoir water for the parameter Sulphate, Silicate, Potassium and Chloride was (33.65 mg/l ±1.65) in July, (11.20 mg/l ±0.28) in April, (1.28 mg/l ±0.03) in June, (32.30 mg/l ±2.10) in June respectively. Fluoride during the whole show no variation and its value found constant (0.30 mg/l ±00).

Table 32.2: Seasonal Variation in Physical Parameter of Moghat Reservoir Water for 2006–2007

Months	*Temperature (ºC)*	*Transparency (cm)*	*Turbidity (JTU)*	*TDS (mg/l)*
July	29.63±0.29	3.88±0.83	12.25±14.38	218.00±9.42
August	26.95±0.13	1.80±0.55	138.50±26.95	190.50±3.70
September	28.63±1.1	1.50±0.99	41.50±11.50	180.00±7.35
October	29.65±0.24	7.77±2.72	38.75±8.50	174.00±6.32
November	21.50±0.48	8.68±1.22	25.25±4.11	69.75±7.90
December	17.93±0.19	8.90±3.10	30.50±8.02	63.50±7.77
January	19.35±1.4	7.23±1.57	22.75±3.86	128.50±1.73
February	22.53±0.48	6.28±1.76	48.25±15.12	105.80±2.20
March	25.68±051	5.84±1.34	71.75±22.29	129.35±0.47
April	27.80±0.18	3.19±1.46	84.75±10.24	149.83±1.46
May	29.50±0.57	3.76±0.91	88.50±6.66	182.50±7.55
June	29.20±0.78	2.98±1.03	105.75±11.32	210.00±12.49
Average±SD	25.69±4.27	5.15±2.63	68.04±39.39	166.81±33.60

The average value during the whole year of study observed for pH, Free CO_2, Total hardness, Ca hardness, Mg hardness, DO, BOD, COD, Sulphate, Fluoride, Silicate, Potassium and Chloride was (8.18±0.20), (2.40 mg/l ±0.78), (121.64 mg/l ±19.23), (29.95 mg/l ±3.53), (15.91 mg/l ±4.98), (6.64 mg/l ±1.37), (3.77 mg/l ±0.59), (10.41 mg/l ± 1.00), (23.66 mg/l ±5.97), (0.30 mg/l ±5.8E-17), (1.17 mg/l±0.09), (23.13 mg/l±4.88) respectively. Results obtained are given in tabular form in Tables 32.1 and 32.2 and their graphical representation are shown in Figures 32.5–32.8.

References

Adefemi, O.S., Asaolu, S.S. and Olaofe, O., 2007. Assessment of the physico-chemical status of water samples from major dams in Ekiti State, Nigeria. *Pakistan Journal of Nutrition*, 6(6): 657–659.

APHA, AWWA and WPCF, 1998. *Standard Methods for the Examination of Water and Wastewater*, 20th edn. American Public Health Association, Washington DC, New York.

Ayoade, A.A., Fagade, S.O. and Adebisi, A.A., 2006. Dynamics of limnological features of two man-made lakes in relation to fish production. *African Journal of Biotechnology*, 5(10): 1013–1021.

Jekatierynczuk-Rudczyk, E., Gorniak, A., Zielinski, P., Dziemian, J., 2002. Daily dynamics of water chemistry in lowland olyhumic dam reservoir. *Polish Journal of Environmental Studies,* 11(5): 521–526.

Khanna, D.R., Gautam, A., Chugh, T. and Sarkar, Praveen, 2000. Impact of abiotic factors on the phytoplanktonic population of a pond at Haridwar. *Environment Conservation Journal,* 1(1): 41–46.

Ngo, L.L., 2006. Optimising reservoir operation: A case study of the Hoa Binh reservoir, Vietnam. *Ph.D. Thesis,* Institute of Environment and Resources Technical University of Denmark.

Polak, J., 2004. Nitrification in the surface water of the Wloclawek Dam Reservoir: The Process Contribution to Biochemical Oxygen Demand (N–BOD). *Polish Journal of Environmental Studies,* 13(4): 415–424.

Van Den Bos, A.C., 2003. A water quality assessment of the Occoquan reservoir and its tributary watershed: 1973–2002. *Thesis of Master of Science in Environmental Engineering,* Graduate Faculty of the Virginia Polytechnic Institute and State University.

Water and Wastewater Management in 2 Vols (2011) *Pages 208–227*
Editors: **D.R. Khanna, R. Bhutiani & Gagan Matta**
Published by: **DAYA PUBLISHING HOUSE, NEW DELHI**

Chapter 33

Seasonal Variation in Physico-chemical and Biological Properties of River Panar, Bihar

B.N. Pandey[1], O.P. Ambasta[2], A.K. Jha[3] and K. Shambhu[3]
[1]*Eco-Genetical Research Laboratory, P.G. Department of Zoology Purnia College, Purnia – 854 301*
[2]*Department of Botany, R.L. College, Madhav Nagar, Purnia – 854 204*
[3]*P.G. Department of Zoology, T.P.College, Madhepura – 852113*

ABSTRACT

A base line study was conducted to determine the impact of anthropogenic activities on the physico-chemical and biological factors of the river Panar. The result revealed well defined seasonal variation.On the basis of physico-chemical and biological parameters the river appears to be polluted. On the basis of Moyle's classification, the river water appeared to be hard type. Likewise on the basis of phosphorous and nitrogen, the investigated water body may be categorized as moderate trophic level. The bulk of phytoplankton was shared by Chlorophyceae (9), Myxophyceae (8), Bacillariophyceae (9) and Euglenophyceae (1). Many of algal species out of the total 27 reported from the river like *Euglena, Microcystis, Anabaena, Chlamydomonas,* and *Spirulina* were recognized as pollution indicator. The zooplankton population comprised of Rotifera (19), Cladocera (5) and Copepoda (3). The zooplankton also comprised some pollution tolerant species like *Brachionus, Keratella, Moina, Mesocyclops, Cypris, Ceriodaphnia, Polyarthra etc.* The study clearly revealed that the aquatic environment of the river has undergone degradation and is not suitable for human use. The main sources of river pollution in the present study were recognized as: (a) discharge of domestic sewage from near by villages; (b) agricultural run off; (c) burning of corpses on the bank of river; (d) bathing habits of human as well as cattles; (e) washing of clothes and utensils; (f) soil erosion; (g) retting of Jute. Proper remedial measures should be taken immediately in order to restore it from further deterioration. There must be alternate waste disposal system away from the river and disposal of solid and liquid waste must be stopped forthwith. Indiscriminate fishing practices and retting of jute should be stopped, Planting of some toxicant reducing macrophytes should be done on both sides of the river. There is urgent need to take up widespread conservation and education programme in this area to highlight the problem of pollution.

Keywords: *Seasonal variation, Physico-chemical characteristics, Phytoplankton, Zooplankton, Water quality.*

Introduction

Freshwater is the most precious resource on the earth, which is becoming scarce. Rivers, streams, ponds and lakes have been the major source of water. But these days, due to industrialization, urbanization and anthropogenic activity, most of the rivers have become a matter of great concern over period of last few decades because of their vital importance in several human uses, all over the world. Due to the unplanned industrialization and population explosion of the country, the rivers have become today one of the most heavily polluted water sources of India.

Of all the earth's ecosystems, rivers are the most dynamic having as their primary function the transportation water. They also carry the load of the dissolved and particulate output of crystal weathering and erosion from land to the sea. Rivers have been studied for many years with inputs from many disciplines of science and humanity. Engineering hydrologists, geographers, economists, biologists, chemists, geologists and social scientists have produced a vast body of literature on regional, hydrographical, chemical and historical aspects (Livingstone, 1963; Barnes and Mann, 1980). From data available with UNESCO (1979) and UN (1981) on twenty-nine of the larger rivers of the world, the inter annual variance of discharge is found to be between 6 and 33 per cent.

In most of the developing countries, the disposal of sewage and industrial wastes is often conducted without critical appraisal of the impact upon their receiving waters. There are numerous sources of domestic and industrial effluents leading to heavy metal enrichment of water and sediments. The rapid urbanization and industrialization, which are, advanced tool and tips of modern civilization have cultivated pollution problems into water, air and land.

The quality of water resources is usually described according to its physical, chemical and biological characteristics. The use of water in any region for different purposes like domestic, industrial, agriculture and fish culture depends on its physico-chemical and biological characteristics. Hence the present work was undertaken which deals with the seasonal variation in physico-chemical and biological properties of river Panar.

Materials and Method

The present investigation was carried out in the river Panar at five sampling stations selected in the stretch of the 15 km. (Figure 33.1) The water samples were collected at the monthly intervals between Jan'2003 to Dec'2003 from each sampling station in a polythene bottle of 3 litres capacity. Water temperature was recorded with an ordinary mercury thermometer graduated from 0° to 50° C. The analysis of pH, dissolved oxygen, free carbon dioxide, BOD,COD, carbonate,bicarbonate, chloride,nitrate, phosphate and sulphate was done according to the methods of APHA (1975) and Trivedy and Goel (1984).

For the planktons, 50 litres of river water was taken and filtered through the bolting silk plankton net (No.30) with 77 mesh/sq cm. The samples were taken from different areas of the stations regularly.

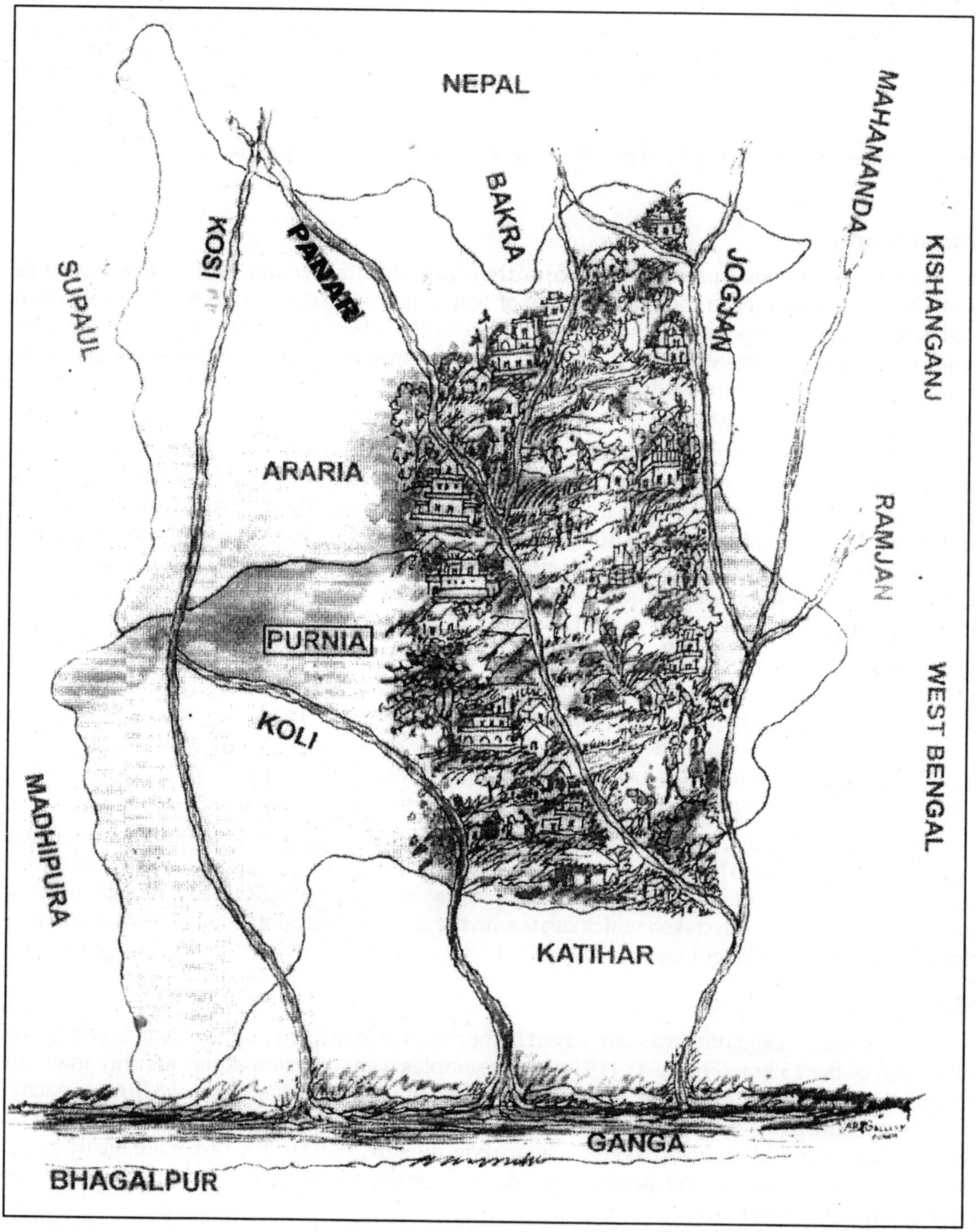

Figure 33.1: River Panar Showing Sampling Sites

The plankton concentrate on the plankton net was preserved with 5 per cent formalin. Separation and counting the plankton was done by taking 1 ml of sub-samples into a Sedgwick Rafter Plankton Counting chamber of 1 ml capacity. All organisms were counted according to the procedure described by Welch (1952).

Results and Discussion

Temperature

It is well known fact that the water temperature exerts direct as well as indirect influence on many abiotic and biotic compositions of aquatic ecosystem. Temperature has variable effects, it may kill some organisms and may stimulate the growth of the other (Kotpal and Bali, 1989). The water temperature regulates various physico-chemical as well as biological activities (Raney and Menzel,1969) and its values and variations have a great bearing upon its productivity (Jhingran, 1982). It also reflects on the dynamics of living organisms (Chandler, 1942). All organisms possess well defined limits of temperature tolerance. All metabolic and physiological activities are greatly influenced by water temperature. Temperature also affects the speed of chemical changes in soil and water. Temperature showed diurnal as well as seasonal variations. In the present investigation, the water temperature followed more or less similar trend of the air temperature. The water temperature varied between 18.9 to 36.9°C (Table 33.1). Dale and Gillsspic (1977) have reported the influence of macrophytes on the water temperature.

Table 33.1: Physico-chemical Characteristics of the River Panar Water

Months	*Temp. °C*		*pH*	*Trans. (cm)*	*COD (mg/l)*	*DO (mg/l)*	*BOD (mg/l)*	*Free CO_2*	*CO_3 (mg/l)*	*HCO_3 (mg/l)*	*PO_4 (mg/l)*	*NO_3 (mg/l)*	*Cl^- (mg/l)*	*Sulphate (mg/l)*
	Air	*Water*												
January	20.1	18.9	8.6	72.8	66.8	9.9	19.9	0.9	0	90.8	0.09	0.38	25.6	55.6
February	22.4	19.5	8.4	703	64.7	9.5	20.4	0.8	0	90.5	0.08	0.41	23.2	66.4
March	30.5	27.9	8.2	58.6	59.5	93	20.6	1.1	0	88.4	0.09	0.43	28.5	73.1
April	34.8	32.5	7.9	50.9	71.6	8.5	21.9	1.9	0	92.1	0.12	0.49	29.4	77.6
May	37.4	35.1	7.5	48.7	74.2	8.3	24.2	1.14	0	99.9	0.14	0.54	32.4	72.9
June	38.2	36.8	7.3	45.1	78.2	8.1	25.4	1.19	0	80.4	0.16	0.56	33.5	71.4
July	38.9	36.9	7.4	38.9	90.4	7.8	25.1	1.04	0	64.6	0.16	0.58	33.9	80.5
August	37.4	35.8	7.3	34.1	92.8	7.6	29.8	1.08	0	55.8	0.18	0.61	32.6	74.2
September	35.9	28.2	7.6	31.5	76.5	6.8	22.9	1.09	0	50.1	0.18	0.64	29.7	81.9
October	33.5	25.4	7.9	35.5	69.8	6.4	21.8	1.04	0	55.2	0.15	0.47	28.2	63.5
November	29.1	22.2	8.0	46.5	66.5	7.9	21.4	1.1	0	59.6	0.12	0.42	26.1	58.3
December	24.5	20.1	8.4	60.1	65.3	8.1	19.6	1.0	0	70.4	0.11	G.40	25.2	60.4

pH

pH is a general physico-chemical parameters and indicates the fate of chemical constituents. It is one of the most important environmental factor of natural water. pH is also reported to play an important role in the formation of algal bloom (Anderson,1961 and King, 1972). Decreasing volume of water often accompanied by progressive change in pH (Adoni, 1975). The variation of pH is often linked with the species compositions and life processes of animal and plant communities inhabiting them. In the present investigation, the pH varied between 7.3–8.6.(Table 33.1). Thus the river water was alkaline throughout the year. The highest pH value during winter appeared to be influenced by water level, large number of phytoplanktons and highest value of dissolved oxygen (Sunder, 1988 and Pandey *et al.*, 1992a). The pH is associated with high photosynthetic activity (Roy,1955). The low pH value during monsoon might be due to high turbidity and elevated water temperature, which stop photosynthetic activity leading to accumulation of free CO_2 which lowers the pH of water during monsoon (Adebisi, 1980).

Transparency

Turbidity restricts the penetration of sunlight and hence reduces photosynthetic activity which in turn is related to the productivity of water mass. The suspended particles causing turbidity may also absorb considerable amount of nutrient element like phosphorus, nitrogen and potassium in their ionic forms making them unavailable for planktonic production. Turbidity makes the water unfit for domestic purpose. In the present study the turbidity varied from 31.5 to 72.8 cm (Table 33.1). The high turbidity during rainy season may be attributed to the land drainage contributing to increase in particulate suspended matter; Roy (1955) reporting on the hydrology of an irrigation tank from Banglore, Karnataka opines that high turbidity adversely affects primary production. The highest transparency during winter season was due to close competition between macrophytic vegetation and phytoplanktons. Kaul *et al.* (1972) and Wetzel (1975) found that macrophytes were capable of storing nutrients, suspended in water in more quantity that the phytoplankton could do. Zutchi and Vass (1973) have similar observations while studying the limnology of Kashmir lakes.

Dissolved Oxygen

The dissolved oxygen content plays a vital role in supporting aquatic life in running waters and is susceptible to environmental changes (Jameson and Rana,1996). Dissolved oxygen affects the solubility and availability of the nutrients and the productivity of the ecosystem (Wetzel, 1975). Dissolved oxygen is one of the most important parameters in the water quality assessment and reflects the physical and biological process prevailing in the water. The volume of oxygen dissolved in water is dependent upon (i) Its temperature (ii) the partial pressure of oxygen in the air in contact with the water at the surface, and (iii) the concentration of dissolved salts. The water receives oxygen mainly through two sources (i) by absorption from the atmosphere at the surface, and (ii) by the photosynthesis of chlorophyll bearing organisms inhabiting the water body. Any water polluted by organic wastes suffers from decline in dissolved oxygen (Butcher, 1947). In the present investigation the dissolved oxygen varied from 6.8–9.9 mg/l (Table 33.1). The higher value of oxygen was observed during winter season when temperature was low. It supports findings of Rao (1975), Bharti and Kori (1975), Mishra and Yadav (1978), Singhal *et al.* 1985), Saxena and Mishra (1991). The low value of oxygen during rainy season might be due to higher growth of bacteria which utilizes oxygen for metabolic activities. However, Madhyastha *et al.* (1999) have reported increased level of dissolved oxygen during monsoon leading to super saturation, which may be due to rainwater.Higher value of O_2 during summer has been reported by Vijaykumar (1992,99) due to increased solar radiation. There should be adequate dissolved oxygen in all the seasons. The river with self-purification capacity could assimilate the existing load (Lester, 1967).

BOD

The Biochemical Oxygen Demand (BOD) is a parameter which enables the determination of relative oxygen requirements especially of wastewaters, polluted waters and effluents. BOD in water showed a very critical situation.The value ranged from 19.6–29.8 mg/l (Table 33.1) Higher value were observed and presented seasonal fluctuation in monsoon and lower values in post monsoon. The overall increase in BOD values can be attributed to the discharge of organic wastes and animal excreta in to the water from domestic sources and animals of the villages located near the rivers. Kaur *et al.* (1996) have reported increase in BOD value in river water due to deposition of animal excretory wastes.

COD

COD values were also very high in the river water, which ranged from 59.5 – 92.8 mg/l (Table 33.1). High COD values were due to algal biomass and other organic matter (Patel and Sinha, (1998). The peak of COD was observed in monsoon. The highest value of BOD and COD in rainy season are due to higher concentration of dissolved solids (Jameel, 1998). High COD value show that a large amount of chemically oxidizable matter is present.

Free CO_2

Free carbon-dioxide is an extremely necessary constituent in an aquatic ecosystem (Welch, 1952). The respiratory activity by aquatic organisms produce considerable amount of CO_2 in the water, which remains dissolved partly in water as carbonic acid and part of it diffuses to atmosphere. The free CO_2 value of the river ranges from 0.8 – 1.2 mg/l (Table 33.1). The low value of CO_2 during winter season might be due to increased photosynthetic activity by phytoplanktons. The high value of CO_2 during summer season suggests active decomposition of organic matter. High carbon dioxide content is indicative of high organic pollution (Todda, 1970 and Cole, 1979).

Vyas and Kumar (1968) showed a direct relationship of free carbon dioxide with transparency, temperature and pH, while an inverse relation with dissolved O_2. According to Bohra and Bharagava (1977) high concentration of CO_2 results in to low pH and high carbonate values. Present observation of high CO_2 in summer are in line with similar studies done earlier by Vyas and Kumar (1968). In summer with an increase in atmospheric temperature there was corresponding rise in water temperature and subsequent increase in biological oxidation of organic matter that might have caused and elevation in the level of CO_2 (Singh and Srivastava, 1988; Michael, 1966).

Chloride

Natural water normally contains low chloride than bicarbonate and sulphates. Large contents of chloride in freshwater is an indication of organic pollution (Thrash *et al.*, 1944). Chloride is generally undertaken as major factor to equalize cation–anion balance of ecosystem. The most important source of chloride in the water is the discharge of domestic sewage. Man and other animals excrete very high quantities of chloride together with nitrogenous compounds. The chloride contents in the river under investigation was found highest during summer and lowest during winter season (Table 33.1). The higher value of chloride during summer months may be due to the gradual decrease in the amount of water. Mishra and Yadav (1978) have also reported increased values of chloride in summer months and have correlated this with gradual decrease in the amount of water. Moreover, the increased amount of excreta laid by various aquatic fauna may also account for the increased values of chloride during summer months. The finding of Cole (1979), Pandey and Mishra (1990) and Swarnalatha

(1994) are also similar with the observations made in the present study in this regard. Munawar (1970) has suggested that high value of chloride is an indication of pollution of animal origin. It has been also suggested that chloride contents also increased with degree of eutrophication (Sinha, 1986).

Alkalinity

Alkalinity of water is the capacity to neutralize a strong acid and is characterized by the presence of all hydroxyl ions capable of combining with hydrogen ion. The role of alkalinity in the determination of productive capacity of aquatic environment has been described by Philipose (1959), according to whom inland waters are categorized into three types, further according to the same author waters with low alkalinity are comparatively low in production. The alkalinity in natural waters is mainly due to the presence of bicarbonates and carbonates. Rai (1980) has reported that waters rich in free CO_2 are comparatively less alkaline whereas waters deficient in this gas are more alkaline. Similar observations have been made by Roy (1955), Zafar (1964) and Munawar (1970). But on the contrary, Singh (1965) has claimed that the alkalinity of water does not seem to be affected either by oxygen or carbon dioxide alkalinity below 500 mg/l indicate low photosynthetic rate (Das, 2002).

In the present investigation carbonate alkalinity was absent throughout the year. Bicarbonate alkalinity ranged from 50.1–99.0 mg/l (Table 33.1). Bicarbonate alkalinity was highest during summer and lowest during rainy season. On the basis of alkalinity values the river water can be classified under high category of nutrient type after the classification of Philipose (1959) as the alkalinity value is > 100 mg/L. On the basis of Moyle's (1946) classification, the river may be termed as hard water type. The low value of alkalinity during monsoon indicated degradation of water quality due to flooding from intensive agricultural zones. The effect of alkalinity in river water is responsible for the increased algal productivity. On the basis of pH, free CO_2 and alkalinity, the river water may be considered moderately alkaline giving support to the observations of Vass *et al.* (1977) of river Jhelum.

Nitrate

Nitrogen occurs in natural waters in elemental state and as organic as well as inorganic nitrogenous compounds. The elemental nitrogen in water is derived mostly from the atmosphere. The other source is the bacterial denitrification of nitrates, nitrites and ammonia. Nitrate parameter is an excellent parameter to judge organic pollution. Thrash *et al.* (1944) have attributed the nitrogen richness of freshwater body to the pollution of animal origin. In aquatic system nitrogen level is regulated through precipitation, atmosphere solution and volatilization under meteorological process; sedimentation influence effluent and groundwater movement under geologic process, nitrogen fixation, denitrification, uptake, growth, decay, hydrophytes pumping and fish and weed removal under biological process (Toetz, 1976). The level of the total nitrogen concentration in an index of carbon budget of freshwater body (Munawar, 1970; Pillai and Sreenivasan, 1975). In the present study the nitrate value ranged from 0.38 to 0.64 mg/l. The low value of nitrate during winter season might be due to high macrophytic and phytoplankton production. Lee *et al.* (1975) have indicated that during winter, the decrease in nitrogen is probably associated with entire uptake of this element by these plant communities. The high value of nitrate during rainy season night be attributed to rain showers, decomposition of organic matters and influx of flood water (Rao and Govind, 1966).

Phosphate

The phosphate in natural water occurs in very small quantities. It is necessary for fertility and is generally recognized as a key nutrient in the productivity. The phosphate more than 2 mg/l in open water gives a sign of organic pollution (Pomeroy *et al.*, 1965). Among the various micro-nutrients phosphorous is found to play an important role in governing the primary productivity of any ecosystem (Hutchinson, 1957). The primary concern to phosphorus lies in its ability to increase the growth of nuisance algal and eutrophication. Butler and Liss (1972) stated that any change in phosphate content may affect the plankton community.

Phosphate is the main nutrient responsible for eutrophication in water bodies. Excess of phosphate is reported to be present in the domestic sewage. According to Arceivala (1981) raw domestic water may contain an average of 10 mg/l of phosphate.

Phosphorus is available in water combined with a number of ions. Phosphate or iron and calcium are very common. In the present investigation the phosphate value was usually low, the minimum being 0.08 mg/l (Table 33.1). The minimum value of phosphate during winter may be due to advance phytoplanktons which utilizes it. Such findings have also been reported by Kataria *et al.* (1996). The maximum value of phosphorous during rainy season might be due to agricultural run off which brings phosphate from the catchment areas. Cattle dung and night soil also add to the phosphorous content of water.

Sulphate

Sulphate, a common anion has a very important role in the soft water systems where certain organic chelate or complex metal ions prevent those ions reacting with other substances (Hutchinson, 1957 and Wetzel, 1975). The increased value of sulphate during monsoon (Table 33.1) might be due to surface run off which brings more suspended solids along with organic and soluble salts (Sinha, 1986). The low value of sulphate during winter might be due to higher phytoplankton population. Sulphate is ecologically important for growth of plants and its short supply may inhibit the development of plankton. Sulphur is also important in protein metabolism.

Phytoplankton

Phytoplankton is an important aquatic flora. They play a key role in maintaining proper equilibrium between abiotic and biotic components of the aquatic ecosystem. Phytoplankton have been regarded as the chief primary producers of natural ecosystems. Being an index of trophic status, phytoplankton reflects the over all environmental condition of the system and its potentiality (Agarwal *et al.*, 1993). Phytoplankton are the fundamental components of aquatic ecosystem as they are the major source of biologically important and labile organic carbon, located at the base of food chain. The magnitude and dynamics of phytoplankton population have became an essential tool to assess the general health of an aquatic ecosystem. Phytoplankton can be used in the treatment of wastewater and either heavy scale deposition of corrosiveness of water and making it potable (Mahadev and Hosmani, 2002). Their density have been reported to be effected by the quality of water (Bilgrami and Dutta, 1979, 1985).

The phytoplankton communities of the river were presented mainly by four groups comprising 38.71 per cent species of Chlorophyceae, 38.96 per cent Myxophyceae, 15.06 per cent Bacillariophyceae and 7.27 per cent Euglenophyceae (Table 33.2 and Figure 33.2).

Chlorophyceae

This group was represented by *Spirogyra* sp., *Volvox* sp,. *Ulothrix* sp,. *Closteridium* sp,. *Chara* sp,. *Pandorina* sp., *Zygnema* sp., *Chlorella* sp. and *Chlamydomonas* sp.

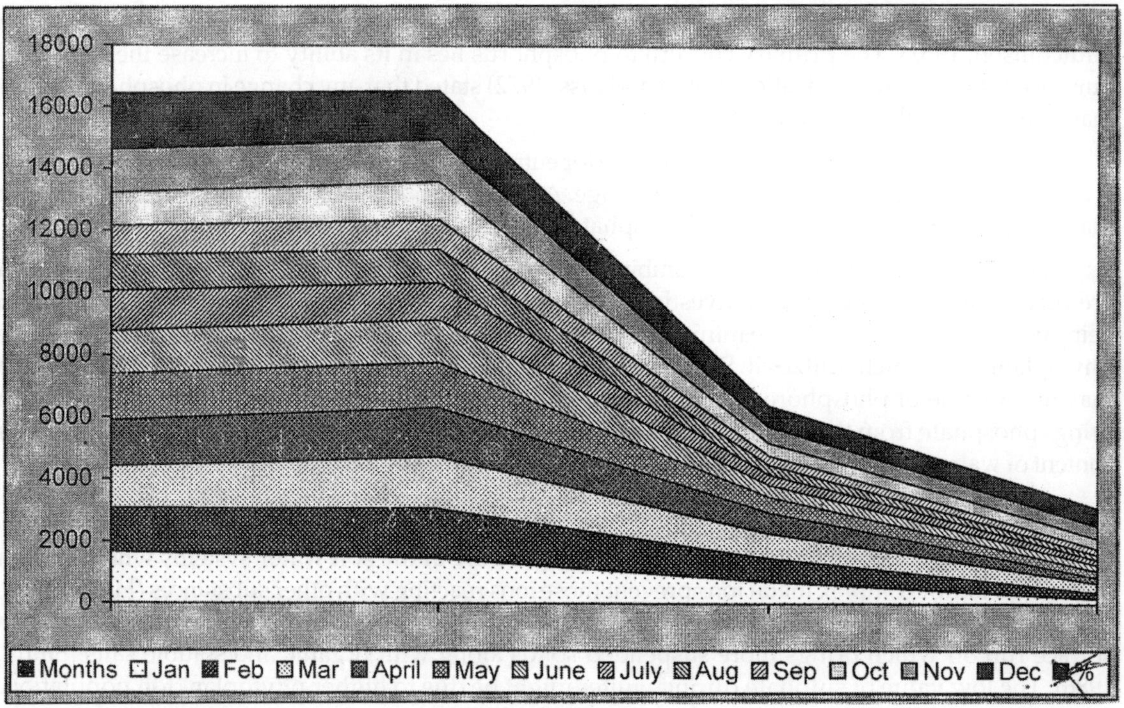

Figure 33.2

Table 33.2: Month-wise Distribution of Phytoplankton Groups in River Panar Water

Months	*Chlorophyceae*	*Myxophyceae*	*Bacillariophyceae*	*Euglenophyceae*
January	1680	1450	720	178
February	1388	1595	758	202
March	1350	1680	790	295
April	1485	1588	670	188
May	1498	1450	498	205
June	1370	1378	390	198
July	1295	1198	340	178
August	1145	1078	320	155
September	950	945	305	195
October	1058	1250	405	325
November	1425	1356	520	395
December	1795	1580	680	575
Per cent	16439 (38.71)	16542 (38.96)	6396 (15.06)	3089 (7.27)

Myxophyceae

This group was represented by *Anabaena spheical, Oscillatoria limnetica, Microcystis robusta, M. aeruginosa, Spirulina* sp., *Merisomopedia* sp., *Nostoc* sp. and *Collastrum* sp.

Bacillariophyceae

This group was represented by *Cymbela, Melosira, Navicula Nitzschia* sp., *Diatoma Syndera, Fragilaria, Pinnularia* and *Cyclotella.*

Euglenophyceae

This group was represented by *Euglena viridis.*

Davis (1955) pointed out that various physical, chemical and biological circumstances must be simultaneously taken into consideration for understanding the fluctuation of plankton population. Temperature, pH alkalinity and phosphate have been emphasized to be significant factors controlling distribution of Cyanophyceae (Singh,1965). George (1976) also reported temperature and pH correlation with phytoplankton. In the present investigation pH, DO and transparency showed a positive relationship with Chlorophyceae. Jackson (1971) has reported alkalinity range of 50 to 110 mg/l as optimum for the Myxophycean where as Sreenivasan (1965) observed 50 mg/l alkalinity for the growth of Myxophyceae. The present study agree with them. The rate of production are closely dependent upon temperature conditions of water body (Mc Camby, 1953). According to Roy (1955) low temperature in Hoogly was favorable for the growth of diatoms resulting in highest plankton peak in the year. The present study shown highest phytoplanktons in winter, when water temperature was low. There was decline in number with increase in temperature suggesting that phytoplankton preferred moderate temperature. Rainfall and high turbidity produced by high wind velocity during rainy season had a direct bearing on phytoplankton population reducing them to minimum numbers.

The river harbours many allergenic algae such as *Chlorella vulgaris, Anabaena, Microcystis, Nostoc* sp., and *Oscillatoria* sp. Allergenic problems caused due to such algal are rhinitis, bronchial asthma, hypersensitivity in fish, cattle and animals couple with symptoms of partial paralysis, loss of balance, hard stool, reduced milk yield, general weakness and photosensitization of skin (Shukla, 1991).

During recent years algae have acquired place of problem organism in eutrophic waters (Trainer, 1978). Toxic characteristic exhibited by certain species *Microcystis acruginosa, Anabaena flosagual* and *Aphanizomenon flosaguae* are the best offenders. Water fowl, fish and livestock are the most frequently affected organism though human deaths have also been reported. *Ulothrix* sp. and *Diatoms* are some of the more troublesome forms (Evans, 1960 and Edmondson, 1974).

The study made on phytoplankton of water bodies all over the world have shown marked differences in the floriotic composition with different levels of pollution which have promoted several workers to use algal composition as the indicator of the level of pollution Hasle, 1947; Rawson, 1956; Hutchinson, 1967 and Palmer, 1969).

Palmer (1969) on the basis of studies carried out at different places in the world has given a list of 60 algal genera and 80 species most tolerant to pollution. The use of diatoms as indicators of pollution has been emphasized by Davis (1964); Patrick (1973); Reynold (1973); Stockner and Benson(1967); Stoermer *et al.* (1978); Rai and Kumar (1976) and Verma *et al.* (1978) have reported *Nitzschia* from highly polluted water in India. Geevarghese and Chandrasekharan (1985) while studying the impact of newsprint factory on Moovattepuzha river found that the diatoms *Nitzschia, Melosira, Diatoma, Pinnularia* and *Navicula* were quite dominating in polluted zone. Some other studies like Raina *et al.* (1982) on river Jhelum, Gunale (1991) on river Mula. Mutha and Adhola (1988) on river Betwa have also reported certain characteristic diatoms in polluted conditions. These authors have also encountered several members of Chlorophyceae, Cyanophyceae and Euglenophyceae like *Pandorina, Scenedesmus, Coelastrum, Chlorella, Chlamydomonas Oocystis, Pediastrm, Microcystis, Oscillatoria, Anabaena,*

Ankistrodesmus and *Euglena* in polluted conditions. Goel and Autade (1995) have reported 61 species of algae during their studies on river Panchganga. Out of these 61 species, 21 are pollution tolerant species. Wani (1998) have reported 91 species of phytoplankton in the lakes of Kashmir. Prasad and Singh (2003) have reported 20 taxa of algae in a tropical water body of Motihari. In the present investigation, 27 species of phytoplanktons were found of which many were pollution tolerant. Algal species can also be used as good indicators of heavy metal pollution in freshwater ecosystems (Forstner and Wittmann, 1979). *Euglena* and *Nitzschia* have been regarded as indicators of Cr and Cu pollution in water (Palmer, 1980). Ramaswamy *et al.* (1982) reported luxuriant growth of *Syndera* and *Nitzschia* and Euglenophyceae in water having high concentration of Cu. However, in present study no such correlation was observed. Franklin (1972) suggested that blue green algae are general indicators of eutrophication of water. Rama Rao *et al.* (1978) have also designated green algae to be the indicators of highly polluted waters.

Zooplankton

Zooplankton forms the vital link between autotorphs and heterotrophs in an aquatic ecosystem. Zooplankton occupy an intermediate position in the food web. Zooplankton, the primary consumers, which play an important role in fish production are being adversely affected. These organisms are highly variable in nature from one water bodies to the other and acts as bioindicators of pollution (Arora, 1966; Sampath *et al.*, 1978; Sharma, 1986 and Saksena, 1987). The Zooplankton forms a link between phytoplankton and macroinvertebrates, which inturn provides food to fishes and aquatic bird. The knowledge of their seasonal qualitative and quantitative fluctuation has been considered always essential for proper understanding of the factors influencing biological productivity and fisheries development. Really, planktonic animals in freshwater are dominated by Rotifers, Cladocera and Copepods. Rotifers are the most sensitive bio-indicators of water quality and their presence may be used as a reference to the physico-chemical characteristics of their habitat. Several zooplankton species have been classified as indicator of polluted conditions. The zooplankton community mainly comprises of Rotifers (46.83 per cent), Cladocera (36.46) and Copepods (16.71 per cent) (Table 33.3 and Figure 33.3). The three groups of zooplankton were:

Rotifera

Brachionus rubens, B. falcatus, B. calyciflorus, B. plicatilis, B.angularis,. B. forticula, B. caudatum, Keratella tropica, K. vulga, K. eanadensei, Monostyla lunaris M. bulla, Horaella sp., *Lepadella* sp., *Cephalodella forticula, Rotaria* sp., *Polyarthra* sp., *Ophryoxus* sp. and *Chydorus gibbus.*

Cladocera

Moina micrura, M. brachiata, Macrothrix sp., *Daphnia* sp., and *Ceriodaphnia* sp.

Copepods

Nauplius sp., *Mesocyclops* sp. and *Heliodaptomus vidunus.*

Species richness is a good indication of the dynamic balance of the communities. The rotifers constitute from 33.33 per cent to 53.84 per cent in total zooplanktonic communities in various waters (Kaushik, 1992). In the present investigation, the percentage of rotifers in total zooplankton is 46.83 per cent The difference in rotifer community structure in the river can be attributed to intense variability in physico-chemical and geographical condition. Other factors responsible can be nutrients, organic matter and phytoplanktonic growth.

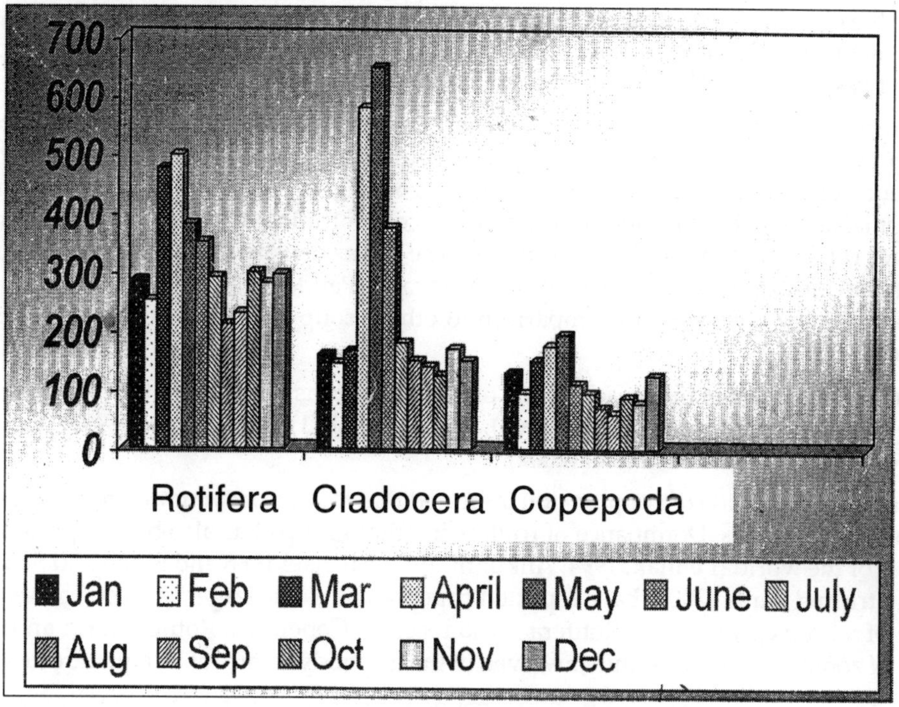

Figure 33.3

Table 33.3: Month-wise Distribution of Zooplankton and Group in River Panar Water

Months	*Rotifera*	*Cladocera*	*Copepods*
January	285	160	130
February	250	145	95
March	475	165	150
April	498	580	175
May	380	650	195
June	350	375	110
July	290	180	95
August	210	150	70
September	230	140	60
October	298	125	88
November	280	170	78
December	295	150	125
Per cent	3841 (46.83)	2990 (36.46)	1371 (16.71)

Lakshminarayana (1965), Chourasia and Adoni (1985) have found a direct relationship between pH and zooplankton production. It is quite evident from the present study that the pH range of 7.3 to

8.2 favoured the growth of zooplankton. Nasar (1977) noted that pH between 7.6 to 7.7 is quite suitable for development of zooplankton. Low value of nutrients act as limiting factor for zooplankton development despite good illumination (Sreenivasan, 1970).

The structure of zooplankton community is characteristic and is the product of the environmental conditions prevailing there. Presence of maximum zooplankton population in summer might be due to the presence of higher population of bacteria. According to Singh (1991), Pandey *et al.* (1995) optimal thermal and nutritional condition and higher concentration of oxygen might be responsible for the higher rotifer population during summer. The lower density of zooplankton during rainy season might be due to flood and fast current of river water (Pandey *et al.*, 1995,2004). The abundance of zooplankton in the ecosystem in comparison to other group is an indication of eutrophication (George, 1966).

Dominance of rotifers in the seasonal data of zooplankton as observed in the present study is in accordance with the findings of Michael (1968); Saha *et al.* (1971); Bahura *et al.* (1993), Pandey *et al.* (1992b, 1994, 1995, 2004) and Hosmani (2002).

The species composition of zooplankton indicated that rotifers occur more predominantly than cladocerans and copepods. Dominance of rotifers in other groups has also been reported in other water bodies of the world (Pennak, 1994; Alikuhni, 1957; Michael 1968 and Singh and Sahai, 1978). However, Patra and Dutta (2004) have reported copepodans as the largest group. The sequence of dominance of various groups was Rotifera > Cladocera > Copepods. Rotifers form an important component of zooplankton community involved in the trophodynamics of an aquatic ecosystem.

The Cladocerans were succeeded by the Copepods, which in turn indicates that Copepods build up their population taking more time than Rotifers and Cladocerans. However, once they become dominant, they continue to dominate the habitat till hydrological conditions favour their existence. The phytoplankton bloom physical and chemical characteristic of water are stated to be greatly responsible for the Copepods population (Patalas, 1975). Copepods were abundant during rainy season followed by winter peak. Winter and rainy peaks have been reported by Maruthanayagm *et al.* (2003) and Pandey *et al.* (2004). Comita and Anderson (1959) found that copepods multiply when that phytoplankton population was abundant. Such an observation was not found in the present investigation. This supports finding of Hosmani (2002). The cladocerans population was scanty in comparison to the rotifers. The main cladoceran peak was observed during summer and rainy seasons. According to Wright (1965) the density of cladocerans is primarily determined by the food supply but in the present investigation decline in the number of cladocerans in the presence of sufficient food may due to fish predation and the active competition between cladocerans and other groups.

The zooplanktonic species have been recognized as pollution indicators by their usual absence or presence in water body. Sampath *et al.* (1978), Verma *et al.* (1984), Sharma (1986), Mishra and Saksena (1990) have commented on the basis of study of zooplankton communities that *Arcella discoides, Difflugia cornata, D. muriformes, Brachionus angularis, B. rubens, B. caudatus, B. falcatus, B. calyciflorus, B. diversicornis, Keratella lenzi, K. tropica, Moina* sp., *Daphnia carinata, D. lumkoltz, Cypris* sp. *Cyprinotus* and *Mesocyclops* are present in mesotrophic waters. *Arcella* sp., *Diffugia* sp. *Centropyxis* sp., *Brachionus angularis, B. calyciflorus, B. rubens, B. caudatus, B. quadridentatus, B. budapestinensis, B. petrodinoides, B. sessilis, Keratella tropica, K. cohlearis, Anuracopysis fissa, Platyias quadricornis, Lecane crenata, L.cornuta, L. lunaris, L. decipience L. pyriformis, L. closterocerca, Colurella bicuspidate, Filinia longiseta, F. terminalis, Asplanchna* sp., *Rotaria neptunia, Cephalodella* sp., *Conchilus* sp., *Polyarthra* sp., *Daphnia parvula, Cerodaphina lacustaris, Chydrous sphaericus, Daphanosoma leuchtenbergianum, Moina* sp., *Mesocyclops edax* and *Diaptomus busiciloides* are representative of eutrophic waters (David and Roy, 1966; Arora, 1966; Sharma, 1986; Bhatti and Rana, 1987; Cap, 1980; Janicki *et al.*, 1979 and Saksena, 1987). In the present study, *B. calyciflorus* sp., *B. falcatus, B. forticula, B. angularis, Cephalodela* sp., *Polyarthra* sp., *Moina, Mesocyclops* sp., *Cypris* sp., *Ceriodaphnia* sp. were present in the river and they are suggestive of pollution.

All forms of organic matters are basically undesirable in the water though to a certain level of is tolerable. This may lead by to serious trouble like eutrophication and depletion of the oxygen content of the system and thus impairing its health. Indian standards set 3 mg/l of DO as the limit. The DO level throughout the year was more than 3 mg/l. This perhaps indicates that the self purification capacity of the river can assimilate the existing load.

The results of physico-chemical study of water quality of the river depict a very critical level of pollution. An over all nutrients status, a low level of dissolved oxygen, high BOD,COD, solids and low oxygen are some of the positive factors favouring the growth of phytoplankton (Palmer, 1969). The growth of algae causes eutrophication, which may enhance the fish production (Lund, 1972).

It is clear from the present findings that the aquatic environment of the river has under gone degradation. The main sources of river pollution in the present study were recognized as:

1. Discharge of domestic sewage from near by villages as well as industrial wastes
2. Agricultural run off
3. Burning of corpses on the bank of river
4. Bathing habits of human as well as cattles.
5. Washing of clothes and utensils and
6. Soil erosion
7. Jute retting

Proper remedial measures should be taken immediately in order to restore it from further deterioration. There must be alternate waste disposal system away from the river and the disposal of solid and liquid wastes must stopped forthwith. Planting some macrophytes like *Polygonum amphibium. Bacopa momiesi Alternanthera sessilis* and *Leonots nepataefolia* on both sides of the river has been reported to reduce toxicants from the polluted river water (Gupta, 2000).

References

Adoni, A.D., 1975. Studies on microbiology of Sagar lake. *Ph.D. Thesis*, Sagar University, Sagar, M.P.

Anderson, G.C., 1961. Recent change in the trophic nature of lake, Washington: A review in algae and metro politancinnati, Ohio. *TRW*, 61(3): 27–33.

Adebisi, A.A., 1980. The physico-chemical hydrology of tropical seasonal upper Ogum river. *Hydrobiol.*, 79: 157–165.

Agarwal, R., Bahura, C.K. and Saxena, M.M., 1933. Planktonic productivity of sewage fish pond and adjoninig oxidation pond at Vallabh Gardens, Bikaner. *Acata. Ecol.*, 15(1): 58–61.

Alikuhni, K.H., 1957. *Fish Culture in the India.* Farm Bulletin No. 20, ICAR, New Delhi.

APHA (American Public Helath Association), American Water Works Association and Pollution Control Federation, 1975. *Standard Methods of Examination of Water and Wastewater*, 14th edn. APHA, Washington, USA.

Arceivala, A.J., 1981. *Wastewater Treatment and Disposal.* Marcel Dekker, New York.

Arora, H.C., 1966. Studies on Indian Rotifera Part III. *J. Zool. Soc.*, 16(1&2):1–6

Bahura, C.K., Bahura, P. and Saxena, M.M., 1993. Zooplanktonic community of Shivbari temple tank Bikaner. *J. Ecobiol.*, 5(1): 5–8.

Barnes, R.K. and Mann, K.H., 1980. *Fundamental of Aquatic Ecosystems.* Blackwell Scientific Pub., Oxford, p. 1–229.

Bharti, S.G. and Kori, A.P., 1975. Limnological studies in ponds and lakes of Dharwad: Diurnal variation in Saydapur Keri. Part I. Karnataka Univ. *J. Sci.*, 20: 157–167.

Bhatti, D.P.S. and Rana, K.S., 1987. Zooplankton in relation to abiotic components in the fort moat of Bharatpur. *Proc. Nat. Acad. Sci. India*, 57(B): 237–242.

Bilgrami, K.S. and Dutta Munshi, J.S., 1979. *Limnological Survey and Impact of Human Activities on the River Ganga (Barauni to Farakka).* Allied Press, Bhagalpur, India.

Bilgrami, K.S. and Dutta Munshi, J.S., 1985. *Ecology of River Ganges: Impact of Human and Conservation of Aquatic Biota (Patna to Farakka).* Allied Press, Bhagalpur, India.

Bohra, O.P. and Bhargava, D.S., 1977. Abiotic factors chlorophyll pigments and primary production in two lakes of Jodhpur. *Geobios*, 4: 215–216.

Butcher, R.W., 1947. Studies on ecology of river. VII. The algae of enriched water. *J. Ecology*, 35: 186–191.

Cap, R.K., 1980. Limnological studies of Western lake Erie. III. Phytoplankton and physical-chemical data from November, 1939 to November, 1940. *Ohico J. Sci.*, 42: 24–44.

Chandler, D.C., 1942. Limnological studies of Western lake Erie. III. Phytoplankton and physical-chemical data from November, 1940. *Ohico. J. Sci.*, 80: 114–118.

Chourasia, S.K. and Adoni, A.D., 1985. Zooplankton dynamics in a shallow eutrophic lake. In: *Proc. Nat. Symp. Pure and Appl. Limnology*, (Ed.) A.D. Adoni. *Bull. Bot. Soc., Sagar*, 32: 30–39.

Cole, G.R., 1979. *A Textbook of Limnology*, 2nd edn. The E.V. Mosley Co., London.

Comita and Anderson, 1959. Cited from phytoplankton, zooplankton relationship in four freshwater bodies of Dharwar.

Butler, E.T. and Liss, S., 1972. Chemical survey of tamar estuary. *J. Mar. Biol.*, 52: 681–699.

Dale, H.M. and Gillsspic, T.J., 1977. The influence of submerged aquatic plants on temperature gradients in shallow water bodies. *Canadian J. Bot.*, 55(16): 2216–2225.

Das, A.K., 2002. Phytoplankton primary production in some selected reservoirs of A.P. *Geobios*, 29: 52–57.

David, A. and Roy, P., 1966. Studies on the pollution of the river Daha (N. Bihar) by sugar mill and distillery wastes. *Environ. Hlth.*, 8(1):6–35.

Davis, C.C., 1955. *The Marine and Freshwater Plankton of Michigan State.* Uni. Press. East Lausing, 562 p.

Edmondson, W.T., 1974. A simplified method for counting phytoplankton. In: *A Manual on Methods for Measuring Primary Productivity in Aquatic Environment*, (Ed.) R.A. Vollenweider. Blackwell Sci. Pub., Oxford.

Evans, G.H., 1960. Algological studies in Ann. Rep. of East African freshwater fisheries. Res. Org. Jinga, Uganda.

Forstner, V. and Wittman, G.T.W., 1979. *Metal Pollution in the Aquatic Environment*. Spring-Verlag, New York.

Franklin, T., 1972. *First Internat. Symp on Taxonomy and Biology of Blue Green Algae.* Univ. Madras, p. 442.

Geevarghese, C. and Chandrashekharan, M.P., 1985. Studies on the ecological aspect of factory waste on river ecosystem. *Poll. Res.*, 4(1): 21–24.

George, M.C., 1966. Comparative plankton ecology of five fish tanks in Delhi, India. *Hydrobiol.*, 27: 81–90.

George, J.P., 1976. Hydrobiological studies on lower lake of Bhopal with special reference to productivity of economic fishes. *Ph.D. Thesis*, Bhopal University.

Goel, P.K. and Autade, V.B., 1995. Ecological studies on the river Panchganga at Kolhapur with emphasis on biological components. In: *Recent Researches in Aquatic Environment*, (Eds.) A. Gautam and N.K. Agarwal. Daya Publishing House, New Delhi, pp. 25–46.

Gunale, V.R., 1991. Algal communities as indicators of pollution. *J. Environ. Biol.*, p. 233–232.

Gupta, A.K., 2000. Sustainable strategies for restoration of degraded Varuna river corridor. *J. Env. and Poll.*, 7(1): 43–46.

Hasler, A.D., 1947. Eutrophication of lakes by domestic drainage. *Ecology*, 28: 288–295.

Hosmani, S.P., 2002. Phytoplankton-zooplankton relationship in four freshwater bodies of Dharwar Ind. *J. Env. and Ecoplan.*, 6(1): 23–28.

Hutchinson, G.E., 1957. *Treatise on Limnology*. John Wiley and Sons, New York.

Hutchinson, G.E., 1967. *A Treatise on Limnology, Vol. III: Introduction to Lake Biology and Limnoplankton.* John Wiley and Sons, N.Y.

Jackson, D.F., 1971. Comparative studies on phytoplankton photosynthesis in relation to total alkalinity *Ver. Limnol.*,14: 125–133.

Jameel, A.A., 1998. Physico-chemical studies in Vyakondan channel water of River Cauvery. *Poll. Res.*, 17: 110–114.

Jameson, J. and Rana, B.C., 1996. Pollution status of river complex Sabarmati at Kheda region of Gujarat. I. Physico-chemical characters. *Poll. Res.*, 15(1): 53–55.

Janicki, A.K., Decosta, J. and Davis, J., 1979. Midsummer crustacean-planktonic communities of seven small compoundments. *Hydrobiol.*, 64: 123–129.

Jhingran, V.G., 1982. *Fish and Fisheries of India*, 2nd edn. Hindustan Publishing Corporation, Delhi, India.

Kataria, H.C., Quereshi, H.A., Iqbal, S.A. and Shandilya, A.K., 1996. Assessment of water quality of Kolar Reservoir in Bhopal, M.P. *Poll.*, 15(2): 191–193.

Kaul, V., Zutshi and Vass, K.K., 1972. Biomass productivity of some macrophytes in Srinagar lakes. In: *Tropical Ecology Emphasizing Organic Production*, (Eds.) P.M. Colley and F.B. Golley, pp. 295–311.

Kaur, H., Dhillon, S.S., Bath, K.S. and Mander, G., 1996. Analysis of the elements polluting River Ghaggara in the region of Punjab. *J. Env. Poll.*, 3(2): 65–68.

Kaushik, S., 1992. Ecological studies of some water bodies at Gwalior, Madhya Pradesh. *Ph.D. Thesis*, Jiwaji University, Gwalior.

King, D.L., 1972. Carbon limitation in sewage lagoons. In: *Nutrients and Eutrophication*. Am. Soc. Limnol. Oceanogr., Kanari, 98: 112.

Kotpal, R.L. and Bali, N.P., 1989. *Concept of Ecology*, 4th edn.Vishal Pub., Jallandhar, India.

Lakshminarayana, J.S.S., 1965. Studies in the phytoplankton of river Ganges, Varanasi, India. I. Physico-chemical characteristics of river Ganges. II. The seasonal algae in the river Ganges. *Hydrobiol.*, 25: 19–165.

Lee, G.F., Bentley, E. and Amundson, R., 1975. Effects of marshes on water quality. In: *Coupling of Land Water System*, (Ed.) A.D. Hasler. 105–126, Spinger-Verlag Berlin, Heidelberg, New York.

Lester, W.F., 1967. Management of river quality. In: *River Management*, (Ed.) P.C.G. Isaac. Maclaren and Son Ltd., London.

Livingstone, D.A., 1963. Chemical composition of rivers and lakes U.S. *G.S.Geological Survey Prof. Pop.* 440G 1–64.

Lund, J.W.J., 1972. Eutrophication Proceedings of the Royal Society of London. *Bio. Sci.*, 180: 371–382.

Mahadev, J. and Hosmani, S.P., 2002. Langlers index and its relation to phytoplankton in two lakes of Mysore city. *Nature Env. and Poll. Tech.*, 1(1): 19–21.

Maruthanayagam, M., Sasikumar, M., and Senthikumar, G.C., 2003. Studies on Zooplankton population in Thirukkulum pond during summer and rainy seasons. *Nature Env. and Poll.Tech.*, 2(1):13–19, 47–60.

Madhyastha, P.D., Rekha, P.D. and Shashikumar, K.C., 1999. Impact of anthropogenic activities on the river quality along pilgrims centres at Dakshina Kannada. *J. Hum. Ecol.*, 10(1): 29–34.

McCamby, A.M., 1953. Factors influencing the growth of phytoplankton. *J. Fish Res. Board. Can.*, 10: 253–282.

Michael, R.G., 1966. Diurnal variation in chemical factors and zooplankton in the surface layers of three freshwater fish ponds. *Ind. J. Fish.*, 13(1 and 2): 48–42.

Michael, R.G., 1968. Studies on zooplankton of a tropical fish pond. Hydrobiol, 2(1–2).

Mishra, G.P. and Yadav, A.K., 1978. A comparative study of physico-chemical characteristics of river and lake water in Central India. *Hydrobiol.*, 59(3): 275–278.

Mishra, S.R. and Saksena, D.N., 1990. Seasonal abundance of the zooplankton of wastewater from the industrial complex at Birla Nagar (Gwalior), India. *Acta. Hydrochem. Hydrobiol.*, 18: 215–220.

Nasar, S.A.K., 1977. Investigations on the seasonal periodicity of zooplankton in freshwater pond in Bhagalpur, India. *Acta. Hydrohem. Hydrobiol.*, 5(6): 577–584.

Palmer, C.M., 1969. Composites rating of algal tolerating organc pollution. *J. Phycol.*, 5: 78–82.

Palmer, C.M., 1980. *Algae and Water Pollution.* Castle House Publication Ltd., U.S.A.

Pandey, B.N., Jha, A.K, Das, P.K.L. and Pandey, K., 1994. Zooplanktonic community in relation to certain physico-chemical factors of konnwamps, Purnea, Bihar. *Env. and Ecol.*, 12(3): 563–567.

Pandey, B.N., Lal, R.N., Mishra, P.K. and Jha, A.K., 1992a. Seasonal rhythms in the physico-chemical properties of river Mahananda (Katihar), Bihar. *Env. and Eco.*, 10(2): 354–357.

Pandey, B.N. and Mishra, R.D., 1990. Hydrobiological features of river Saura. *Bio. J.*, 2(2): 342–354.

Pandey, B.N., Mishra, A.K., Jha, A.K., and Lal, R.N., 1992b. Studies on qualitative composition and seasonal fluctuation in plankton composition of river Mahananda, Katihar (Bihar). *J. Ecotoxicol. Environ. Monit.*, 2(2): 93–97.

Pandey, B.N., Mishra, A.K., Das, P.K.L and Jha, A.K., 1995. Studies on hydrobiological condition of river Saura in relation to its impact on biological health. In: *Recent Researches in Aquatic Environment*, (Eds.) Ashutosh Gautam and N.K. Agrawal. Daya Publishing House, New Delhi.

Pandey, B.N., Hussain, S., Jha, A.K. and Shyamanand, 2004. Seasonal fluctuation of zooplanktonic community in relation to certain physico-chemical parameters of river Ramjan of Kishanganj, Bihar. *Nature. Env. and Poll. Tech.*, 3: 325–330.

Patel, N.K. and Sinha, B.K., 1998. Study of the pollution load in the ponds of Burla area near Hirakund Dam of Orissa. *J. Env. and Poll.*, 5(2): 57–160.

Patalas, R., 1975. The crustacean plankton communities of fourteen North American great lakes. *Ver. Intern. Ver. Limnol.*, 19: 504 – 511.

Patra, S.B., and Datta, N.C., 2004. Seasonal fluctuation of different zooplanktonic groups of a rainfed wetland in relation to some abiotic factors. Insd. *J. Environ. and Ecoplan.*, 8(1): 7–12.

Patrick, R., 1973. Use of algae, especially diatoms in assessment of water quality. In: *Biological Method for Assessment of Water Quality*, ASIM/STP, pp. 76–95.

Pennak, R.W., 1944. Diurnal movements of zooplankton organisms in some Colorado mountain lakes. *Ecology*, 25: 387–403.

Philipose, M.T., 1959. Freshwater phytoplankton of inland fisheries. *Proc. Symp. Algology*, 4: 275–291.

Pillai, K.V. and Sreenivasan, A., 1975. A note on the carbon and nitrogen status of some lakes, reservoir and pond of Tamil Nadu. *J. Inland. Fish. Soc.*, 2(195): 65–97.

Pomeroy, L.R., Smith, E.E. and Grant, C.M., 1965. The exchange of phosphate between estuarine water and sediments. *Limnol. Oceanor.*, 10(2): 167–172.

Prasad, B.B. and Singh, R.B., 2003. Composition abundance and distribution of phytoplankton and zoobenthos in a tropical water body. *Nature Environ. and Poll.*, 2(3): 255–258.

Rai, D.N., 1980. Ecological structures of certain Swamps of Darbhanga (Bihar). *Ph.D. Thesis*, Bhagalpur University, Bhagalpur.

Rai, L.C. and Kumar, H.D., 1976. Systematic and ecological studies on algae of some habitats polluted with fertilizer factory effluents near Shapuri, Varanasi, India. *Nova Hedwigia*, 27: 805–811.

Rama Rao, S., Singh, U.P. and Mail, L.P., 1978. Pollution of river Kaun (Indore). India. I. Biological assessment of pollution. *Water*, 12: 555–559.

Ramaswamy, S.N., Somashekhar, R.K. and Areka, G.D., 1982. Ecological studies on algae in wastewater from rubber tyre factory. *Indian J. Environ. Hlth.*, 24: 1–7.

Raina, R., Subla, B.A. and Zutchi, D.P., 1982. Water quality and plankton of Jhelum River. *Int. J. Ecol. Environ. Sci.*, 8: 11–17.

Raney, E.D. and Menzel, B.W., 1969. *Icthyological Association Bull*. Ithaca. BYZ.

Rao, C.B., 1955. On the distribution of algae in a group of six small ponds. II. Algal periodicity. *J. Ecol.*, 43: 291–308.

Rao, D.S. and Govind, B.V., 1966. Hydrobiological studies of Tungbhadra reservoir. *Indian J. Fish*, 11: 321–344.

Rawson, O.S., 1956. Algal indicators of trophic lake types. *Limnol. Oceanoger.*, 1: 18–25.

Roy, H., 1955. Plankton ecology of River Hooglly (West Bengal). *Ecology*, 36: 169–175.

Reynolds, C.S., 1973. The seasonal periodicity of planktonic diatoms in a shallow eutrophic lake. *Freshwater Biol.*, 3: 89–100.

Saha, G.N., Sehgal, K.L., Mitra, E.and Nandy, A.C., 1971. Studies on the seasonal and diurnal variation in physico-chemical and biological conditions of a perennial freshwater pond. *J. Inland Soc. India*, 3: 79–102.

Sampath, V., Sreenivasan, A. and Ananathanarayanam, R., 1978. Rotifers as biological indicators of water quality in Cauvery river. *Abst. Symp Environ. Biol.*, p. 5.

Saksena, D.N., 1987. Rotifers as indicator of water quality. *Acta Hydrochem., Hydrobiol.*, 15: 169–175.

Saxena, D.N. and Mishra, B.R., 1991. The water quality index self purification capacity of sewage collecting channel, Morar (Kalpi) River at Gwalior, Madhaya Pradesh. *J. Freshwater Biol.*, 3(2): 169–175.

Sharma, B.K., 1986. Assessment of pollution indicators in Indian Rotatoria. *J.Meghalaya Sci. Soc.*, p. 47–49.

Shyam Sunder, 1988. Monitoring the water quality in a stretch of river Jhelum, Kashmir. In: *Ecology and Pollution of Indian Rivers*, Ashish Publishing House, New Delhi, p. 131–162.

Shukla, A.C., 1991. Biological studies of the Ganga ecosystem between Kannauj and Shuklaganj. In: *The Ganga: A Scientific Study*. Northern Book Centre, New Delhi.

Singh, V.P., 1965. Phytoplankton periodicity in a small lake near Delhi. *J. Phykos*, 4: 61–68.

Singh, R., 1991. Rotifer fauna of Jamalpur, Munger, Bihar. *J. Curr. Bio. Sci.*, 8: 45–48.

Singh, S.B. and Sahay, R., 1978. Seasonal fluctuation of zooplankton population in relation to certain physico-chemical factors of a pond. *Geobios*, 5: 228–230.

Sinha, M.P., 1986. Limno-biotic study of trophic status of a polluted freshwater reservoir of coal field area. *Poll. Res.*, 5: 13–17.

Singh, S.R. and Srivastava, V.K., 1988. Variation of water quality of Ganga river between Buxer and Ballia. *Poll. Res.*, 7(3–4): 85–92.

Singhal, R.N., Jeet, S. and Deviesh, R.W., 1985. The relationship among physical, chemical and plankton characteristics of unregulated rural ponds in Haryana, India. *Trop. Eco.*, 26: 43–53.

Sreenivasan, A., 1965. Limnology of tropical Impoundments. III. Limnology and productivity of Amaravati reservoir (Madras state), India. *Hydrobiol.*, 26: 501–516.

Sreenivasan, A., 1970. Limnology of tropical impoundment: A comparative study of the major reservoirs in Madras state (India). *Hydrobiol.*, 36: 443–469.

Swarnalatha, N., 1994. Seasonal rhytms of various physico-chemical characteristics of a lake. *J. Mendel.*, 11(1): 65–66.

Stoermer, E.L., Ladonski, B.G. and Schelske, C.L., 1978. Population response of lake Michigan phytoplankton to nitrogen and phosphorous enrichment. *Hydrobiol.*, 57: 249–265.

Stockner, J.G. and Benson, W.W., 1967. The succession of diatom assemblages in the recent sediments of lake Washington. *Limnol. Oceanoger.*, 12(5): 513–532.

Thrash, J.C., Beale, J.F. and Suckling, E.V., 1944. *The Examination of Water Supplies*, (Ed.) E.W. Taylor, 1949.

Todda, B.K., 1970. *Water Encyclopedia*. Water Information Centre, Port Washington, New York.

Toetz, D.W., 1976. Mineral cycling in reservoirs. *Ann. Okla. Acad. Sci.*, 5: 21–28.

Trainer, F.R., 1978. *Inductory Phycology*. John Wiley and Sons, p. 1–525.

Trivedy, R.K. and Goel, P.K., 1984. *Chemical and Biological Methods for Water Pollution Studies*. Environmental Publications.

United Nations, 1981. *River Inputs to Ocean Systems*. New York, 384 pp.

UNESCO, 1979. *Discharge of Selected Rivers of the World*. UNESCO, Paris.

Vass, K.K., Raina, H.S., Zutchi, D.P. and Khan, M.A., 1977. Hydrobiological studies of river Jhelum. *Geobios*, 4: 238–242.

Verma, S.R., Shukla, G.R. and Ahuja, S.R., 1978. Pollution studies of few rivers of western Uttar Pradesh with reference to biological indicators. *Ind. J. Env. Health*, 20: 87–99.

Verma, A.M., Dudani, V.K., Kumar, B., Kumar, S. and Pandey, S., 1988. *Proc. Nat. Acad. Sci., India*, 58(B): 11–259.

Verma, S.R., Sharma, P., Tyagi, A., Rani, S., Gupta, A.K. and Dalela, R.C., 1984. Pollution and saprobic status of eastern Kalnadi. *Limonologica (Berlin)*, 15: 69–133.

Vijay Kumar, K., 1992. Limnological studies of perennial and seasonal standing water bodies of Gulbarga area. *Ph.D. Thesis*, Gulbarga University, Gulbarga.

Vijay Kumar, K., Holkar, D. and Kaur, D., 1999. Limonological studies on Chandrampalli reservoir, Gulbarga. In: *Freshwater Ecosystem*, (Ed.) K. Vijay Kumar. Daya Publishing House, Delhi.

Vyas, L.N. and Kumar, H.D., 1968. Studies on the phytoplankton and other algae of India Sagar tank, Udaypur, India. *Hydrobiol.*, 31: 421–434.

Wani, I.A., 1998. Lentic aquatic resources of Himalaya: The case of Kashmir Himalaya, India. In: *Conservation and Management of Aquatic Resources*, Daya Publishing House, Delhi–35.

Welch, P.S., 1952. *Limnology*, 2nd edn. McGraw Hill Book Company, New York. Tronto and London.

Wetzel, R.G., 1975. *Limnology*. W.B. Sunders Co. Philadelphia, USA, p. 742.

Wright, J.C., 1965. The population dynamics and production of Daphnia in Canyon Ferry Reservior, Mentana. *Limnol, Oceanoger.*, 10: 583–590.

Zafar, A.R., 1964. On the ecology of algae in certain fish ponds of Hyderabad, India. II. Distribution of unicellular and colonial forms. *Hydrobiol.*, 24(4): 556–566.

Zutchi, D.P. and Vass, K.K., 1973. Variation in the water quality of some Kashmir lakes. *Trop. Ecol.*, 14(2): 182–196.

Water and Wastewater Management in 2 Vols (2011) *Pages 228–235*
Editors: **D.R. Khanna, R. Bhutiani & Gagan Matta**
Published by: **DAYA PUBLISHING HOUSE, NEW DELHI**

Chapter 34

Limnological Studies of Gudavi Wetlands, Sorab, Shimoga, Karnataka

G.Y. Dayananda, A. Naveed, K.L. Naik and B.B. Hosetti
Department of Applied Zoology, Bioscience Complex,
Jnana Sahyadri, Kuvempu University, Shankaraghatta, Shimoga

ABSTRACT

A study on the water quality of three freshwater ponds connected in series at Gudavi village near Shimoga is studied for physico-chemical factors. The DO was 8.81 mg/l in Kallambi, 3.84 mg/l in Vaddekere and 7.78 mg/l in Gudavi pond respectively and such a variation is attributed to the fecal (guano) droppings discharged by about 12,000 resident birds which include Egrets, Cormorants, Ibises, Herons, water fowls and other wetland dependent species in the Vaddekere of the sanctuary. A comparative study of nutrients and BOD levels in the three consecutive ponds revealed that the Vaddekere pond is most polluted. Due to this reason the water quality in the Vaddekere was altered and the emergent trees on the pond which were used for roosting and nesting by the birds are affected and evidenced by dying of trees. The death of island trees will lead to loss of roosting and nesting sites for Gudavi birds. Hence an urgent action plan to protect the trees or replanting new trees in the water bodies by creating islands every year is suggested.

Keywords: *Limnology, Guano droppings, Waterfowls, Wetlands and Gudavi.*

Introduction

The natural freshwater which is available to man in the form of reservoirs, ponds and lakes is spread in an area of 0.72 million ha and was reduced to about 0.65 million ha in the country (Saxena, 1996). It means there is a decline of 0.07 million ha of water spread area. According to a survey conducted by NEERI revealed that about 70 per cent of the available water in India is polluted (Agarwal *et al.*, 1982). For limnological study evaluation of physico-chemical factor is a basic step in that each factor contributes in making up of specific ecosystem, which determines the trophic dynamics of the water body. The management of any aquatic ecosystem is a means of conservation of freshwater habitat with an aim to maintain the water quality or to rehabilitate the physico-chemical characteristics of water.

The lentic water systems are popularly referred as wetlands. These wetlands should be scientifically studied as potential productive areas. In a meeting on wetlands held at Ramsar, Iran in 1971 major emphasis was given to water fowls. It was well established that wetlands are very essential for migratory birds and other aquatic biota (Mahajan, 1981, 1988). Lentic wetlands are important not only for water fowls but also for various purposes *viz*, their role in flood control, reducing sediment load, recycling of biogenic salts, source of irrigation water, animal husbandry, aquaculture refuge for rare and endangered fauna and as agents for recharging groundwater. In view of the above importance some ecological aspects of Gudavi wetlands is carried out and discussed in the present chapter.

Materials and Method

Study Area

Gudavi wetlands are located in the Sorab taluk of Shimoga district, at about 13 km away from Sorab city and 0.5 km from Gudavi village. These wetlands occupies the water spread area of about 33 hectares. Remaining area is moist deciduous, forest species interspread with grassy patches.

Gudavi wetlands lies between latitude of 14° 25' 59'' to 14° 26' 41'' and longitude 75° 6' 43'' to 75° 1' 28''. These are connected in series, they are Kallambi, Vaddekere and Gudavi ponds. The rainfed water from Kallambi flows for a distance of 2 km through irrigation channel and enters into the Vaddekere. The outflow of water from Vaddekere enters into Gudavi pond, from there it is used for irrigation at the downstream croplands. Vaddekere and Gudavi ponds are separated by a common bund.

Sampling and Analysis

The water sampling has been carried out during breeding season with an interval of 15 days from June 2000 to February 2001. The parameters like air and water temperature, dissolved oxygen, biological oxygen demand, carbon dioxide, chlorides, calcium, phosphates and sulphates are determined according to standard methods (APHA, 2000 and Trivedi and Goel, 1984).

Results and Discussion

The important factor which influence the plankton dynamics in the pond water are temperature and light intensity (Hosmani, 2002). The change in physico-chemical characteristics are brought about by the fluctuation in climate conditions. The water temperature during the study period was around 24.96 °C in Kallambi, 24.57 °C in Vaddekere and 25.02 °C in Gudavi ponds respectively. It was always 1 °C below the ambient temperature. The temperature in the range 15 °C to 30 °C is found to be suitable for algal productivity, Hence the existing temperature triggered the algal photosynthesis inturn enhanced the fish productivity which is one of the main source of food for birds living in the wetlands of the sanctuary. One of the very interesting subject to note that always the water temperature remained lower than air temperature in all ponds at all seasons. The pH values varied around 7.63 ± 0.52 in Kallambi, 6.77 ± 0.94 in Vaddekere and 7.35 ± 0.85 in Gudavi water samples respectively (Tables 34.1–34.3). In Kallambi pond the water pH was always alkaline during study period from

Table 34.1: The Physico-chemical Characteristics of Water from Kallambi Pond

Month of Sampling 2000–01	Air Temp. (°C)	Water Temp. (°C)	pH	DO (mg/l)	BOD (mg/l)	CO_2 (mg/l)	Cl (mg/l)	Ca (mg/l)	TA (mg/l)	PO_4 (mg/l)	SO_4 (mg/l)
June	25.5	23.0	7.5	8.04	6.76	36.44	48.96	32.00	84.0	0.32	0.01
July	23.6	21.7	7.0	6.86	6.00	29.80	66.03	26.70	67.0	1.02	0.03
August	22.1	21.0	7.5	7.48	5.25	32.0	54.40	25.06	78.0	0.48	0.05
September	25.0	23.0	7.0	9.66	4.80	35.2	65.32	22.40	85.0	0.15	0.08
October	27.0	26.2	7.2	8.26	3.90	28.0	17.04	19.23	50.0	0.40	0.08
November	28.6	26.5	8.0	10.11	5.25	34.12	61.33	20.00	82.0	0.15	0.04
December	28.5	27.0	8.0	10.0	7.00	36.02	64.00	21.00	23.0	0.12	0.06
January	30.2	27.5	8.0	10.11	7.50	35.0	61.00	18.00	21.0	0.50	0.10
February	30.0	28.8	8.5	8.75	5.92	36.02	50.15	20.00	90.0	0.22	0.10
Mean±SD	26.72± 2.85	24.96± 2.81	7.63± 0.52	8.81± 1.22	5.82± 1.14	33.62± 3.01	54.24± 15.33	22.71± 4.46	64.44± 26.85	0.37± 0.28	0.06± 0.03

Table 34.2: The Physico-chemical Characteristics of Water from Vaddekere

Month of Sampling 2000–01	Air Temp. (°C)	Water Temp. (°C)	pH	DO (mg/l)	BOD (mg/l)	CO_2 (mg/l)	Cl (mg/l)	Ca (mg/l)	TA (mg/l)	PO_4 (mg/l)	SO_4 (mg/l)
June	26.0	23.8	6.0	5.04	22.34	66.60	41.20	27.03	98.00	1.78	0.07
July	23.6	21.5	6.0	5.28	26.04	70.00	53.00	22.20	88.00	1.84	0.04
August	23.1	22.3	6.0	4.62	20.20	62.00	68.00	21.08	69.00	2.06	0.08
September	25.0	21.0	6.5	2.40	28.00	61.60	71.00	22.14	90.00	2.13	0.12
October	28.4	27.0	6.0	1.72	24.00	72.00	36.92	24.34	75.00	2.18	0.00
November	28.6	27.0	7.2	2.20	24.00	70.00	68.16	24.04	100.0	2.20	0.10
December	29.5	28.0	8.0	3.00	18.00	48.00	63.00	21.50	81.00	2.00	0.06
January	29.0	26.0	8.5	6.53	12.00	42.00	48.13	25.00	21.00	1.86	0.06
February	–	–	–	–	–	–	–	–	–	–	–
Mean±SD	26.65± 2.38	24.57± 2.58	6.77± 0.94	3.84± 1.63	21.82± 4.73	61.52± 10.26	56.17± 12.34	23.41± 1.90	77.75± 23.65	2.00± 0.15	0.07± 0.74

Table 34.3: The Physico-chemical Characteristics of Water from Gudavi Pond

Month of Sampling 2000–01	*Air Temp. (°C)*	*Water Temp. (°C)*	*pH*	*DO (mg/l)*	*BOD (mg/l)*	*CO_2 (mg/l)*	*Cl (mg/l)*	*Ca (mg/l)*	*TA (mg/l)*	*PO_4 (mg/l)*	*SO_4 (mg/l)*
June	26.0	23.5	7.0	8.20	12.11	17.10	28.68	22.08	80.0	2.50	0.05
July	23.6	22.0	7.5	8.75	12.73	22.00	34.20	18.33	67.0	1.86	0.10
August	23.1	22.2	7.0	7.83	10.15	24.00	36.00	20.68	64.0	2.02	0.09
September	25.0	21.0	6.4	11.20	17.20	26.40	76.68	17.63	85.0	1.18	0.07
October	27.0	27.0	6.0	6.80	10.80	52.00	25.56	24.04	40.0	1.12	0.10
November	28.6	27.0	7.8	8.13	14.13	44.00	66.61	18.63	65.0	1.18	0.09
December	28.5	28.0	8.5	5.00	6.15	51.00	67.61	20.63	67.0	1.50	0.07
January	29.0	26.0	8.5	6.40	5.20	34.00	65.00	21.00	60.0	0.86	0.05
February	29.2	27.5	7.5	5.28	6.93	44.00	43.02	24.04	80.0	1.12	0.10
Mean±SD	26.66± 2.35	25.02± 2.67	7.35± 0.85	7.78± 1.71	10.6± 3.96	34.94± 13.19	49.26± 19.57	20.78± 2.33	67.55± 13.46	1.48± 0.53	0.08± 0.02

June 2000 to February 2001. Where as it was acidic during September and October in the other two ponds and it was 6.5 to 6.0 in Vaddekere and 6.4 and 6.0 in Gudavi ponds respectively. However, it remained alkaline in later part of the season. Since all enzymes are pH dependent and will directly enhance metabolic activities (Hosetti, 1987). Hence, it is essential to maintain alkaline pH or around neutral to support sustainable fish and algal productivity in these ponds.

The DO is a vital factor used in qualifying water samples. Most of the desired fishes may not survive if the DO falls below 4 mg/l. In the present study DO was always above 5 mg/l in Kallambi and Gudavi tanks. It was recorded an average 8.81 mg/l in Kallambi and 7.78 mg/l in Gudavi tanks, but in contrast to the two ponds, Vaddekere pond always showed low levels of oxygen except during January 2001. It was below 3 mg/l in September 2000 to December 2000 and a maximum of 6.5 mg/l in January 2001, when there was no bird activity. Such a drastic low levels of DO in Vaddekere during breeding seasons are attributed to biodegradation of guano dropped by birds (Dayananda, 2001).

The minimum DO which supports most of the desired fish species is above 4 mg/l. The reduced DO level in Vaddekere pond revealed that water is polluted by oxygen demanding materials (Hosetti and Patil, 1998; Hosetti and Kumar, 2001) added through fecal pollution. This might be attributed to the reason that in Vaddekere pond there were about 12,000 birds residing on emergent trees and discharged excreta. The uric acid and other organic matter which was present in excreta needed more oxygen to stabilize the waste biologically.

A humus material can be utilized by living organisms. But, once, it is discharged into an aquatic ecosystem, it will be consumed and oxidized by a variety of facultative organisms. The facultative organisms utilize the dissolved oxygen for the process of biodegradation. If all the oxygen is depleted then facultative aerobes also disappear. The quality of water will be further degraded, it will change into a saprobic system. Then water is considered to be polluted and evidenced by presence of oxygen demanding materials. Thus, the BOD is the amount of Oxygen required to oxidize biodegradable substance in an aquatic ecosystem under saprobic situations.

Biological oxygen demand is an indirect measure of organic load present in any aquatic systems. In the present study the 5 day BOD values 5.82, 21.82 and 10.6 mg/l recorded for Kallambi, Vaddekere and Gudavi ponds respectively. According to drinking water standard BOD should be less than 6 mg/l, but it was above the drinking water standard in Vaddekere and Gudavi ponds. The relatively high BOD levels in Vaddekere pond was due to the discharge of fecal matter by large number of birds. The high BOD in Gudavi tank attributed to two reasons. The water from Vaddekere entering into Gudavi tank gets diluted and hence low BOD values were recorded in Gudavi pond.

In tropical slow flowing polluted rivers and in stagnant water bodies a group of microbes dominated by algae and bacteria play a key role in process of purification in a symbiotic manner. That is bacteria degrade organic matter into simpler molecules of nutrients and release oxygen available back to the bacteria for use, ultimately leading to up gradation of water quality evidenced through reduced BOD and increased DO levels (Hosetti *et al.*, 1995). Another reasons are human activities like washing and bathing in Gudavi pond which are also responsible for high phosphates and BOD. The study on BOD level of the water sample of the three ponds revealed that, it is necessary to reduce the BOD level by artificial management, such as removing of sediment in summer and managing the water retention time at least in Vaddekere pond where birds reside during breeding seasons in large numbers.

The CO_2 level was 33.6, 61.5 and 34.9 mg/l recorded in the ponds, Kallambi, Vaddekere and Gudavi respectively. The higher amount of CO_2 was recorded and may affect the pH levels. Hence the water was acidic. The increase in the amount of CO_2 indicates that the ponds are inadequately aerated (Kumar, 1996; Singh, 1997). It also indicates that these ponds are filled with large amount of silt and sediments. The bacteria and fungi living in their sediments undertake decomposition of organic substances and also release CO_2 during the respiration processes.

Chlorides are gradually considered as nutrients. Presence of high levels of chlorides indicates that the water is polluted. The chlorine level was 54.2, 56.1 and 49.2 mg/l in the water samples from Kallambi, Vaddekere and Gudavi ponds respectively. In drinking water samples usually chlorine is around 20 mg/l. In contrast to this, the three ponds showed chloride three times more than that of drinking water standards. Such high levels of chloride perhaps originated from agricultural activity. Large amount of fertilizer and manures used in agriculture might have percolated in these ponds. The fecal discharge of birds is also responsible for increase in chlorides in Vaddekere (Hosetti *et al.*, 1995). According to Thresh *et al.* (1944) it is advocated that the presence of high chloride is an indication of pollution of animal origin. Chlorides are antibiotic in nature and the presence of large amount of chloride is lethal to pathogenic bacteria and chloride around 29 mg/l is considered to be favorable for freshwater community. However, chloride itself along with phosphate and nitrates makes the water eutrophic (Schmitz, 1996).

Calcium was 22.7, 23.4 and 20.7 mg/l in respective ponds. The levels of calcium in all these three ponds are comparable. Under high chloride levels the impact of calcium ions is yet to be studied. In freshwater ponds relatively high levels of calcium may be originated from the soil. Calcium is a micronutrient required for all organisms for metabolic activities (Gautam, 1995).

Phosphates are important macronutrients essential for plant growth. It was 0.37, 2.00 and 1.48 mg/l recorded in Kallambi, Vaddekere and Gudavi ponds respectively. Phosphate contents considered to be nutrients of major importance in the production process. It was within irrigation standards in both Kallambi and Gudavi ponds and it was very high in Vaddekere pond. Higher levels of phosphates are originated from the guano discharged by the birds. Due to this particular reason the Vaddekere pond became eutrophic (Kumar, 1996). The eutrophic nature of pond is evidenced by large growth of floating hydrophytes which include Lemna major, Lemna minor, Pistia, Salvinia, Trapa, *etc.* Due to availability plenty of phosphates and other nutrients the hydrophytes grown and occupied large surface area in the second and third ponds, the light penetration was checked and the algal photosynthesis also hampered (Hosetti and Latha, 1996) and low level of oxygen was recorded in such places as we found in Vaddekere pond samples. This change has also checked the free movements of the Ducks *viz.* Coot, Grey duck, Moorhen and Jacanas in these ponds.

The major sources of sulphate in natural water are rocks, fertilizers and waste discharges from industries. The amount of sulphate in water is a factor of concern in determining the magnitude of problems that can arise from reduction it into hydrogen sulphides. In the present study, the sulphate varied around 0.06, 0.07 and 0.08 mg/l recorded in Kallambi, Vaddekere and Gudavi ponds respectively. In this study sulphate recorded a highest concentration of 0.08 mg/l in Gudavi pond, which is well within the tolerance limit prescribed by ISI standards. Hence, it can be concluded that the ponds under study are oligotrophic (based on sulphate content), However the Vaddekere was always eutrophic with reference to other nutrients *viz.* phosphates, nitrates and chlorides.

The data on the evaluation of water quality revealed that due to guano droppings by large number of birds in small area of Vaddekere the water quality was deteriorated. Due to changes water quality in the Vaddekere the roosting trees viz, *Vitex leucoxylon*, *Kirganelia reticulate* and *Phyllanthus polyphyllus* are also affected. In order to avoid this there is need to plant more trees so that the birds aggregation can be spread and thereby pollution of water in Vaddekere may also be controlled. Now, it is good to exploit the water from the Vaddekere and the Gudavi for irrigation purpose for the crop land situated in the downstream regions.

References

Agarwal, A., Sharma, K. and Sharma, R., 1982. *State of India's Environment: A Citizens Report*. Centre for Science Environ., New Delhi.

APHA, 2000. *Standard Methods for the Examination of Water and Wastewater*, 16th edn. Washington D.C.

Dayananda, G.Y., 2001. Water quality evaluation with reference to Gudavi wetlands, Sorab, Shimoga, Karnataka. *M.Phil. Dissertation*, Kuvempu University, Shimoga.

Gautam, A., 1995. *Recent Researches in Aquatic Environment*. Daya Publishing House, Delhi.

Hosetti, B.B., 1987. Studies on use of some important groups of microbes in the treatment of wastewaters. *Ph.D. Thesis*, Karnataka University, Dharwad.

Hosetti, B.B. and Patil, H.S., 1998. Evaluation of catalase activity in relation to changes in physico-chemical parameters in a polluted river. In: *Threatened Habitats*, (Eds.), V.P. Agarwal and I.D. Chaturvedi. Society of Biosciences, India.

Hosetti, B.B. and Latha, P., 1996. Impact of plankton and weeds on distillary effluents meandering through natural ponds. *Advances in Biosciences*, 15: 63–70.

Hosetti, B.B. and Kumar, Arvind, 2001. *A Textbook of Applied Aquatic Biology*. Daya Publishing House, Delhi, India.

Hosetti, B.B., Shivaprasad, R.S. and Natraja, S., 1995. Limnology of a eutrophic pond with reference to physico-chemical factors at Navile, Shimoga. In: *Recent Researches in Aquatic Environment*, (Eds.) G. Asuthosh and N.K. Agrawal. Daya Publishing House, Delhi, pp. 99–105.

Hosmani, P.S., 2002. Ecological diversity of algae in freshwaters. In: *Wetlands Conservation and Management*, (Ed.) B.B. Hosetti. Pointer Publishers, Jaipur, pp. 65–87.

Kotpal, R.L. and Bali, M.P., 1989. *Concepts in Ecology*. Vishal Publishers, New Delhi, India.

Kumar, Arvind, 1997. Comparative hydrological studies of tropical waterbodies with special reference to sewage pollution in South Bihar. *J. Ecobiol.*, 9(4): 255–262.

Kumar, A., 1996. The limnological profiles of a tropical fish pond at Dumka (Bihar.). *J. Ecobiol.*, 8: 117–122.

Kumar, Arvind, 1997. Comparative hydrological studies of tropical waterbodies with special reference to sewage pollution in South Bihar. *J. Ecobiol.*, 9(4): 255–262.

Mahajan, K.K., 1981. Wetlands of India: A survey report and suggestions for their optimum utilization. In: *Proc. Int. Symp. Water Resources Conservation, Pollution and Abatement*, Roorkee, December, 11–14, pp. 371–378.

Mahajan, K.K., 1988. Dangerously threatened habitats: The wetlands–An overview and suggestions for their conservation as harbours of living natural resources. In: *Threatened Habitats*. Jagmandir Book Agency, New Delhi, pp. 241–268.

Saxena, S., 1996. Ecological studies of Gauri tank at with special reference to organic productivity *Ph.D. Thesis*, Jiwaji University, Gwalior.

Schimitz, R.J., 1996. *Introduction to Water Pollution Biology*. Asian Book Pvt. Ltd., New Delhi, India.

Singh., 1997. Limnological studies of River Mayurakshi in Santal Pargana Bihar. *Ph.D. Thesis*, S.K. University, Dumka.

Thresh, J.C., Suckling, E.V. and Deadle, J.F., 1944. *The Examination of Water and Wastewater Supplies*, (Ed.) Taylor.

Trivedi, R.K. and Goel, P.K., 1984. *Chemical and Biological Methods for Water Pollution Studies*. Env. Pub. Karad, India.

Water and Wastewater Management in 2 Vols (2011) *Pages* ***236–239***
Editors: **D.R. Khanna, R. Bhutiani & Gagan Matta**
Published by: **DAYA PUBLISHING HOUSE, NEW DELHI**

Chapter 35

Water Quality of River Ganga in Respect of Physico-chemical Characteristics at Kangri Village, District Haridwar

Unnati Vishnoi[1], Jagjit Kaur[2] and S.K. Pathak[3]
[1]Department of Chemistry, K.L.D.A.V. (P.G.) College, Roorkee, U.K.
[2]M.S. College, Saharanpur, U.P.
[3]Govt. Girls P.G. College, Mhow, M.P.

ABSTRACT

This paper deals with the water quality status of river Ganga at Kangri village with respect to physico-chemical characteristics. Parameters studied was pH, total solids, total dissolved solid, TSS, conductivity, alkalinity, hardness, DO, BOD, COD. With respect to pH the water of river Ganga was found alkaline.

Keywords: *Water quality, Physico-chemical, River, DO, BOD, COD.*

Introduction

Kangri village (District Haridwar) is situated at the east bank of river Ganga where number of people and cattle take bath daily. Kangri village is connected by Haridwar-Najibabad road. Ganga river passes through the Haridwar, Shyampur and Gendi Khata villages. Number of few workers have carried out their investigations on water quality of river Ganga with respect to different physico-chemical characterstics from its origin to merging point (Singh *et al.*, 1988, Khanna *et al.*, 2003). But no study has been conducted to assess the water quality of river Ganga at Kangri village. The present chapter includes one year study (Jan.–Dec. 2005).

Materials and Method

Two sampling stations were selected upstream Kangri village denoted as sampling station A and downstream Kangri village denoted as sampling station B situated 1 km in downstream of Kangri village.

Water samples of river Ganga were collected in neat and clean two litre white plastic Jercanes for physico-chemical parameters. Sample preservation and analytical methods were adopted as per APHA, AWWA, WPCF-(1992), Khanna and Bhutiani (2004). All chemicals, reagents and solvents were used for the analysis of analytical (AR) grade. Fine chemicals and glass wares were used of Borosil made.

Results and Discussion

The physico-chemical parameters are given in Tables 35.1–35.3. The values of studied parameters pH, TS, TSS, alkalinity, total hardness, chloride, free CO_2, conductivity, water temperature varied between 7.00-8.50, 119.00-229.00 mg/l, 28.90-51.20 mg/l, 52.00-92.00 mg/l, 64.00-98.00 mg/l, 6.20-27.20 mg/l, 1.40-3.60 mg/l, 0.10-0.58 µmho/cm, 11.2-21.5 °C, at sampling station-A. Where at sampling station-B the values of these parameters ranged from 6.80-8.20, 130.00-262.00 mg/l, 30.00-79.00 mg/l, 58.00-102.00 mg/l, 60.00-98.00 mg/l, 6.90-58.10 mg/l, 1.20-3.90 mg/l, 0.12-0.65 µmho/cm, 12.4-20.5 °C respectively and of DO, COD, BOD ranged between 9.0-11.5, 10.0-29.0, 0.4-1.5 mg/l at sampling station-A whereas at sampling station-B the values of these parameters ranged from 8.10-11.00, 9.00-27.20, 0.60-3.00 mg/l respectively. The average value of pH, TS, TDS, TSS, conductivity, DO, COD, alkalinity, total hardness and chloride were obtained within the tolerance limit of drinking purpose whereas average values of BOD is 0.94 mg/l at sampling station-A and 2.0 mg/l at sampling station-B (Table 35.3).

Table 35.1: Variation in Physico-chemical Characteristics of River Ganga at Sampling Station-A of Kangri Vllage (Haridwar), during Jan. to Dec., 2005

Parameters	*Month*											
	Jan	*Feb*	*Mar*	*Apr*	*May*	*Jun*	*Jul*	*Aug*	*Sept*	*Oct*	*Nov*	*Dec*
Water temp. (°C)	11.20	13.00	18.40	18.60	20.30	21.50	18.00	17.70	17.60	17.30	19.00	13.80
TS (mg/l)	122.00	119.00	128.00	157.00	174.10	210.00	221.20	229.00	220.00	221.90	200.00	185.00
TDS (mg/l)	92.00	88.00	89.00	118.00	125.00	158.00	169.00	178.00	175.00	180.00	157.00	145.00
TSS (mg/l)	29.00	28.90	35.00	37.00	48.00	49.00	51.20	51.00	42.00	40.50	38.00	33.00
Conductivity	0.10	0.14	0.19	0.18	0.20	0.28	0.58	0.43	0.45	0.39	0.20	0.21
pH	7.00	7.30	7.20	7.80	7.80	8.00	8.50	8.10	8.40	7.90	7.00	7.50
DO (mg/l)	11.50	10.80	10.00	10.10	10.20	9.00	9.10	9.30	9.50	10.00	10.10	11.00
Free CO_2 (mg/l)	1.40	1.90	2.00	2.20	2.40	2.60	3.60	3.00	2.60	2.40	2.00	1.80
COD (mg/l)	10.00	11.20	11.5	12.00	16.00	20.00	24.00	29.00	23.00	12.20	19.00	14.50
BOD (mg/l)	0.50	0.80	1.10	0.60	0.90	1.40	1.50	1.20	1.20	1.00	0.70	0.40
Alkalinity (mg/l)	52.00	56.00	59.00	61.00	70.00	72.00	74.00	79.00	71.00	92.00	88.00	74.00
Total Hardness (mg/l)	76.00	64.00	69.00	70.00	74.00	84.00	98.00	75.00	80.00	81.00	80.00	73.00
Chloride (mg/l)	6.20	6.90	7.20	11.50	17.00	19.00	27.20	23.00	17.60	9.80	7.20	7.00

Table 35.2: Variation in Physico-chemical Characteristics of River Ganga at Sampling Station-B of Kangri Vllage (Haridwar), during Jan. to Dec., 2005

Parameters	*Month*											
	Jan	*Feb*	*Mar*	*Apr*	*May*	*Jun*	*Jul*	*Aug*	*Sept*	*Oct*	*Nov*	*Dec*
Water temp. (°C)	12.40	12.60	18.50	19.00	20.00	20.50	20.00	18.60	17.90	19.30	18.10	15.00
TS (mg/l)	130.00	140.00	144.00	170.00	183.00	229.00	259.00	262.00	253.00	230.00	223.00	195.00
TDS (mg/l)	100.00	104.00	92.00	133.00	128.00	173.00	189.00	181.00	179.00	185.00	160.00	135.00
TSS (mg/l)	30.00	33.00	48.00	35.00	54.00	49.00	69.00	79.00	72.00	46.00	63.00	61.00
Conductivity	0.13	0.19	0.20	0.12	0.14	0.34	0.65	0.58	0.59	0.58	0.51	0.37
pH	6.80	6.90	7.00	7.10	7.20	7.30	7.50	7.80	7.50	7.60	8.20	7.60
DO (mg/l)	10.00	9.10	8.10	9.50	8.90	8.40	9.50	10.00	10.10	9.70	10.00	11.00
Free CO_2 (mg/l)	1.20	2.80	2.00	2.90	2.30	3.50	3.90	2.60	207.00	3.30	209.00	2.60
COD (mg/l)	9.00	10.20	10.50	11.00	14.50	19.20	22.50	27.20	22.40	10.20	16.50	13.50
BOD (mg/l)	2.20	0.60	106.00	2.00	2.50	2.90	3.00	1.50	106.00	1.50	2.00	2.60
Alkalinity (mg/l)	58.00	65.00	59.00	78.00	95.00	100.00	102.00	74.00	79.00	94.00	95.00	80.00
Total Hardness (mg/l)	60.00	64.00	76.00	79.00	90.00	96.00	98.00	79.00	93.00	89.00	97.00	89.00
Chloride (mg/l)	6.90	7.50	9.00	12.00	21.00	33.00	58.10	24.50	20.00	16.00	14.00	12.00

Table 35.3: Average Value of Physico-chemical Parameters at Sampling Stations-A and B at Kangri Village during Jan. to Dec., 2005

Parameters	*Jan–Dec 2005*			*Jan–Dec 2005*		
	Sampling Station A			*Sampling Station B*		
	Min.	*Max.*	*Average*	*Min.*	*Max.*	*Average*
Water temp.	11.20	21.50	17.20	12.40	20.50	31.10
TS	119.00	220.90	182.10	130.00	262.00	201.50
TDS	88.00	180.00	139.50	92.00	189.00	136.90
TSS	29.00	51.20	40.20	30.00	79.00	53.30
Conductivity	0.10	0.58	0.28	0.12	0.65	0.37
pH	7.00	8.50	7.70	6.80	8.20	7.40
DO	9.00	11.50	10.1	8.10	11.00	8.80
Free CO_2	1.40	3.60	2.30	1.20	3.90	2.70
COD	10.00	29.00	15.10	9.00	27.20	13.90
BOD	0.40	1.50	0.94	0.60	3.00	2.00
Alkalinity	52.00	92.00	70.70	58.00	102.00	81.60
Total Hardness	64.00	98.00	77.00	60.00	98.00	84.20
Chloride	6.20	27.20	13.30	6.90	58.10	19.50

Maximum values of the studied parameters were observed during monsoon season except DO and minimum values of these parameters obtained during winter, lower values of DO were observed during rainy period at both sampling station which may be due to dilution in rainy season and super saturation of oxygen at lower temperature in winter. After winter period as temperature of water rises than the free CO_2, COD, BOD was increased and DO decreased of river Ganga, Khanna *et al.*, 2003 had also found the similar trend. Water quality of river Ganga with respect to pH is alkaline similar trends was also found by Singh, 1988.

References

APHA, AWWA, WPCF, 1992. *Standard Methods for the Examination of Water and Wastewater*, 18th edn. American Public Health Association, Washington DC.

Khanna, D.R., Singh, Shakun, Gautam, Ashutosh and Singh, J.P., 2003. Assessment of water quality of river Ganga in district Bulandshahar (UP) India. *J. Nat. Con.*, 15(1): 165-167.

Khanna, D.R. and Bhutiani, R., 2004. *Water Analysis at a Glance*. Pub. ASEA, pp. 1–116.

Khanna, D.R. and Chugh, Tarun, 2004. *Microbial Ecology*. Discovery Publishing House, Delhi, pp. 1–277.

Singh, J.P., Yadav, P.K. and Singh, L., 1988. Pollution status of Sangam and its adjoining rivers before the Kumbh Mela at Allahabad. *Indian J. Environment Protection*, 8(11): 839–842.

Water and Wastewater Management in 2 Vols (2011) *Pages* **240–243**
Editors: **D.R. Khanna, R. Bhutiani & Gagan Matta**
Published by: **DAYA PUBLISHING HOUSE, NEW DELHI**

Chapter 36

Hydrobiological Studies in the Upstream of River Kunda at Khargone, Madhya Pradesh

Bharti Khare
Botany Department, Govt. P.G.College, Barwani

ABSTRACT

The present paper deals with the hydro-biological observation on the upstream water of river Kunda at Khargone, M.P. The data revealed that 4 major groups of algae *i.e.*, Chlorophyceae, Bacillariophyceae, Cyanophyceae and Euglenophyceae are present in the upstream of Kunda river.

Keywords: *Phycology, Freshwater biology, Algal taxonomy, Limnology, Water analysis.*

Introduction

Algae plays an important role in all types of water bodies. Polluted water harbour characteristic types of algal taxa. Algae in general and blue green algae in particular cause additional problems in water supplying. Kunda river, a tributory of Narmada is sufficiently rich in various types of algae. Earlier works on the algal flora of M.P. by Desikachary and Mall (1955), Agarkar and Agarkar (1962), and others reveal that our knowledge regarding the algal flora of M.P. is scanty. Seerwani (1963) and Mahajan (1987 and 1991) have published some research papers on the algae of Khargone. Recently physico-chemical and biological characterization of the river Kunda at downstream of Khargone has been reported (Mahajan *et al.*, 2002 and Mahajan, 2005) but till now no such type of work has been done on the upstream of this river. Therefore the present investigation has been taken up.

Materials and Method

Water samples were collected during 2005-06 from the Dargah site of the river Kunda near water filteration and purification plant of Khargone city. This place is only 3 km away from the city. Algal taxa were identified after consulting the standard literature and monographs (Desikachary, 1959; Randhawa, 1959; Prescott, 1964 and Phillipose, 1967) and every attempt has been made to bring the nomenclature up-todate. Enumeration of algal members of different classes is shown in Table 36.1. Physico-chemical characteristics of water samples were determined following Golterman and Clymo, 1969; Trivedi and Goel (1986) and APHA (1989) and the data are shown in Table 36.2.

Table 36.1: Algal Taxa Reported in the Upstream Water of River Kunda at Khargone, M.P.

Chlorophyceae	
1.	*Chlorella* sp.
2.	*Cosmarium* sp.
3.	*Mougeotia abnormis*
4.	*Scenedesmus armatus*
5.	*S. brasiliensis*
6.	*Schizomeris* sp.
7.	*Spirogyra condensata*
8.	*S. sinensis*
9.	*Staurastrum* sp.
10.	*Zygnema himalayansis*
Bacillariophyceae	
1.	*Fragillaria rumpens*
2.	*G. sphaeroperum*
3.	*Navicula cuspidata*
4.	*N. pupula*
5.	*Nitzschia*
Cyanophyceae	
1.	*Anacystis nidulens*
2.	*Gomphosphaeria* sp.
3.	*Oscillatoria princeps*
4.	*O. sancta*
Euglenophyceae	
1.	*Euglena*
2.	*Phacus*

Results and Discussion

It is revealed from Table 36.1 that altogether 21 algal taxa are reported, out of which 10 taxa belong to Chlorophyceae, 5 to Bacillariophyceae, 4 to Cyanophyceae and 2 to Euglenophyceae. Important members reported are *Zygnema* and *Scenedesmus*. Occurrence of Anacystis and *Schizomeris* is significant because these taxa are new records for West Nimar district. Fertile stages of *Zygnema* and

Spirogyra were also reported. Apart from this, 3 members of zooplankton are also reported *i.e., Daphnia, Cyclops* and *Rhabditis.* From Table 36.2 the water analysis shows that pH of water is alkaline (7.7), DO is maximum in winter months while minimum during monsoon months. The values of Conductivity, Total alkalinity, Hardness, Magnesium, Calcium and Chlorides are maximum in summer while minimum in winter months.

Table 36.2: Physico-chemical Parameters of Upstream Water of River Kunda at Khargone, M.P.

Parameters	*Upstream*
Physical	
Air temperature (°C)	28.4
Water temperature (°C)	23.5
Turbidity (NTU)	17.5
Conductivity (μmhos)	125
pH	7.9
Chemical	
Dissolved Oxygen (mg/l)	15
BOD (mg/l)	1.5
Total Solids (mg/l)	51
Total Hardness (mg/l)	180
Chlorides (mg/l)	128
Calcium (mg/l)	26.05
Magnesium (mg/l)	7.62
Total alkalinity (mg/l)	85

Acknowledgements

The author is grateful to Dr. S.L. Yadav, Principal, Govt. P.G. College, Barwani for laboratory facilities and encouragements and Dr. S.K.Mahajan, Ex-Professor of Botany, Govt. P.G.College, Khargone for going through the manuscript and helpful suggestions.

References

Agarkar, D.S. and Agarkar, M.S., 1972. Contributions to demids of M.P. (India), Port. *Acta Biol.*, 12 : 159–178.

APHA (American Public Health Association), 1989. *Standard Methods for Examination of Water and Wastewater*, 17th edn. Washington DC, USA, 1452 pp.

Desikachary, T.V. and Mall, L.P., 1955. On the occurrence of *Lyengariella tirupatiensis* Des. *Curr. Sci.*, 24: 245–246 (from Pachmarhi).

Desikachary, T.V., 1959. *Cyanophyta*. ICAR, NewDelhi, p. 700.

Golterman, H.L. and Clymo, R.S., 1969. *Methods for the Chemical Analysis of Freshwater*. B.P. Handbook No. 8, Blackwell Sci. Pub., Oxford, 166 pp.

Mahajan, S.K., 1987. Algal flora of Pawagiri Oon, Madhya Pradesh. *Phykos*, 26(1 and 2): 61–62.

Mahajan, S.K., 1991. Algal flora of Choolgiri Hills, Madhya Pradesh. *Phykos*, 30(1 and 2): 87–89.

Mahajan, S.K., Garde, Yogesh, Bhawsar, Rakesh and Sharma, Sharad, 2002. Physico-chemical and biological characterization of the river Kunda at downstream of Khargone (Madhya Pradesh), India. *Environment Conservation Journal*, 3(2): 53–55.

Mahajan, S.K., 2005. Algal flora of a recently constructed dam on river Dalki in West Nimar district of M.P. *Indian Hydrobiology*, 8(2): 113–116.

Philipose, M.T., *1967.Chlorococcales.* ICAR, New Delhi.

Prescott, G.W., 1964. *The Freshwater Algae.* WMC Brown Co., Publ., Dubuque, pp. 258.

Randhawa, M.S., 1959. *Zygnemaceae.* ICAR, New Delhi.

Seerwani, A.B., 1963. Occurrence of *Enteromorpha* at Khargone, Madhya Pradesh. *Curr. Sci.*, 32: 182.

Trivedy, R.K. and Goel, P.K., 1986. *Chemical and Biological Methods for Water Pollution Studies*, 2nd edn. Environmental Publishers, Karad, M.S., p. 250.

Water and Wastewater Management in 2 Vols (2011) *Pages* **244–247**
Editors: **D.R. Khanna, R. Bhutiani & Gagan Matta**
Published by: **DAYA PUBLISHING HOUSE, NEW DELHI**

Chapter 37

Abiotic Status of Song River and its Relations to Zoo and Phytoplankton at Nepali Farm, Dehradun

Unnati Vishnoi[1], Kumar S. Chandra[2] and Gagan Matta[3]
[1]*Department of Chemistry, K.L.D.A.V. (P.G.) College, Roorkee*
[2]*Department of Zoology and Environmental Science, Gurukula Kangari Vishwavidyalya, Haridwar*

ABSTRACT

Water samples were collected from Song River during pre-autumn to post-autumn 2006. This study indicates the direct relationship among various physico-chemical parameters like water temperature, free carbon dioxide, transparency and DO, BOD, COD, hardness and phosphate and inverse relation was found between, DO and BOD, DO and COD during the course of study.

Keywords: *Abiotic, Zooplankton, Phytoplankton, Biological.*

Introduction

In the lotic system, the significance of physico-chemical and biological data for the assessment of water quality has been recognized. The studies consider water quality usually involve physico-chemical and biological variables.

The Song River is a tributary of River Ganga. It is a spring fed River. It originates from near Dehradun. The river ultimately mixed in to Ganga near Raiwala. The present investigation was carried out the study of the abiotic characteristics.

Number of workers have carried out their work on aquatic quality of different Rivers. Khanna (1993), Khanna (2003) of River Ganga and Badola and Singh (1982), studied about the abiotic and biotic factors of River Alaknanda. Whereas the study about the Song River is scanty therefore the study will be helpful for other workers.

Materials and Method

Water samples were collected fortnightly during paddy in different season (pre-autumn, post-autumn) in plastic containers and analyzed for abiotic characteristics by standard methods (APHA, 1998). Plankton were collected through a filtering the water through a planktonic net and identified by with the help of Jhingran (1975), Khanna (1993).

Results and Discussion

The physico-chemical parameter and planktonic fluctuation obtained during the study period are tabulated in Tables 37.1–37.3. The temperature showed a negative relationship with the dissolved oxygen. Water is often attributed to the fact that the oxygen is dissolved more during the period of active photosynthesis. Hutchinson (1957) concluded that cold water has greater capacity for holding dissolved gases. The maximum dissolved oxygen was recorded (9.13 mg/l±0.12) in pre-autumn and minimum (8.68 mg/l±0.30) in post-autumn. It was also reported by Badola and Singh (1982). The maximum total solids was observed (210 mg/l pre-autumn 10) in post-autumn and minimum(155 mg/l 5) in pre-autumn. The value of the total solids is incresed from pre-autumn post-autumn. as reported by David (1956) in Bhadra River (Mysore).

The maximum value of pH was observed (8.33 mg/l ± 0.16) in pre-autumn and minimum (7.91 mg/l ±0.01) in post-autumn. Maximum value of pH in winter might to be due to increase in algal population in the river. Khanna (1993) found in the river Ganga at Haridwar. pH and DO showed a positive relationship to one another as also found by Ali *et al.* (1988) in a eutrophic lake.

Table 37.1: Seasonal Variations in Physico-chemical Parameters of Song River

Parameters	*Seasons*			
	Pre-Autumn	*Autumn*	*Post-Autumn*	*Average*
Temperature (°C)	19.0±0.57	19.9±0.16	22.0±0.50	20.3±1.70
Total solid (mg/l)	155.0±5.0	185.0±15.0	210.0±10.0	183.3±22.4
pH	8.33±0.16	8.10±0.10	7.91±0.01	8.11±0.12
Free CO_2 (mg/l)	1.33±0.16	1.39±0.07	1.50±0.14	1.40±0.07
Dissolved Oxygen (mg/l)	9.13±0.12	8.72±0.22	8.80±0.30	8.88±0.30
Alkalinity (mg/l)	168.33±0.66	142.50±8.30	140.81±9.16	150.55±12.58
Chloride (ppm)	16.68±0.12	17.33±1.41	17.80±1.47	17.27±0.45
Sulphate (mg/l)	1.07±0.01	1.26±0.02	1.29±0.02	1.20±0.09
BOD (mg/l)	2.15±0.35	2.13±0.25	3.33±0.37	2.53±0.56
COD (mg/l)	2.15±0.17	2.54±0.27	3.45±0.05	2.71±0.54

The minimum value of free carbon dioxide was found (1.33 mg/l ± 0.16) in pre-autumn and maximum(1.50 mg/l ± 0.14) in post-autumn. Free carbon di oxide and DO showed a negative relationship to one another.

Table 37.2: Seasonal Quantitative Analysis of the Plankton of the River Song

Seasons	*Plankton*		
	Phytoplankton (µ/l)	*Zooplankton (µ/l)*	*Total Plankton (µ/l)*
Pre-autumn	929.5±61.50	98.0±6.32	1027.5±96.17
Autumn	832.5±53.5	83.5±2.5	916.0±56.0
Post-autumn	719.5±27.5	67.5±2.5	787.0±19.7
Average	827.1±85.81	83.0±12.45	910.1±85.81

Table 37.3: Number of Different Groups among Phytoplankton of the River Song during Different Seasons

Seasons	*Phytoplankton*		
	Bacillariophyceae (µ/l)	*Chlorophyceae (µ/l)*	*Myxophyceae (µ/l)*
Pre-autumn	598.0±42	226.5±18.5	105.0±1.0
Autumn	528.0±42	204.5±10.5	100.0±1.0
Post-autumn	437.5±21.5	186.0±4.0	96.0±1.4
Average	521.16±6.80	205.66±16.55	100.33±6.37

The maximum value of alkalinity was (168.33 mg/l ± 6.66) in pre-autumn and minimum (140.83 g/lm ± 9.16) in post-autumn. Increasing alkalinity was decomposed the organic matter. It was also observed by Venkateswarlu and Jayanti (1968) in the river Sabarmati.

The chloride value was (16.68 mg/l ± 0.12) in pre-autumn which are lowest and highest in post-autumn (17.80 mg/l ± 1.47). It show the positive relationship with temperature (Mohanty, 1981 in Bhubaneswar).

The sulphate value was minimum (1.07 mg/l ± 0.07) in pre-autumn and maximum (1.20 mg/l ± 0.02) in post-autumn (Singh, 1988). The maximum value of biochemical oxygen demand was recorded (3.33 mg/l ± 0.37) in post-autumn and minimum (2.15 mg/l ± 0.35) in pre-autumn. Biochemical oxygen demand and DO show negative relationship which is according to Verma *et al.* (1984) in eastern Kalinadi.

The COD was highest (3.45 mg/l ± 0.5) in post-autumn and lowest (2.15 mg/l ± 0.17) in pre-autumn. Biochemical oxygen demand and COD showed a positive relationship with one another like Chopra and Patrick (1994) in the river Ganga at Rishikesh. COD showed a negative relationship with DO as Verma *et al.*(1984) in eastern Kalinadi. The total planktonic concentration was maximum (1027.50 µ/l± 96.77) in pre-autumn and minimum (787.0 µ/l± 19.76) in post-autumn. There is a inverse co-relation in the temperature and the phytoplankton and zooplankton production. When the increase in temperature causes decreased in plankton production. Badola and Singh (1982) reported high value of plankton during pre-autumn in the river Ganga of Garhwal Himalaya.

References

Ali, A., Reddy, K.R. and Debusk, W.F., 1988. Seasonal changes in sediment and water chemistry of a subtropical shallow eutrophic lake. *Hydrobiologia*, 159: 159–167.

APHA, 1998. *Standard Methods for the Examination of Water and Wastewater*, 20th edn. American Public Health Association, 1015, 15th street, New Washington, 15: 1–1134.

Badola, S.P. and Singh H.R., 1982. Hydrobiology of river Alaknanda of the Garhwal Himalayas. *Indian Eco.*, 8: 269–276.

Chopra, A.K. and Patric, Nirmal I 1994. Effect of domestic sewage on self-purification of Ganga Water at Rishikesh. I. Physico-chemical parameters. *Ad. Bios.*, 13(11): 75–82.

David, A., 1956. Studies on the pollution of the Bhadra river at Bhadrawati Fisheries effluents. *Proc. Nat. Inst. Set. India*, 93(3): 132–160.

Hutchinson, G.C., 1957. *A Treatise on Limnology, Geography, Physics and Chemistry, Vol. 1.* John Wiley and Sons, Inc., New York, p. 1–1015.

Jhingran, V.G., 1975. *Fish and Fisheries of India*. Hindustan Publishing Corporation, India, p. 1–954.

Khanna, D.R., 1993. *Ecology and Pollution of Ganga River*. Ashish Publishing House, Delhi, p. 1–241.

Mohanty, R.C., 1981. Water quality studies of some water bodies of Bhubaneswar. *Ph.D. Thesis*, Utkal University, pp. 1–240.

Singh, H.R., 1988. Pollution studies on upper Ganga and its tributaries at upper Himalaya. *Project Report.*

Venkateswartu, T. and Jayanti, T.V.C., 1968. Hydrobiological studies of the river Sabarmati to evaluate water quality. *Hydrobiologia*, 31: 442–448.

Verma, S.R., Sharma, P., Tyagi, A., Rani, S., Gupta, A.K. and Daleta, R.G., 1984. Pollution and saprobic status of eastern Kalinadi. *Limnologica* (Berlin), 15(1): 69–133.

Water and Wastewater Management in 2 Vols (2011) *Pages 248–261*
Editors: D.R. Khanna, R. Bhutiani & Gagan Matta
Published by: DAYA PUBLISHING HOUSE, NEW DELHI

Chapter 38

Impact of Sugar Industries of Kushinagar District, U.P. on Water Quality: Using Remote Sensing and Geographical Information System (GIS)

Gyanendra Awasthi[1], Vinay Kumar[1], Harish Chandra[2], Ajay Singh[3] and Arvind Mathur[4]
[1]Department of Biochemistry, [2]Department of Microbiology, [3]Department of Chemistry, Dolphin (P.G.) Institute of Bio-Medical and Natural Sciences, Dehradun
[4]Remote Sensing Application Centre, Jankipuram, Lucknow

ABSTRACT

Quality of water is an important aspect for drinking and other related requirement. Water has been the most exploited natural system due to ever increasing demand of man for food, cloth, industrialization; water obtained from open dung wells and hand-pumps is used for domestic purpose. In the present study effect of sugar effluent on water quality at given region in two seasonal conditions pre-rainy season and post-rainy season and the sampled data was analyzed separately for pH, Total Dissolved Solid (TDS), Nitrate, Phosphate, Dissolved Oxygen (DO), Biochemical Oxygen Demand (BOD) and Chemical Oxygen Demand (COD) were estimated and followed by G.I.S. to produce respective maps. It was found that the pH value was alkaline in pre-rainy season and slightly acidic in post-rainy season. TDS was found high in 66 per cent area in pre-rainy season and 74.6 per cent in post-rainy season above permissible limit. In pre-rainy season 83.4 per cent area and in post-rainy season 98.4 per cent area contains high electrical conductivity. In pre-rainy season BOD was found more than 6mg/dl in 80.9 per cent in pre-rainy season and in 84.2 per cent area in post-rainy season COD was found above prescribed limit (>10mg/dl) in 59.5 per cent area in pre-rainy season and in 93.6 per cent area in post-rainy season 33.3 per cent area contain high phosphate in pre-rainy season 46.03 per cent area contains high phosphate in post-rainy season. Nitrate was found in permissible limit in pre-rainy season

46.6 per cent area was found to contain nitrate in post-rainy season 7.1 per cent of the geographical area is having fluoride concentration more than 1.5mg/dl and 28.5 per cent area is having fluoride concentration below the normal limit. Thus, 35.6 per cent of total area is not suitable for drinking purpose 6.34 per cent area contain fluoride above prescribed limit.

Keywords: *Dissolved oxygen, BOD, COD, GIS.*

Introduction

Groundwater is important source of freshwater for agriculture, drinking domestic uses in many regions of the world and also in India. Demand of groundwater has been increasing day by day for irrigation by bringing area under cultivation. Water generally contains different species of cations and anions in varying amounts. The concentration of dissolved constituents in water is an important determinant for its quality (Islam, *et al.*, 2003). Rapid growth of population, urbanization, coupled with industrialization pose serious concern to the vulnerability of water resources owing to regional imbalance of chemical parameters and eventual safe use of groundwater for drinking purpose (Mathur and Priyakant, 2005).

In general most of sugar industries don't process the effluent for removal of organic and inorganic pollutants and release the entire effluent outside the premises which eventually spreads throughout the low lying area adjacent to the sugar factory or discharges into natural course. This in turn can migrate and mix with the shallow water aquifer system through seepage and therefore can cause detrimental effect on groundwater quality. Total estimated amount of production of sugar industries in India is $7.5x10^6$ tonnes/year. Sugar mill consume 1500-2000 liters of water and generate about 1000 liters of wastewater ton of crane crushed. The effluent is mainly floor washing wastewater and condensate water, leakage in valves and glands of pipeline add sugarcane juice syrup, molasses in the effluent. The sugarcane effluent has B.O.D. 1000–1500mg/l, but appears relatively clean initially, however after stagnating for some time it turns black and starts emitting foul odor if untreated effluent is discharged in water courses it depletes dissolved oxygen in water and make environment unfit for aquatic life. If untreated effluent is discharged on land decaying organic solids, oil and grease clog the soil pores. Most of the used bagasse as sugar mill fuel in boilers produces particulate matter, oxides of nitrogen, carbon,sulphur and water vapors. These causes increase in nitrate, phosphate and sulphate content (Central Water Pollution Control Board, Annual report, 1996-1997).

The district Kushinagar lies between latitudes 26° 50' and 27° 0' N and longitude 83° 40' and 84° 0' E and fall in the survey of India toposheet No. 63N/9 and 63 N/13. Topographically the area exhibits a flat terrain with a gentle slop from north to south. It is situated in the doab of river (Gandak, Rapti and Ghagra). It comprises three tehsils of *viz.*, Padrauna, Ramkola and Kaptanganj. The main surface water source is Banari nala, which traverses the entire area in a zigzag way flowing in North West direction. Western Gandak network canal is the main irrigation canal system. The other canals of the area include Kurmai minor, Barhara distributory, Padrauna minor, Khasai Branch and Sepaha distributory. A Number of lakes namely Kamaini Taal, and Balkariyal Taal are also located in the area climate of this area is moderate as compared to western part of state. The maximum temperature in May and June seldom exceed 41°C and generally it remains around 36°C during winter, the average minimum temperature for December and January drops to about 10°C. Average annual rainfall in the area is about 1145 mm. Bulk of the rainfall occurs during June to September due to South-west monsoon.

The area under reference is a fertile alluvial plain with gentle slopes towards South-east. Major amount of sugar industries located in area of Padrauna and Ramkola Tehsils.

Materials and Method

Water Analysis

Water samples were collected from 3 areas Kaptanganj, Ramkola, Padrauna for period of pre-rainy season [April to May] and for period of post-rainy season [Oct to Nov]. Samples are collected from different point and non point sources *i.e.* River, tube wells, canals, hand pumps *etc.* From each sampling site two samples were collected in chemical resistant pre cleaned plastic bottles, which were thoroughly washed twice with the some water required to be analyzed. The concentration of total dissolved solids was measured on the spot itself with the help of portable electric conductivity meter. TDS was determined by gravimetric technique or method. Dissolved Oxygen, Biological Oxygen Demand, Chemical Oxygen Demand measured by Winkler's-Iodometric Titration method, 5-days BOD test, closed reflex colorimetric method respectively. Fluoride, Nitrate, Phosphate measured by UV-Vis. spectrophotometric method (APHA, 1992; Abbasi, 1998). The entire standard used for calibration was ISO 9001 certified reference material procured.

Result and Discussion

The monitoring parameters are measured on the basis of pre-rainy season and post-rainy season effect on groundwater and surface water. The analytical result of 42 water samples in pre-rainy season and 63 water samples in post-rainy season, collected from ISI land pumps, river, tube wells, canals, drains *etc.* All the eight parameters *i.e.*, pH, TDS, Nitrate, phosphate, D.O., B.O.D., C.O.D., fluoride in ground and surface water were compared with the permissible/tolerable limit laid down by B.I.S. [Bureau of Indian Standards I.S.:10500,1993] for drinking purpose.

A pH range of 6.5-8.5 is normally acceptable as per guide line suggested by BIS and WHO (BIS, 1991; WHO, 1993) pH values at Kaptanganj, Padrauna and Ramkola varies from 6.83-10.27, 7.2-8.12 and 7.14-8.71 indicating alkaline nature of water at all stations. The pH range observed at, Padrauna and Ramkola were well among prescribed range and at Kaptanganj were on alkaline side. But in post-rainy season the pH value of these stations Kaptanganj, Padrauna and Ramkola ranges from 7.24-8.51, 6.8-8.2 and 4.4-8.5.

The electrical conductivity limit is 300 µS/c.m., normally acceptable as per guideline suggested by BIS and WHO (BIS; 1991, WHO, 1993).The conductivity value at Kaptanganj, Padrauna and Ramkola ranges from 189-1324.8, 484-1901and 690-1885 µS/c.m.respectively in pre-rainy season. High conductivity values are observed in low flow periods at the above three stations. It is due to industrial waste disposal at water bodies and dissolved solids from water shed. Figure 38.1 shows spatial and percentage distribution of electrical conductivity in district. It clearly shows 83.4 per cent of electrical conductivity in excess of prescribed limit only 16.6 per cent suitable with electrical conductivity for drinking purpose. Electrical conductivity at Kushinagar district increases during post-rainy season. High conductivity was observed in high flow periods at above three stations. The conductivity value at Kaptanganj, Padrauna and Ramkola ranges from 155-1480,600-2030 and 541-2000 µS/c.m.respectively (Table 38.1).

GIS analysis (Figure 38.1) revealed that 1.6 per cent area of district Kushinagar contains less than 300 µS/c.m and safe for drinking purpose. Total dissolved solids (TDS) indicate general water quality or salinity and is usually related to conductivity. Water containing more than 500 mg/dl of TDS is

Table 38.1: Comparative Data of Various Parameters in Kushinagar District during Pre-Rainy and Post-Rainy Season

Season	Site	Parameters								
		TDS (mg/dl)	pH	Electrical Conductivity (μS/cm)	DO (mg/dl)	BOD (mg/dl)	COD (mg/dl)	NO_3^- (mg/dl)	PO_4^{3-} (mg/dl)	Fl (mg/dl)
Pre rainy	Kaptanganj	122-8611	6.83-10.27	189-1324.8	5.2-7.8	2.0-55.0	7.2-82.0	0.09-3.95	0.01-0.63	0.35-1.50
	Padrauna	305-1157	7.2-8.12	484-1901	5.3-8.2	4.0-34.0	5.8-66.5	0.09-8.78	0.01-2.85	0.17-0.96
	Ramkola	414-6329	7.14-8.71	690-1885	4.9-6.3	2.8-20.8	5.2-24.6	0.29-1.4	0.01-0.66	0.34-1.56
Post rainy	Kaptanganj	88.35-8379	7.24-8.51	155-1480	5.4-7.6	4.2-10.4	8.42-22.4	4.8-48	0.0-0.98	0.37-1.78
	Padrauna	342-6441	6.8-8.2	600-2030	0.0-6.4	8.2-88	19.42-52.2	11.09-84.79	0.02-0.91	0.41-1.03
	Ramkola	308.37-1140	4.4-8.5	541-2000	5.1-7.0	3.0-14.6	7.42-28.9	3.38-72.96	0.0-0.045	0.30-1.57

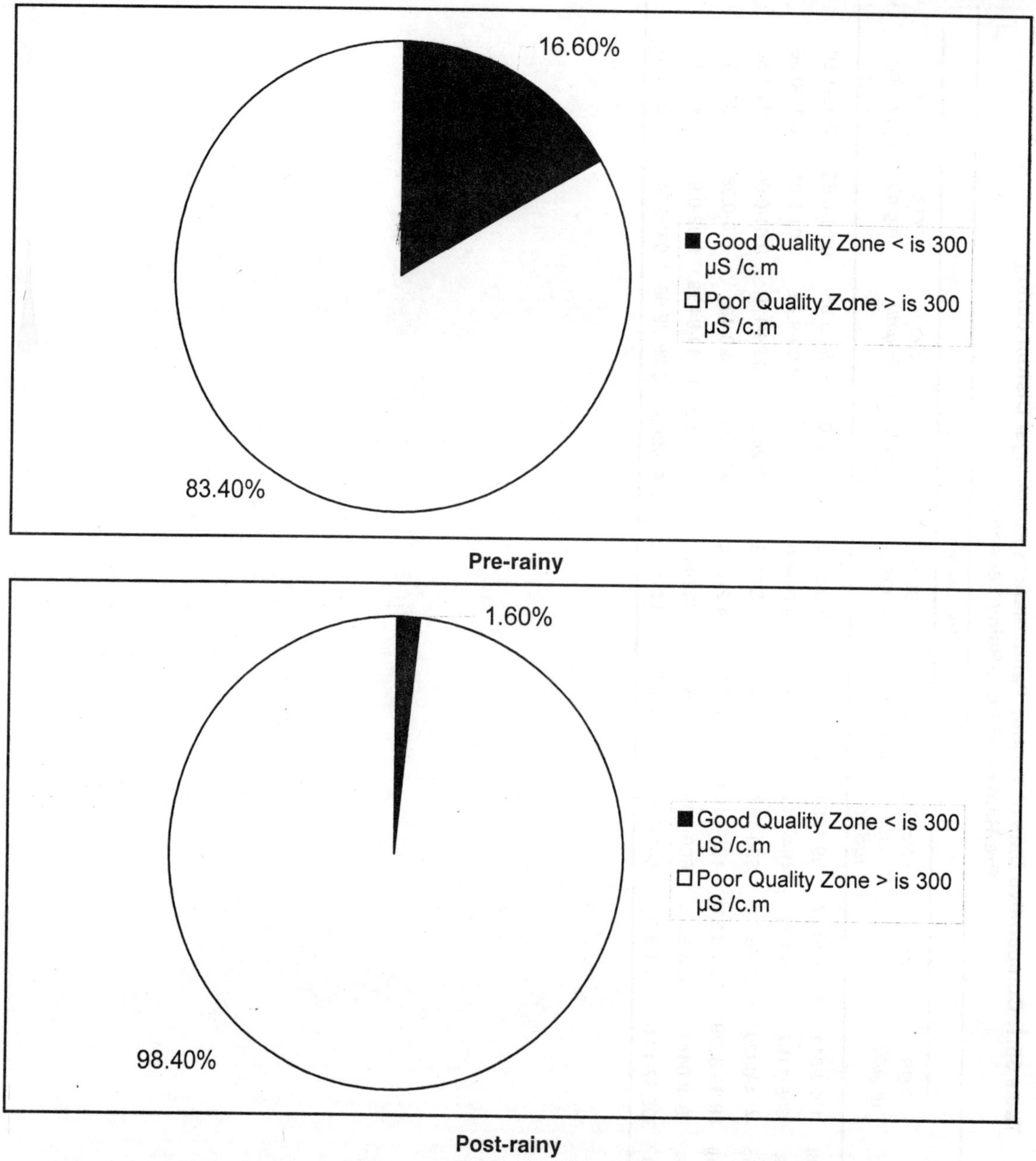

Figure 38.1: Percentage Distribution of Electrical Conductivity in Kushinagar District

considered suitable for drinking 500 mg/dl is the desirable limit and 2000 mg/dl is the permissible limit (BIS1991;WHO, 1993).The TDS Kaptanganj, Padrauna and Ramkola vary from 122-8611mg/dl, 305-1157mg/dl and 414–6329mg/dl respectively. High TDS at these three stations is due to industrial waste percolation or seepage in groundwater. The G.I.S. analysis reveal that 66.6 per cent area of

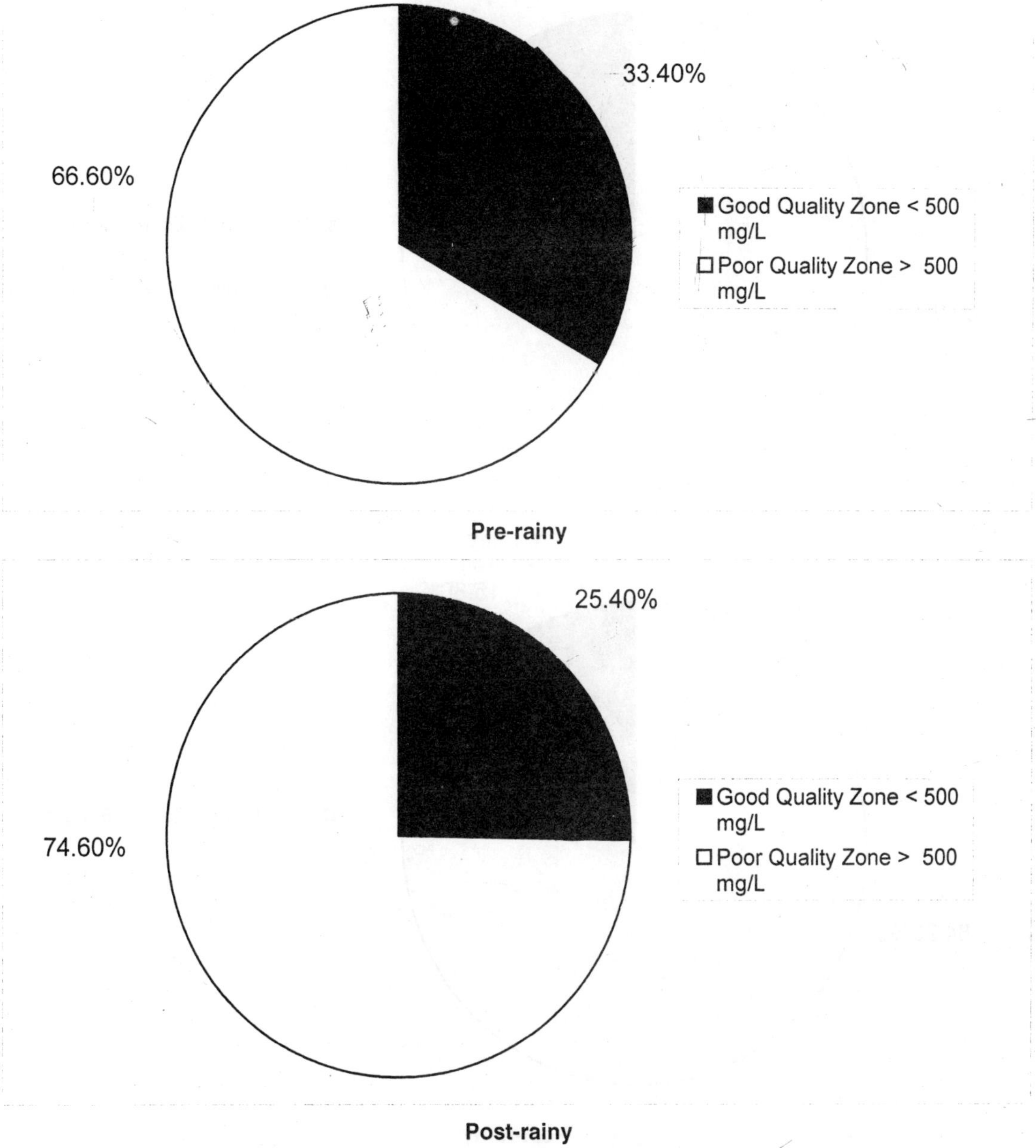

Figure 38.2: Percentage Distribution of TDS in Groundwater in District Kushinagar

district Kushinagar contains TDS>500mg/dl and is not safe for drinking purpose (Figure 38.2). TDS in post-rainy season in Kaptanganj, Padrauna and Ramkola vary from 88.35–837.9, 342–6441and 308.37 – 1140 (Table 38.1).

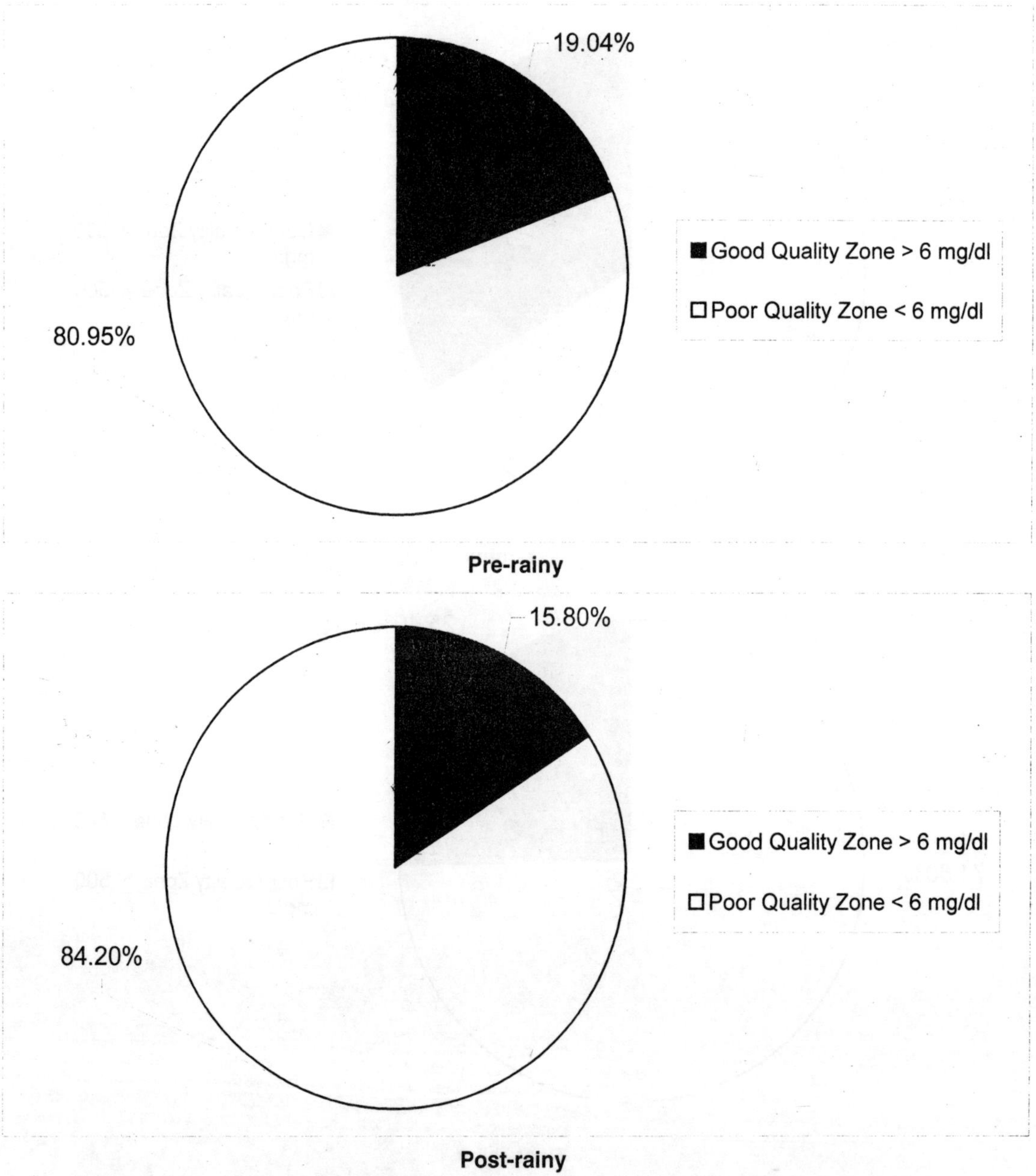

Figure 38.3: Percentage Distribution of DO in Groundwater in District Kushinagar

On the basis of spatial and percentage distribution of TDS in Kushinagar in post-rainy season 74.60 per cent area of Kushinagar contains TDS > 500 mg/L and is not safe for drinking purpose (Figure 38.2). In pre-rainy season 83.40 per cent area and in post-rainy season 98.40 per cent area of Kushinagar contains E. conductivity > 300 μS/c.m. Higher values of TDS at Kushinagar appear due

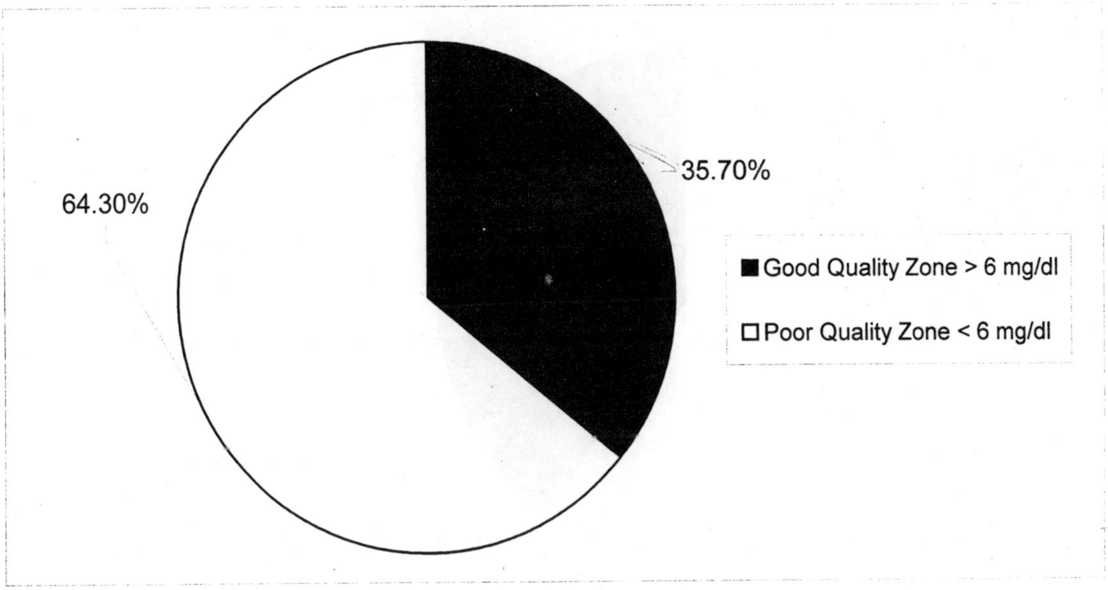

Pre-rainy

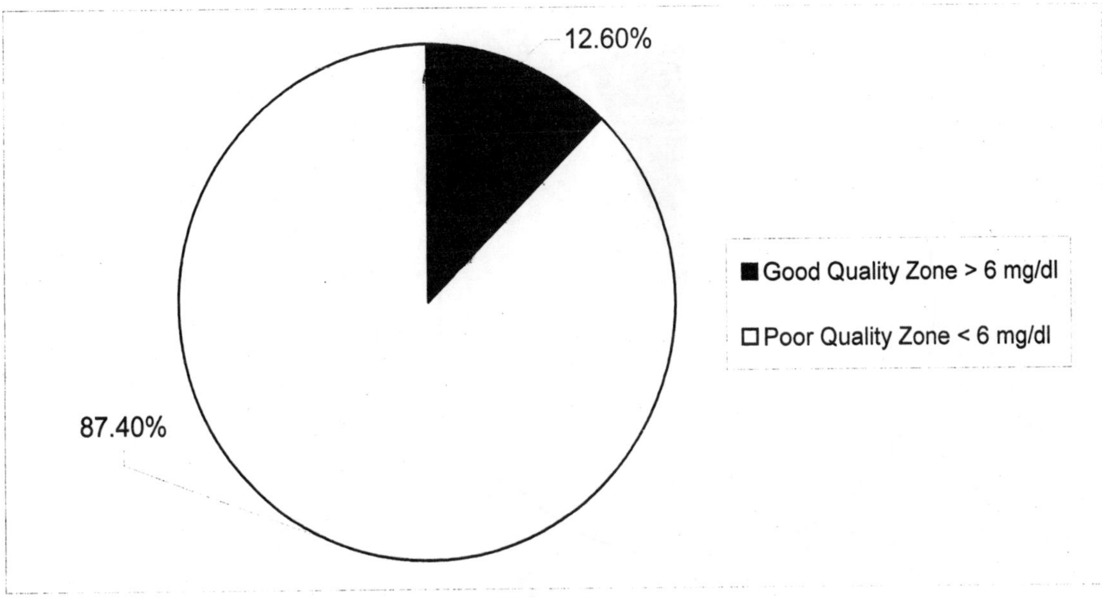

Post-rainy

Figure 38.4: Percentage Distribution of BOD in Groundwater in District Kushinagar

to the contribution of dissolved solid from watershed and disposable of Industrial effluent on water bodies; one of the most important reasons are high amount of agricultural runoff and discharge of effluent on water bodies (Mathur, 2005). Contribution of watershed due to dissolution or weathering of rocks and soil as water passes over or percolates through them (Decoursey, 1985). A large amount

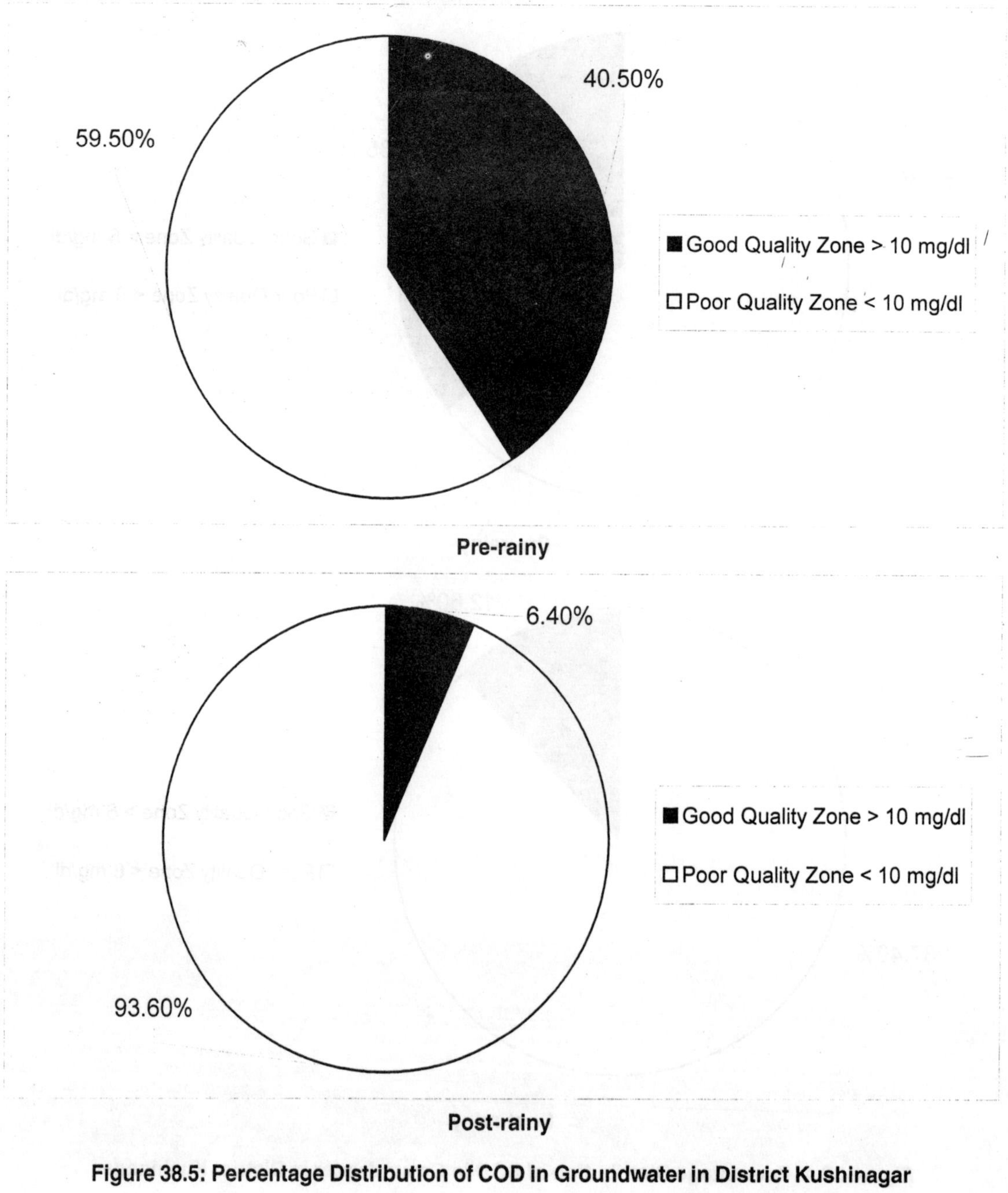

Figure 38.5: Percentage Distribution of COD in Groundwater in District Kushinagar

of sediment load is transported from the watershed during the steep slopes of the land (Rai and Sharma, 1998).

Dissolved Oxygen (DO) is important parameter in the drinking water and is inversely related to the Biological Oxygen Demand (BOD) and Chemical Oxygen Demand (COD). DO, BOD and COD

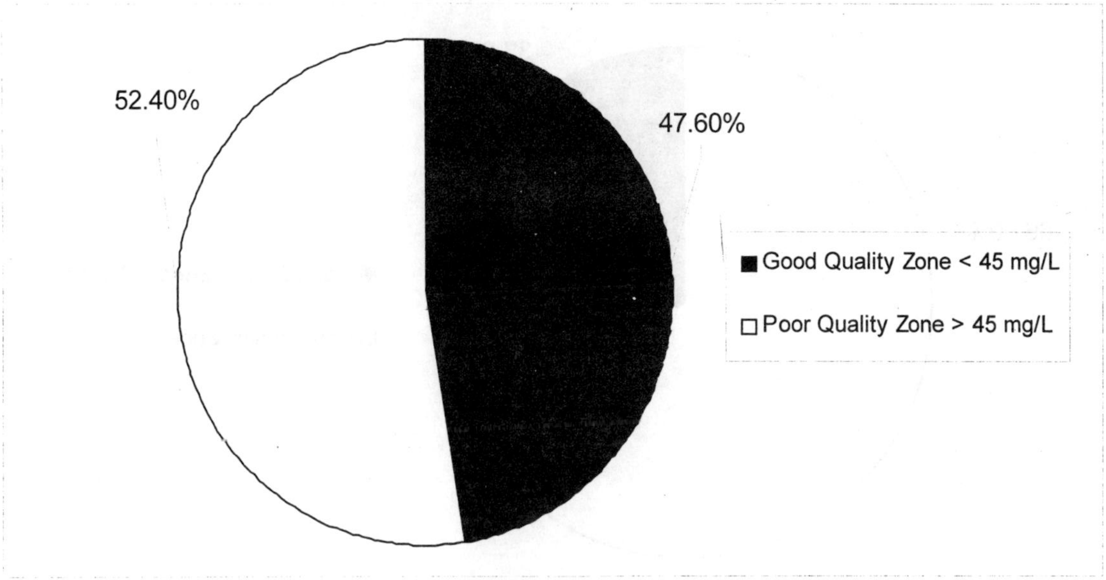

Figure 38.6: Percentage Distribution of Nitrate in Groundwater on Post-rainy Season, District Kushinagar

values at Kaptanjanj range from 5.2–7.8 mg/dl, 2-55.0 mg/dl and 7.2–82.0 respectively. At Padrauna values of DO, BOD and COD ranges from 5.3–8.2 mg/dl, 4.0-34.0 mg/dl and 5.8–66.5 and at Ramkola values of DO, BOD and COD 4.9–6.3 mg/dl, 2.8–20.8 mg/dl and 5.2-24.6 mg/dl.These values are much higher than the permissible limits prescribed by WHO and BIS (WHO, 1993; BIS, 1991), D.O. 4.0-6.0 mg/dl BOD < 6 mg/dl and COD < 10 mg/dl. At Kaptanganj these values are much higher than Ramkola and Padarauna due to positive loading of industrial waste at water bodies. In the post-rainy season values of DO, BOD and COD 5.4–7.6 mg/dl, 4.2-10.4 mg/dl and 8.42–22.4 2mg/dl. In Padrauna values of DO, BOD and COD was 0.0-6.4 mg/dl, 8.2-88 mg/dl and 19.2-52.2 mg/dl at Ramkola 5.1-7.0mg/dl, 3.0-14.6 mg/dl and 7.42–28.9 mg/dl (Table 38.1). The condition in post-rainy season of water quality at Kushinagar showed strong positive loading of BOD and COD and strong negative loading on pH and DO. This represent anthropogenic pollution sources and can be explained on the basis that high level of dissolved organic matter consume large amount of oxygen, which undergoes anaerobic fermentation processes leading to formation of ammonia and organic acids. Hydrolysis of this acidic material causes decrease of water pH values (Singh *et al.*, 2004).

In the pre-rainy season nitrate value in Kaptanganj, Padrauna and Ramkola ranges from 0.09-3.95 mg/dl, 0.09-8.78 mg/dl and 0.29-1.4 mg/dl respectively. None of the samples contain nitrate beyond the permissible limit of 40.0 mg/dl prescribed by BIS (BIS, 1991) for drinking water supplies. Concentration of nitrate in post-rainy season at Kaptanganj, Padrauna and Ramkola 4.8-48 mg/dl, 11.09-84.79 mg/dl and 3.38-72.96 mg/dl (Table 38.1).

Nitrate concentration in groundwater is enhanced by shallow groundwater table and excessive application of nitrogenous fertilizer (Steversen, 1996). In an around area of high urbanization, industrialization, municipal, industrial water may contribute high level of nitrate to groundwater (Handa *et al.*, 1982). Nitrate concentration showed a clear seasonal pattern.The seasonable difference

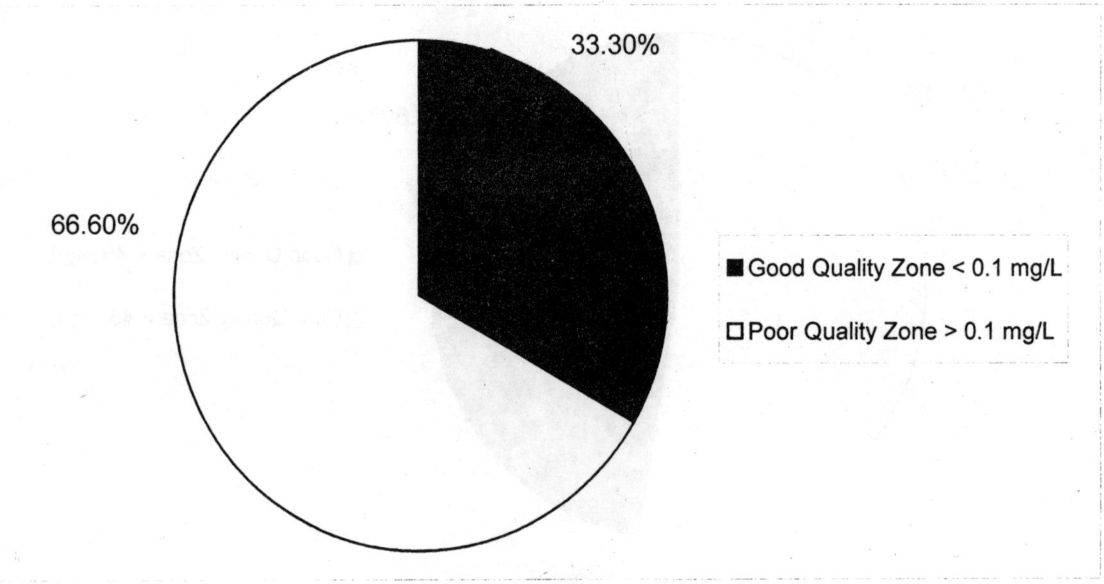

Pre-rainy

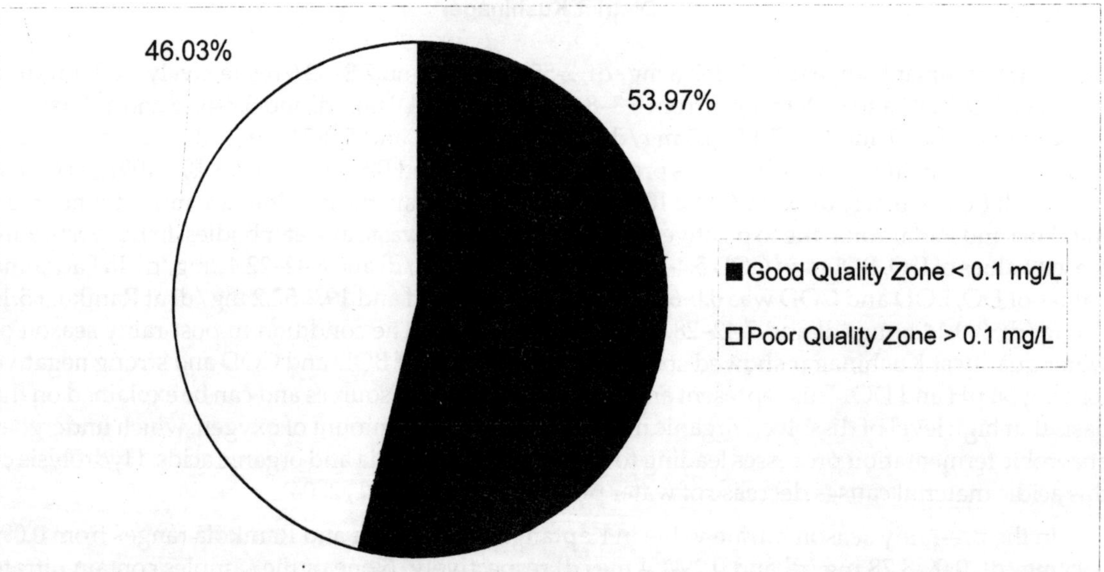

Post-rainy

Figure 35.7: Percentage Distribution of Phosphate in Groundwater District Kushinagar

can be account for the variation in the amount of readily leachable nitrate within soil profile (Powlson D.S., 1993). Autumn sown crops utilize only small amount of nitrate the large quantity being available for leaching. During summer month (pre-rainy season) plant uptake is high, available nitrate is low

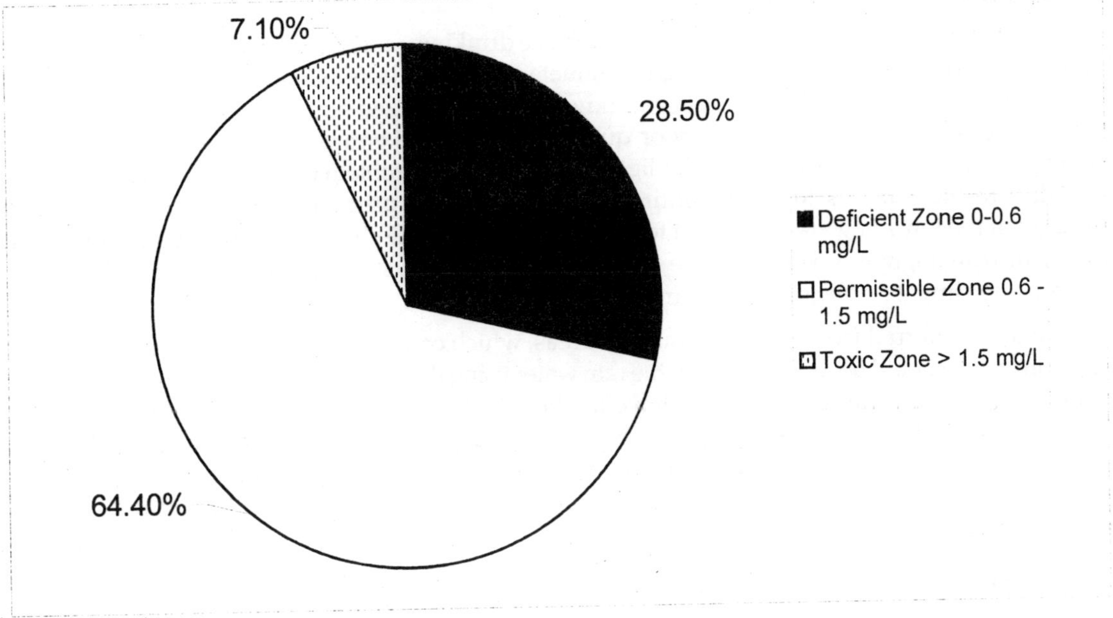

Figure 38.8: Percentage Distribution of Fluoride in Groundwater on District Kushinagar

on the basis of percentage distribution of groundwater in Kushinagar, 47.6 per cent of groundwater contain nitrate above 45mg/dl and is not safe for drinking.

The phosphate concentration at Kaptanganj, Padrauna and Ramkola ranges from 0.01- 0.63 mg/dl, 0.01- 2.85mg/dl and 0.01-2.85 mg/dl respectively high as compared to 0.1 mg/dl limit on the basis of percentage distribution of water quality 33.3 per cent of area has poor water quality and 66.6 per cent area has good quality. Phosphate concentration was high in post-rainy season. The phosphate concentration at Kaptanganj, Padrauna and Ramkola ranges from 0.09-0.98 mg/dl,0.02-0.91 mg/dl and 0.0-0.045 mg/dl respectively. On the basis of percentage distribution of groundwater in Kushinagar 46.03 per cent area contains phosphate above 0.1 mg/dl and is not safe for drinking purpose and remaining 53.97 per cent area contains phosphate above permissible limit. The high amount of phosphate in groundwater in post-rainy season due to increased weathering and break down of the soil structure and addition of mineral fertilizers (Jenkins, 1995).

On the basis of spatial and percentage distribution of groundwater in Kushinagar district 6.4 per cent area contain fluoride >1.5 mg/dl and is not safe for drinking purpose. Fluoride has considerable physiological significance for man and animal. Evidences so far accumulated shows that the upward adjustment of fluoride in drinking water to 1.0 mg/l result in a great reduction in the incidence of dental carries (Handa, 1975). As such the problem of high level of fluoride is encountered in many geographical zones of globe. The globe prevalence of fluorisis is reported to be about 32 per cent (Srivastava, 1997). It is not surprising that 17 of the states and union territories are endemic for fluorosis (Ministry of Health, 1962 and Srinivasan, 1959)

Conclusion

The main objective of this study was to undertake drinking water quality survey of India in sugar industrial area of district Kushinagar. Eight parameters *viz.* pH, Electric Conductivity, TDS, DO, BOD, COD, Fluoride, Nitrate and Phosphate in groundwater were analyzed separately and further combined to delineate, suitable moderate and poor quality zones in the entire district. Areas having high concentrate of these parameters were highlighted and nature of problem was classified using chemical and GIS based techniques. Release of entire effluent outside the premises which eventually spreads throughout the low lying areas adjacent to the sugar factory or discharges into natural water course. This in turn can migrate and mix with the shallow water aquifer system through seepage and therefore can cause detrimental effects on the groundwater quality.

It may be inferred from the study that the areas, which contains the high Total Dissolved Solid values in water, should be recharged with the rain water using the percolation wells. Since the untreated water is being used, the consumer will have the health effect due to high degree of residue. High concentration of fluoride content in the study area need urgent attention to prevent possible health hazards by way of getting defluoridation plant and to inhibit discharge of sugar industry waste in water bodies. Other remedial measures like use of lime, alum, bone charcoal filters may be used on domestic scale (Valdiya, 1987).

References

APHA, 1992. *Standard Methods for Examination of Water and Wastewater*, 18th edn. American Public Health Association, Washington, DC.

BIS, 1991. *Drinking Water Specification*. IS : 10500 : 1991. Bureau of Indian Standards, New Delhi, India. Central Water Pollution Control Board, *Annual Report*, 1996–1997.

DeCoursey, D.G., 1985. Mathematical models for nonpoint water pollution control. *J. Soil Water Conservation*, 40: 408–413.

Handa, B.K., 1975. Geochemistry and gensis of fluoride containing groundwater in India. *Groundwater*, 13(3): 25–31.

Handa, B.K., Goel, D.K., Kumar, A., Singh, T.N. and Sodhi, T.N., 1982. Pollution of groundwater by nitrate in Uttar Pradesh. *IAWPC Tech. Annual Review*, 1: 95–103.

Islam, M.J., Zaman, M.W., Mahmood, R.I., and Rahman, M.S., 2003.Toxicity assessment of groundwater in different aquifers of Khagrachari in Bangladesh. *Asian Journal of Plant Sciences*, 2(3): 257–260.

Jenkins, A., Sloan, W.T. and Cosby, B.J., 1995. Stream chemistry in the middle hills and high mountains of the Himalayas, Nepal. *Journal of Hydrology*, 166: 61–79.

Singh, Kunwar P., Malik, Amrita, Mohan, Dinesh and Sinha, Sarita, 2004. Multivariate statistical techniques for the evaluation of spatial and temporal variations in water quality of Gomti River (India): A case study. *Water Research*, 38(18): 3980–3992.

Mathur, A. and Priyakant, 2005. GIS based delineation of suitable groundwater quality zones for drinking purpose: A case study of district Mainpuri, U.P. *Poll. Res.*, 24(1): 59–68.

Ministry of Health, GOI, New Dehli, 1962. *Manual on Water Supply*.

Powlson, D.S., 1993. Understanding the soil nitrogen cycle. *Soil Uses Manage.*, 9: 86–94.

Rai, S.C. and Sharma, E., 1998. Hydrology and Nutrient flux in an agrarial watershed of the Sikkim Himalaya. *J. Soil Water Conservation*, 53(2): 125–132.

Srinivasan, T.G., 1959. Removal of fluoride from water by alkali digested paddy husk carbon. Central Public Health Engineering and Research Institute, 1: 1022–1039.

Sterversen, F.S., 1986. The internal cycle of nitrogen of soil: carbon, nitrogen, phosphorus, sulphur, and micro-nutrients. John Wiley and Sons, New York.

Valdiya, K.S., 1987. *Environmental Geology: India Context*. Tata McGraw Hill, New Delhi, p. 479–481.

WHO, 1993. *Guidelines for Drinking Water Quality, Vol. 1: Recommendations,* 2nd edn. World Health Organisation, Geneva.

Water and Wastewater Management in 2 Vols (2011) *Pages 262–265*
Editors: **D.R. Khanna, R. Bhutiani & Gagan Matta**
Published by: **DAYA PUBLISHING HOUSE, NEW DELHI**

Chapter 39

Alteration in Surface Water Quality Near Textile Industrial Area in Panipat Region

D.S. Malik, P.K. Bharti and Sumit Grover
Department of Zoology and Environmental Science,
Gurukula Kangri University, Haridwar

ABSTRACT

Mostly the water is used for industrial, municipal and agricultural purposes. In the last two decades, the rapid growth of industrialization and urbanization have created negative impacts on the environment. The industrial, municipal and agricultural wastes containing pesticides, insecticides, fertilizers residues, organic pollutants and heavy metals in their effluents have polluted surface and groundwater. So, water resources must be managed and conserve on sustainable basis for the most beneficial uses in the future. The present study is aimed for an assessment of water quality affected by textile industrial effluents.

Keywords: *Surface water quality, Industrial pollution, Textile effluent, Water characteristics.*

Introduction

In the modern industrialization period, the most of water resources have been affected enormously by seepage, leaching and intermixing of industrial effluents in maximum metropolitan cities and industrial townships. The textile effluents consists of high concentrations of heavy metals, organic pollutants and toxic colours that may alter the surface water quality of the region and ultimately causes the health hazards among livestock and human being.

In India, the industrial effluents have contributed a major source of pollution. Different industries of the country contribute to generate about 16 per cent of the total wastewater. The treatment facilities have been installed in less than one third of the polluting industries (Gopal, 1994). Most of effluent treatment plants (ETP) have not performed physical and chemical treatment processes satisfactory due to economic reasons on commercial scales.

About 6,34,900 metric tons of dyes are used worldwide each year in textile dyeing industries and nearly 10-15 per cent of them are discharged as effluents. The colours affect the nature of water, inhibit sunlight penetration and reduce the photosynthetic action and may cause toxic effect to living organisms. The contaminated wastewater without any treatment may cause adverse effect on the health of human, domestic animals, wildlife and environment (Sharma *et al.*, 1999). Thus, contaminated groundwater have deteriorated immensely the post agriculture irrigation, drinking utilities and impacts on soil systems. So, it is very much essential to treat dye contaminated wastewater before discharging it to the surface water and to develop simple and economical methods to treat the waste at point sources suitable for developing countries.

Materials and Method

Panipat city is situated on GT road in Haryana, which is well known for textile handloom and dye industries. The effluent samples were collected from the common effluent outlet of dye houses (S1) and large pond (S2), where the effluent drain ends up and analyzed by the Standard methods (APHA, 1995).

Results and Discussion

Colours of the wastewater were found varying from pinkish red to brownish yellow. Hence, colour from the textile waste can cause significant health problem among public living in surrounding. Odour is often associated with oil mists or solvent vapours and is removed by removal of the mist or by recovery of vapours. Sources of odour are Sulphur dyeing on cotton. Odour of the textile effluent was existed from Threshold level to Strong level of the odour table. Velocity of the effluents were totally dependent on the production output, amount of residual solids and amount of water used by the industry. The velocity was found low 0.35 m/s in effluent drain because of the solids load. And it was found nil at the pond site because ponds are itself the stagnant ecosystems.

Temperature may vary from 22°C to 24°C in the common effluent drain and polluted pond, having not much fluctuation in industrial area. In this study it was recorded that the pH of the effluent of all the sampling station were found alkaline (7.2-8.2). Alkalinity is caused by strong bases and the salts of strong alkalies and weak acids. Extreme pH or rapid pH change can exert stress conditions or kill biotic life in aquatic system. Malik *et al.* (2004) observed the alkaline value of pH in textile effluent and found creating the sedative nature in groundwater as well as agricultural soil nearby the textile industrial area of Panipat.

The dissolved solids in wastewater are mainly inorganic salts. They are particularly important as they are relatively unaffected by biological treatment processes and can accumulate in water recirculation system. Failure to remove them may lead to an increase in total solids content of surface water sources (341-370 mg/l). The dissolved solids in wastewater, if not controlled, may be harmful to vegetation and also preclude use in various irrigation processes. T.D.S. in the effluent were observed (308-336.3 mg/l) very high than the prescribed limits of C.P.C.B. and B.I.S. Residual is the solid matter present in water, with high solids is harmful to organism. Welch (1952) stated that dissolved solids vary qualitatively and quantitatively in different water depending upon the season, location and other factors.

Table 39.1: Quality Characteristics of Effluent and Pond Water in Panipat Region

Parameters	*Unit*	*Effluent Drain (S1)*	*Polluted Pond (S2)*	*Standards (BIS)*
Odour	–	Threshold	Strong	–
Colour	–	Pinkish red	Brownish yellow	0.0
Velocity	m/s	0.35±0.005	Stagnant	–
pH	–	8.2±0.14	7.2±0.60	5.5–9.0
Temperature	°C	22.5±1.5	20.0±2.0	40.0
TS	mg/l	370.0±90.0	341.0±81.0	–
TDS	mg/l	336.3±81.89	308.0±73.8	–
TSS	mg/l	33.6±8.21	30.9±7.28	100
DO	mg/l	1.64±0 .21	1.28±0.0 I	–
BOD	mg/l	272.5±42.5	465.0±55.0	30.0
COD	mg/l	791.5±51.5	1245.0±65.0	250
Cl	mg/l	340.8±28.4	369.2±28.4	600
Alkalinity	mg/l	590±60	880±20	–

Total suspended solids include both organic and inorganic materials. They adversely affect fishes and bottom fauna of pond ecosystem. They produce carbon dioxide, hydrogen sulphide, methane and other noxious gases. They increase the turbidity of water, reduce light penetration and impair photosynthetic activity of aquatic plants. T.S.S. in the effluent of all the sampling station was (30.9-33.6 mg/l). Similar observation has been recorded by Chavan and Wagh (2005) in the industrial effluent near Sukhna river at Midc area, Aurangabad.

The value of dissolved oxygen in the sampling stations were found fluctuating from 1.28-1.64 mg/l. As the matter of the fact, the effluents were highly toxic in nature which, could not allow any fauna to survive in this condition. The value of dissolved oxygen was even less in the summer month this trend was stated by Khanna *et al.* (1992).

Wastewater with high BOD/COD indicates presence of decomposing organic matter and subsequent high bacterial counts that degrade the quality and potential uses in terms of domestic and even agriculture irrigation. BOD of the effluent of the sampling station is high (272.5-465.0 mg/l). COD of the effluents and pond water were found very high (791.5-1245.0 mg/l). Sevimli and Sarikaya (2005) also observed the similar trend while performing an experiment on decolorization of textile effluent and dye solution by ozonation. Senthinathan and Azeez (1999) and Groff (1993) reported that textile wastewater may cause problem if used as untreated effluent that consist of higher COD, pH, temperature, phosphorous and absorbable organic halogen. COD shows a positive relationship with BOD as observed by Chopra and Patric (1994) in Ganga river of Rishikesh.

The value of Alkalinity in the effluent is very high (550-880 mg/l) as compared to the standards of C.P.C.B. Bhat and Kulkarni (2005) also observed the increase in alkalinity of textile effluent and concluded the research on reducing the COD of textile wastewater.

Chloride is one of the most conservative constituents in wastewater, being affected very little either by vegetative growth or other biological processes, or by precipitation or anion exchange reaction.

Chloride concentration in the effluent was also observed high (369.2-340.8 mg/l). Similar trend of increased chloride concentration was stated by Aboulhassan *et al.* (2005), while conducting a work on the effect of natural flocculants on textile wastewater. In area where discharge of industrial wastes containing high concentration of chlorides must be controlled to safeguard the receiving waters.

Conclusion

Having discussed all the physico-chemical characteristics of effluent drain and pond water, we conclude that the pond water became very toxic to deteriorate the groundwater quality of area. The flora and fauna of this pond is almost in dying stage due to the toxic pollutants of textile industries. The effluent may also cause various health problems such as skin allergies, lung infection, carcino-mutagenic disorders among the human being through contaminated groundwater supply by municipal tube wells water scheme. In order to achieve the standards it is suggested to install the effluent treatment plants (ETP) by the small scale textile industries.

References

Aboulhassan, M.A., Souabi, S., Yaacoubi, A. and Baudu, M., 2005. Treatment of textile wastewater using a natural flocculants. *Environmental Technology*, 26: 705–711.

APHA, 1995. *Standard Methods for Examination of Water and Wastewater*, 19th edn. American Public Health Association, Inc., New York, pp. 1170.

Bhat, Ramananda and Kulkarni, Raghavendra, 2005. COD reduction of dye industry effluent. *J. Industrial Pollution Control*, 21(1): 147–150.

Chavan, T.P. and Wagh, S.B., 2005. Physico-chemical characteristics of industrial effluent near Sukhna river at mid area, Chikalthana, Aurangabad. *J. Industrial Pollution Control*, 21(1): 23–26.

Chopra, A.K., Patrik and Nirmal, J., 1994. Effect of domestic sewage on self-purification of Ganga water at Rishikesh. *Bio Science*, 13(11): 75–82.

Groff, K.A., 1993. Textile waste. *J. Water Environmental Research*, 65: 421.

Gopal, B., 1994. Conservation of inland water in India: An overview. Verh. Internet. Verein. *Hydrobiologia*, 384: 267.

Khanna, D.R., Badola, S.P., Singh, H.R. and Dobriyal, Anoop K., 1992. Observation on seasonal trends in diatomic diversity in the river Ganga at Sapt-Sarovar, Haridwar. In: *Recent Research in Coldwater Fisheries*, (Ed.) K.L. Sehgal. Today and Tomorrow Printer and Publisher, New Delhi, p. 99–107.

Malik, D.S., Yadav, Rashmi and Bharti, P.K., 2004. Accumulation of heavy metals in crop plants through irrigation of contaminated groundwater in panipat region. *Environmental Conservation Journal*, 5(1–3): 101–104.

Senthinathan, S. and Azeez, P.A., 1999. Influence of dyeing and bleaching industries an groundwater of Tripur, Tamil Nadu, India. *Bull. Environ. Contam. Toxicol.*, 62: 330–334.

Sevimli, M.F. and Sarikaya, H.Z., 2005. Effect of some operational parameters on the decolourization of textile effluent and dye solution by ozonation. *Environmental Technology*, 26: 135–143.

Sharma, K.P., Sharma, K., Bharadwaj, S.M. and Chaturvedi, R.K., 1999: Environmental impact assessment of textile printing industries in Sanagar, Jaipur: A case study. *J. Int. Bot. Soc.*, 78: 71–74.

Welch, P.S., 1952. *Limnology*, 2nd edn. McGraw Hill Book Co., New York, Toronto and London, pp. 538.

Water and Wastewater Management in 2 Vols (2011) *Pages 266–270*
Editors: **D.R. Khanna, R. Bhutiani & Gagan Matta**
Published by: **DAYA PUBLISHING HOUSE, NEW DELHI**

Chapter 40

Limnology and Biodiversity of Fish Fauna in Virla Reservoir, M.P.

S.K. Pathak[1] and L.K. Mudgal[2]
[1]*Department of Zoology, Govt. Autonomous Holkar Science College, Indore*
[2]*Govt. Girls P.G. College, Moti Tabela, Indore*

ABSTRACT

The limnological factors in Virla reservoir of west Nimar district (Khargone) that were monitored for two years (July 2001 to June 2003) show high positive correlations among different factors such as alkalinity and phosphates (0.893),transparency and dissolved oxygen (0.885), pH and alkalinity (0.873), alkalinity and BOD (0.859), pH and phosphate (0.826), BOD and phosphate (0.764) (Table 40.2). The total number of 29 species of fishes were recorded in Virla reservoir with order Cypriniformes contributing maximum of 19 members (65.51 per cent) of species followed by four members (13.79 per cent) of order Ophiocephaliformes.

Keywords: *Limnology, Biodiversity, Fish, Fauna.*

Introduction

All organisms have unique characteristics and maintain ecological balances (equilibrium).For enhancement of fish production and suitable management, it is necessary to interact fish species with abiotic and biotic factors. Probable threats and measures to fish conservation were also studied. Some of the important contributions in this regard are those of Menon (1949), Singh and Shromany (1964), Pandey (1999) and Rema *et al.* (1999). The present study deals with the ecology and fish diversity in this reservoir which receives water from the local nallah and the adjacent agricultural field during rainy season.

Material and Methods

The Virla reservoir is an irrigation water body having a total length of about 270 m, a catchment area of about 50 sq. kms. The gross storage capacity is about 525 hectare. It is located near Virla village of Khargone district headquarter that falls under 21°- 50′-30″ latitude and 75 °-23′-30″ longitude and is 309 meters above the mean sea level.

The physico-chemical parameters such as water temperature, transparency, pH and dissolved oxygen were measured at site with mercury thermometer, Secchi disc, digital pH meter and Winkler's titration method. Besides this total alkalinity, BOD, total hardness, free carbondioxide, phosphates were analysed in laboratory by following the standard methods of APHA (1989). Fishes were caught with the help of local fishermans during course of study by operating cast net and drag nets, collected and identified according to Jayaram (1994), Srivastava (1980) and Day (1958).

Results and Discussion

Water temperature ranged from 19.20°C to 34.20 °C. The minimum value was recorded during the month of January and the maximum during June. The trend of variation is mainly affected by seasons and contribute to biodiversity of fish in aquatic system. Transparency values ranged from 57.0 to 170.0 cms. The maximum value was observed in the month of January when the water of reservoir was most clear with high transparency. The water was dirty with the minimum transparency during the rainy season when sand and silt came with runoff water to the reservoir. Transparency is an important factor which affects survival and growth of fish and biodiversity of fish in aquatic habitat.

pH of the reservoir fluctuated from 7.38 to 9.26 with the minimum value during December and the maximum during June. It shows direct significant correlation (r= 0.873) with total alkalinity. Same trend was also reported by Bhutiani (2004). Total alkalinity varied from 133 mg/l to 256 mg/l. The highest value was observed in the month of June while the lowest in the month of January. Dissolved oxygen fluctuated from 5.80 mg/l to 9.84 mg/l. The highest concentration that was recorded in winter month shows inverse correlation with BOD, total alkalinity, water temperature and pH. It depicts positive correlation with transparency which enhances the photosynthesis activity. The same type of results were also recorded by Khanna (1993).

The value of BOD ranged between 3.26 mg/l to 5.53 mg/l. The lowest value was recorded in month of January, while the maximum in the month of August. It shows negative correlation with transparency and dissolved oxygen. Free carbon dioxide value ranged between 3.86 mg/l to 4.55 mg/l. It was observed in the month of December and January, which shows more respiratory activities and incomplete utilization in photosynthesis. Similar trend was obtained by Khanna and Chugh (2004).

The value of total hardness varied from 150 mg/l to 233 mg/l. It was low during winter season.Phosphate varied from 0.20 mg/l to 0.50 mg/l. The maximum value was found during the rainy season which may be due to the influx of the rainwater.The same observation was recorded by Sharma and Hussain (1999), Das *et al.*(2000) and Das and Chand (2003). The phosphate is one of the important nutrients which influences the productivity. This nutrient play an important role in food chain and ichthyobiodiversity in aquatic ecosystem.

Various species of fishes contributes the biodiversity of the reservoir under investigation included the fishes that came to the reservoir through local nallah, adjacent agriculture fields and those of reservoir itself. Subsistence dominant fishery of reservoir are categorised as major carp, cat fishes, murrel and miscellaneous. The observed fishes in reservoir have been tabulated in Table 40.1.

It was found that the fish biodiversity of Virla reservoir contributed by Cypriniformes with maximum of 19 numbers of species (65.51 per cent) followed by four numbers (13.79 per cent) of Ophiocephaliformes, two numbers (6.90 per cent) of Perciformes and Mastacembeleformes and one number (3.45 per cent) of Clupeiformes and Beloniformes. Due to more fecundity of major carp and suitable environmental condition there exists a relatively higher number of Cypriniformes. Such type of observation was also recorded by Talwar and Jhingran (1991) and Das and Chand (2003) in inland fishes.

Table 40.1: List of Fishes Recorded in Virla Reservoir during July 2001 to June 2003

Order	Family	Genera
Cypriniformes	Cyprinidae	1. *Catla catla*
		2. *Cirrhinus mrigala* (Ham)
		3. *Cirrhinus reba* (Ham.)
		4. *Labeo rohita* (Ham)
		5. *Labeo calbasu* (Ham)
		6. *Labeo bata* (Ham)
		7. *Puntius ticto* (Ham)
		8. *Puntius sophore* (Ham)
		9. *Rasbora daniconius* (Ham)
		10. *Nemachelius botia* (Ham)
		11. *Nemachelius beavani* (Ham)
		12. *Nemachelius aurius* (Ham)
	Siluridae	1. *Ompok bimaculatus* (Bloch)
		2. *Wallago attu* (Sclin)
	Bagridae	1. *Mystus seenghala* (Skyes)
		2. *Mystus teengara* (Ham)
		3. *Mystus aor* (Ham)
	Schielbeidae	1. *Eutropiichthys vacha* (Ham)
	Saccobranchidae	1. *Heteropneustes fossilis* (Bloch)
Clupeiformes	Notopteroidae	1. *Notopterus notopterlls* (Pallas)
Beloniformes	Belonidae	1. *Xenentodon cancila* (Ham)
Ophiocephaliformes	Ophiocephalidae	1. *Channa marulius* (Ham)
		2. *Channa gachua* (Ham)
		3. *Channa striatus* (Bloch)
		4. *Channa punctatus* (Bloch)
Perciformes	Centropomidae	1. *Ambasis nama* (Ham)
		2. *Ambasis ranga* (Ham)
Mastacembeleformes	Mastacembelidae	1. *Mastacembelus armatus* (Lac)
		2. *Mastacembellis pancalus* (Ham)

Table 40.2: Correlation Matrix of Physico-chemical Factors at Virla Reservoir During July, 2001 to June, 2003

Parameters	*Water Temp.*	*Trans-parency*	*pH*	*CO_2*	*Alkalinity*	*DO*	*BOD*	*Hardness*	*Phos-phates*
Water Temp.	1	–0.651	0.879	–0.605	0.885	–0.706	0.652	0.762	0.821
Transparency		1	–0.718	0.651	–0.827	0.885	–0.839	–0.259	–0.784
pH			1	–0.562	0.873	–0.786	0.747	0.726	0.826
Carbondioxide				1	–0.592	0.518	–0.450	–0.332	–0.511
Alkalinity					1	–0.895	0.859	0.497	0.893
DO						1	–0.988	–0.238	–0.794
BOD							1	0.190	0.764
Hardness								1	0.579
Phosphates									1

Acknowledgement

The authors are thankful to the principal, Govt. P.G. College, Khargone for providing laboratory facilities.

References

APHA, 1989. *Standard Methods for the Examination of Water and Wastewater*, 19th edn. American Public Health Association, New York.

Bhutiani, R., 2004. Limnological status of river Suswa with reference to its mathematical modeling. *Ph.D. Thesis*, G.K.V., Haridwar.

Das, S.K. and Chand, B.K., 2003. Limnology and biodiversity of ichthyofauna in a pond of southern Orissa, India. *J. Ecotox. Env. Monit.*, 13(2): 97–102.

Das, S.K., Sahoo, J.K. and Saksena, D.N., 2000. Dynamic of abiotic factors in relation to growth of tiger shrimp *Penaeus monodon* in traditional shrimp culture system. *J. Inland Fish Soc.*, India, 32(1): 28–34.

Day, F., 1958. *The Fishes of India*. William Dawson and Sons, Ltd. London, Vol. I, p. 777and II, p. 198.

Jayaram, K.C., 1994. *The Freshwater Fishes of India, Pakistan, Bangladesh, Burma and Sri Lanka*. Zoological Survey of India, Calcutta, p. 475.

Khanna, D.R., 1993. *Ecology and Pollution of Ganga River*. Ashish Publishing House, New Delhi.

Khanna, D.R. and Chugh, Tarun, 2003. *Microbial Ecology: A Study of River Ganga*. Discovery Publications, New Delhi.

Menon, A.G.K., 1949. Fishes of Kumaon Himalayas. *J. Bombay Nat. Hist. Soc.*, 43: 40–454.

Pandey, A.C., 1999. An annotated list of fish genetic resources of Faizabad district, U.P. *Indian J. Fish.*, 46(1): 95–100.

Rema Devi, K., Indra, T.J., Raghunathan, M.B. and Mary, Bai M., 1999. On collection of fish fauna from Chennai, Chengleput and Thiruvallur districts of Tamil Nadu. *Rec. Zool. Survey India*, 97(4): 151–166.

Sharma, B.K. and Hussain, M.D., 1999. Temporal variations in abiotic factors of a tropical flood plain lake, Upper Assam (N.E. India). *Rec. Zool. Surv. India*, 97(4): 145–150.

Singh, B.M. and Shiromany, P.P., 1964. The fishes of Doon Valley. *Ichthyologica*, 3: 85–92.

Srivastava, G.J., 1980. *Fishes of Eastern Uttar Pradesh*. Vishwavidyalaya Prakashan, Varanasi, p. 207.

Talwar, P.K. and Jhingran, A.G., 1991. *Inland Fishes of India and Adjacent Countries*, Vols. I and II. Oxford and IBH Publishing Company Pvt. Ltd., New Delhi, p. 1158.

Water and Wastewater Management in 2 Vols (2011) *Pages* **271–275**
Editors: **D.R. Khanna, R. Bhutiani & Gagan Matta**
Published by: **DAYA PUBLISHING HOUSE, NEW DELHI**

Chapter 41

Pond and Riverine Algae of Khargone, Madhya Pradesh

Pramod Kumar Verma
Botany Department, Govt. Girls College, Khargone – 451 001, M.P.

ABSTRACT

Algae were collected from temporary and permanent water accumulations and banks of river Kunda at Khargone during 2005. In all 68 algal taxa were recorded, out of these 34 taxa belong to ponds and 18 to the river Kunda while remaining 16 algal taxa are common to both Pond and river.

Keywords: *Phycology, Lentic, Lotic systems.*

Introduction

In past few decades various workers have reported a large number of algae from different parts of India. Phycological studies of Khargone and its neighbourhood areas have earlier been done by Seerwani (1963) and Mahajan (1987, 1988, 2004 and 2005) but a detailed comparative study of Ponds and riverine algae is still lacking. In the present investigation an attempt has been made to study the distributional pattern of algal flora of Khargone in lentic and lotic water bodies.

Khargone (21°45′ N, 75°30' E and 250.38 m above MSL) is the district headquarter of West Nimar of M.P. It is 89 km from Khandwa Central Railway Station. The average rainfall is 709.20 mm. The algal flora of West Nimar is very rich and there is much scope for algalogical studies in future.

Materials and Method

For the present study two different types of water bodies *i.e.* settling tank, Adampur dam, temporary and permanent water accumulations of Khargone representing the lentic system and Sarkari Ghat, Navgrah Ghat and Meldera Ghat of river Kunda representing the lotic system, were selected. All the sites are situated in different parts of Khargone city.

The algal samples were collected at monthly intervals during the period January to December 2005. Identification of algal forms was done with the help of relevant literature and monographs (Desikachary, 1959; Randhawa, 1959; Phillipose, 1967; Ahmed, 1967 and Venkateswarlu, 1970).

Enumeration of Algal Flora

Only in Pond

Cyanophyceae

Anabaena ambigua Rao forma

Arthrospira martinii

Cylindrospermum spirale kg

Gloeocapsa stegophila (Itz,) Rabenh

Gloeotrichia raciboraskii Wolosz. forma

Lyngbya dendrobia Bruhl et. Biswas

Microcystis aeruginosa Kutz

Nostoc linckia (Roth.) Born et. Flab fonna

Oscillatoria sancta Kutz

Rivularia bacearíana (De Not) Born et. Flah

Spirulina major Kutz

S. *mahajanii* Mahajan

Bacillariophyceae

Cymbella cistula (Hempr.) Grwn

Navicula viridis Kutz

Chlorophyceae

Chara zeylanica Wild

Chlorella vulgaris Beyernick

Cladophora glomerata (Linn.) kutz

Closterium monoliferum (Bory.) Ehrenb

Coleochaete orbicularis Pring

Cosmarium granatum Breb

Desmidium swartzii Ag

Edorina elegans Ehrenb

Hydrodictyon reticulatum (L.) Lagerh (Planktonic)

Mougeotia affinis Kg

Oedocladium indicum Singh

Oedogonium sylvaticum Halles fonna

Pandorina morum (Muell.) Bory

Pleurococcus naegelii (Cohdat.)

Sirogonium indicum Singh

Spirogyra elongata Kg

S. neglecta (Hass.) Kutz

Vaucheria terrestris Lyng. em. Walz

Stigeolonium tenue Kutz. (Ag.) and *Englena viridis* Ehrenb

Euglenophyceae

Euglena uiridis Ehrenb (Planktonic)

Only in River

Cyanophyceae

Anabaena variabilis Kuetz.ex Born.et Flab. (Periphytic)

Lyngbya lutea (Agar.) Gom. (Planktonic)

Oscillatoria formosa Bory ex Gom. (Planktonic)

O. ornata Kuetz. var. crassa Rao, C.B. (Planktonic)

O. princeps Vauch. ex Gom. (Planktonic)

Bacillariophyceae

Cymbella tumida (Breb.) Gandhi (Planktonic)

Gomphonema lanceolatum var. affine (Kuetz.) A CL (Planktonic)

Navicula constans Hust var. symmetrica Hust. (Periphytic)

N. minuta (Cleve) ACl. (Benthic)

Nitzschia kittonii Smith (Planktonic)

N microcephala Grun. var. elengantula Grun. (Periphytic)

Pinnularia balatonis (Pant.) Gandhi (Planktonic)

Synedra ulna var.subaequalis Grun (Planktonic)

Chlorophyceae

Closterium tumidum John. (Planktonic)

Euastrum spinulosum Delp. var. *inermius* Nordst. Benthic

Scenedesmus armatus Chod. var. *bicaudatus* (Gugi-Printz.) Chodat (Planktonic).

Spirogyra setiformis (Roth.) Kuetz. (Periphytic)

Euglenophyceae

Euglena polymorpha Dong (Planktonic)

Common in Pond and River

Cyanophyceae

Merismopedia punctata Meyen (Planktonic)

Nostoc calcicola Breb. ex Born. et Flah. (Planktonic)

Bacillariophyceae

Achnanthes exigua Grun. (Planktonic)

Cymbella cymbiformis (Kutez.) V.H. var. *Jimboii* (Pant.) A Cl. (Planktonic)

Gomphonema constrictum Ehr. var *capitata* (Ehr.) Cl. (Planktonic)

Navicula injlaia Donk (Planktonic)

Nitzschia umbilicata Hust. (Planktonic)

Chlorophyceae

Coelastrum microporum Naeg. (Planktonic)

Cosmarium auriculatum Reinsch (Planktonic)

Euastrum binale (Turb.) Ehr. *var.juvae* croas. (Benthic)

Pediastrum tetras (Ehr.) Ralfs.var. *excisum* (Rabenh.) Hans. (Planktonic)

Scenedesmus acuminatu (Lager) Chod. (Periphytic)

S. quadricauda (Trup.) Breb. (Planktonic)

S. quadricauda var. *bicaudatus* Hans. (Periphytic)

Spirogyra elliptica Jao (Benthic)

Euglenophyceae

Euglena acus Ehr. (Planktonic)

Results and Discussion

The present study revealed that out of 68 algal taxa, 19 taxa-belong to-Cyanophyceae, 15 to Bacillariophyceae, 31 to Chlorophyceae and only 3 taxa to Euglenopyceae. Thus green algae are in maximum number followed by blue green and Diatoms. The representation of euglenophyceae is least in this survey. The taxa mentioned in section' C' indicate that they able to tolerate water pollution condition to some extent.

Acknowledgements

The author is thankful to the Principal, Govt. P.G. College, Khargone for laboratory facilities.

References

Ahmed, M.R., 1967. Algal Flora of some ponds of Kanpur. *Hydrobiol.*, 29(1–2): 156–164.

Desikachary, T.V., 1959. *Cyanophyta.* ICAR, New Delhi.

Mahajan, S.K., 1987. Algal flora of Pawagiri, Oon, Madhya Pradesh. *Phykos*, 26: 61–62.

Mahajan, S.K., 1988. Phycological studies of Aurangpura, Madhya Pradesh. In: *75th Proc. Ind. Sci. Cong.*, Part III, Abst. p. 16.

Mahajan, S.K., 2004. Validity and affinities of *Spirulina mahajanii* (=*Arthrospira mahajanii*) with existing species of Spirulina (= *Arthrospira*). *Geobios*, 32: 96.

Mahajan, S.K., 2005. Algal flora of a recently constructed dam on river Balki in West Nimar district of M.P., India. *Hydrobiology*, 8(2): 113–116.

Phillipose, M.T., 1967. *Chlorococcales*. ICAR, New Delhi.

Randhawa, M.S., 1959. *Zygnemaceae*. ICAR, New Delhi.

Seerwani, A.B., 1963. Occurrence of *Enteromorhpa* at Khargone, Madhya Pradesh, *Curr. Sci.*, 32: 182.

Venkateshwarlu, V., 1970. An ecological study of the algae of the river Moosi, Hyderabad (India) with special reference to water pollution. IV. Periodicity of some common species of algae. *Hydrobiologia*, 35(1): 45–64.

Water and Wastewater Management in 2 Vols (2011) *Pages 276–280*
Editors: **D.R. Khanna, R. Bhutiani & Gagan Matta**
Published by: **DAYA PUBLISHING HOUSE, NEW DELHI**

Chapter 42

Primary Production Efficiency in Different Environmental Settings of Sahstradhara Hill-stream, Dehradun

D.S. Malik and P.K. Bharti
Department of Zoology and Environmental Science,
Gurukul Kangri University, Haridwar – 249 404

ABSTRACT

An investigation was conducted on primary productivity and water quality of Sahastradhra hill-stream for evaluation of aquatic ecosystem status in different environmental settings along with the stream. The factors such as temperature, pH, carbon dioxide and plankton on which primary productivity depends were studied. It was pointed out that the Gross primary productivity showed a negative relationship with temperature, pH and carbon dioxide.

The present chapter deals with the production efficiency of Sahastradhara hill stream, Dehradun and maximum production efficiency of Sahastradhara stream was found 66.66 per cent at sampling site- III in the month of December 2003, while minimum production efficiency was recorded 20.79 per cent at site- IV during May 2004.

Keywords: *Production efficiency, Primary productivity, Photosynthesis, Hill-stream ecology.*

Introduction

The rate of production of organic matter per unit time is termed as productivity with the two levels; primary and secondary in producers and consumers respectively. The amount of new organic matter as plant tissue built up by photosynthesis is termed as primary production. Primary productivity is the rate at which the sun's radiation energy is stored by photosynthetic activity of producer organisms.

The animal and plant communities inhabiting water bodies are considered as the direct manifestation of the productivity. It is well established that the primary productivity is controlled by several physico-chemical and biotic interactions (Pandey and Mishra, 2000).

There are few successive steps of production process; Gross Primary Productivity (GPP) rate of photosynthesis and includes the organic matter used up in the respiration; Net primary productivity (NPP) the rate of storage of organic matter in plant tissues in excess of the respiratory utilization by the producer; Net community productivity (NCP) or community respiration (CR) the rate of organic matter not used by heterotrophs. Production effeciency depends upon net primary productivity. If NPP is relatively high, than production efficiency will be automatically high in the particular ecosystem.

Materials and Method

Five sampling site were selected for the measurement of primary production and efficiency in Sahastradhara river. We choose the *light and dark bottle method* of Gaarder and Gran for estimation of primary productivity. The water samples were collected monthly from five sampling stations of 'Sahastradhara stream' during December 2003 to May 2004 in morning period 9:00 hrs. to 10:00 hrs. The samples for plankton and physico-chemical parameters were collected and analyzed by using rinsed borosil glassware, with the help of the procedure described by APHA (1995), Trivedi and Goel (1984), Santhanam *et al.* (1989) and Bharti (2004).

Results and Discussion

It is well established that the primary productivity is controlled by several physico-chemical and biotic interactions. The biotic communities inhabiting water bodies are considered as the direct manifestation of the productivity. The maximum gross primary productivity of Sahastradhara stream was observed during winter season. Kar *et al.* (1978) recorded high productivity also in winter season.

Table 42.1: Primary Productivity (Mean values in mg/l) of Sahastradhara Stream

Parameters	*Site–I*	*Site–II*	*Site–III*	*Site–IV*	*Site–V*
NPP	(0.200–0.355)	(0.200–0.255)	(0.140–0.500)	(0.105–0.320)	(0.200–0.330)
Mean	0.271±0.069	0.220±0.0215	0.336±0.108	0.231±0.086	0.265±0.0522
GPP	(0.605–0.805)	(0.600–0.705)	(0.550–1.01)	(0.505–0.900)	(0.580–0.810)
Mean	0.6925±0.074	0.640±0.0776	0.716±0.1.344	0.729±0.0975	0.676±0.826
CR	(0.405–0.450)	(0.400–0.450)	(0.390–0.480)	(0.400–0.580)	(0.380–0.480)
Mean	0.420±0.0184	0.420±0.0635	0.421±0.0394	0.498±0.156	0.411±0.0375
NPE (per cent)	(33.05–44.09)	(33.33–36.17)	(25.45–66.66)	(20.79–37.03)	(34.48–42.55)
Mean	38.77±4.7	34.29±1.101	44.15±15.56	30.485±6.445	38.87±3.069

Temperature plays a vital role in all physico-biochemical reactions and self-purification power of aquatic system. Temperature fluctuation measurement is useful to determine the trends of bio-chemical and biological activities. Temperature was recorded maximum (17.7°C ± 0.4) in the month of May and minimum (13.1°C±0.66) in January. The temperature of water decreases form September and increases from January onward. Badola and Singh (1981) observed similar trend in the river Alaknanda in the Garhwal region. Joshi and Kathait (1998) also reported similar trend in three Tributaries of river Ganga. The higher productivity in the present studies was concomitant with water temperature range from 14°C to 18°C. Singh and Swaroop (1980) observed positive correlation with productivity and

Table 42.2: Planktonic Population (Mean Values) of Sahastradhara Stream

Planktonic Diversity	Site–I	Site–II	Site–III	Site–IV	Site–V
Total plankton (n/l)	(700–1540)	(725–1525)	(693–1480)	(700–1490)	(730–1583)
Mean	1130.25±316.8	1133±316	1101.5±323.9	1063.75±346.9	1163.25±339.4
Phytoplankton (n/l)	(625–1380)	(635–1380)	(615–1345)	(635–1340)	(640–1410)
Mean	1018.25±286.1	1023.75±291.5	992.5±296.3	951.25±318.2	1037.5±259.5
Zooplankton (n/l)	(75–160)	(77–145)	(78–140)	(65–150)	(90–173)
Mean	112±31.61	109.25±27.09	109±28.6	112.5±31.72	125.75±36.66

Table 42.3: Physico-chemical Parameters (Mean Values) of Sahastradhara Stream

Parameters	Site–I	Site–II	Site–III	Site–IV	Site–V
Temperature (°C)	(14–18)	(13 17.5)	(13–17)	(13.5–18)	(12–18)
Mean	15.75±1.14	15.37±1.63	15.0±1.41	15.75±1.6	15.75±2.277
Velocity (m/s)	(0.3–0.4)	(0.28–0.32)	(0.8–1.0)	(0.78–0.82)	(0.6–0.7)
Mean	0.3±0.14	0.9±0.1	0.35±0.05	00.8±0.02	0.65±0.05
Turbidity (JTU)	(1.0–5.0)	(2.0–5.0)	(3.0–6.0)	(2.0–6.0)	(3.0–7.0)
Mean	2.75±1.47	3.25±1.30	4.5±1.15	3.75±1.48	4.75±1.48
Total solids (mg/l)	(955–1427)	(1090–1332)	(1115–1443)	(1135–1520)	(1148–1535)
Mean	1218±183.24	1234.2±89.8	1277.25±122.6	1305.7±166.1	1290.75±146.1
TDS (mg/l)	(840–1278)	(895–1217)	(1010–1322)	(1039–1380)	(965–1420)
Mean	1093.25±174.9	1094.25±120.5	1140.5±134.6	1189±158.01	1134.25±176.1
TSS (mg/l)	(113–149)	(119–195)	(105–207)	(105–140)	(115-211)
Mean	124.75±14.39	140.0±28.16	136.75±40.95	116.75±13.8	156.5±41.69
pH	(7.2–8.1)	(7.3–8.1)	(7.5–7.9)	(7.6–8.0)	(7.2–8.1)
Mean.	7.625±0.43	7.6±0.28	7.75±0.15	7.75±1.66	7.6±0.339
Free CO_2 (mg/l)	(1.40–1.69)	(1.62–1.92)	(1.62–2.20)	(1.65–1.85)	(1.62–2.20)
Mean	1.6525±0.39	1.75±.017	1.825±0.25	1.74±0.09	1.87±0.21
DO (mg/l)	(7.04–11.08)	(6.84–10.07)	(7.04–9.47)	(6.04–9.06)	(6.04–9.47)
Mean	8.41±1.62	8.46±1.18	8.65±0.95	7.55±0.87	7.49±1.36
BOD (mg/l)	(1.60–2.0)	(1.7–2.2)	(2.0–2.5)	(1.8–2.2)	(1.9–2.3)
Mean	1.875±0.162	1.98±0.23	2.25±.18	2.0±0.14	2.05±0.15
COD (mg/l)	(2.1–3.0)	(2.2–2.9)	(2.6–3.2)	(2.4–3.0)	(2.4–2.9)
Mean	2.55±0.32	2.6±0.25	2.9±00.22	2.68±00.22	2.7±0.18
Hardness (mg/l)	(190-210)	(185–220)	(200–250)	(180–230)	(185–235)
Mean	196.25±8.2	200.0±15.41	222.5±19.2	200.0±18.71	198.75±21.02
Alkalinity (mg/l)	(150'–480)	(140–500)	(170–600)	(130–570)	(120–540)
Mean	250.0±134.7	262.5±140.8	305±172.1	292.5±168.8	317.5±175.2

water temperature. A positive correlation existed between gross primary productively and calcium in Cauvery (Lakshminarayan, 2001). Vashney *et al.* (1983) recorded a positive correlation between gross primary productivity and dissolved oxygen while the negative correlation between gross primary productivity and temperature and free carbon dioxide. Sumithra (1971) reported a distinct correlation between productivity and alkalinity, we have also find the similar on individual values but there may be a standard error in correlations due to the mean value of parameters during the different months.

Table 42.4: Correlation (=CORREL) Between Primary Productivity (Mean values) and Physico-chemical Parameters (Mean values) of Sahastradhara Stream

Parameters	*Net Primary Productivity*	*Gross Primary Productivity*	*Respiration*	*Net Production Efficiency*
Temperature (°C)	–0.59945	0.006189	0.311429	–0.57232
Velocity (m/s)	–0.82147	–0.40554	0.387568	–0.78591
Turbidity (JTU)	0.459379	0.237629	–0.09751	0.389256
Total Solids (mg/l)	0.086601	0.565255	0.544701	–0.16675
TDS (mg/l)	0.01787	0.761621	0.801179	–0.33031
TSS (mg/l)	0.165469	–0.58282	–0.73453	0.444083
pH	0.409892	0.87509	0.647625	0.023325
Free CO_2 (mg/l)	0.335796	–0.03623	–0.24934	0.360837
DO (mg/l)	0.43596	–0.20694	–0.47748	0.522691
BOD (mg/l)	0.715985	0.330189	–0.11921	0.583037
COD (mg/l)	0.757185	0.484589	–0.01968	0.572202
Hardness (mg/l)	0.811178	0.38677	–0.13668	0.659382
Alkalinity (mg/l)	0.380434	0.349649	0.067672	0.259953

In this study it was recorded that the pH was always slightly alkaline. The pH of natural water was controlled in a greater extent by the interaction of hydroxyl ions, resulting form the dissociation of carbonic acid, and form hydroxyl ion arising from the hydrolysis of bicarbonate, was recorded highest (8.02 ± 0.1166) in December and lowest (7.4 ± 0.1414) in the month of April. The higher rate of photosynthesis decreases free carbon dioxide content and increases dissolved oxygen (Denham, 1938) while higher rate of decomposition increases free carbon dioxide. Similar observations were found in Sahastradhara stream during investigating period. Varshney *et al.* (1983) observed negative correlation between productivity and turbidity, total dissolved solids and suspended solids because these factors deceases the light penetration and further affects the rate of photosynthesis in aquatic ecosystem (Khanna, 1993), while negative correlation between GPP and pH, temperature and free carbon dioxide.

The maximum value of free carbon dioxide was observed (2.2 mg/l ±0.155) in May and minimum value in January (1.70 mg/l ±.066) free carbon dioxide and dissolved oxygen showed a negative relationship to one another. The highest rate of photosynthesis decreases free carbon-dioxide content and higher rate of decomposition increases free carbon dioxide.

The maximum net primary productivity was observed (0.335 ±.093) in the month of December and minimum (0.170 ±.04) in the month of May. The maximum gross primary productivity was observed (0.796 ±.123) in the month of January and minimum (0.569 ±.0379) in the month of May. The maximum net community respiration was observed (0.479 ±.0575) in the month of January and minimum (0.399 ±.0102) in the month of April. Net production efficiency shows the positive correlation with dissolved oxygen, while the negative correlation between net production efficiency and carbon dioxide. The maximum net production efficiency was observed (43.79 per cent) in the month of December and minimum (29.88 per cent) in the month of May. Similar trend was observed by Yeragi and Shaikh (2003) on the primary productivity of Tansa River. Kar *et al.* (1978) however, recorded low and high productivity in summer and winter respectively.

References

APHA, 1995. *Standard Methods for Examination of Water and Wastewater*, 19th edn. American Public Health Association, Inc., New York, pp. 1970.

Badola, S.P. and Singh, H.R., 1981. Hydrobiology of the river Alaknanda of Garhwal Himalaya. *Indian J. Ecol.*, 8(2): 269–276.

Bharti, P.K., 2004. Planktonic population in Sahastradhara hill stream at Dehradun. *M.Sc. Dissertation*, pp. 102.

Denham, S.C., 1938. A limnological investigation of the West Fork and common branch of White river, Invest. *Indian Lakes and Streams*, 1: 17–71.

Joshi, B.D. and Kathait, R.S., 1998. Limnological characteristics of three tributaries of river Ganga between Devprayag and Rishikesh. *Him. J. Env. Zool.*, 12: 225–243.

Kar, O.K., Mishra, P.C., Das, M.C. and Das, R.C., 1978. Pollution studies in river Ib. III Plankton. *Pollution and Primary Productivity*, p. 175.

Khanna, D.R., 1993. *Ecology and Pollution of Ganga River*. Ashish Publishing House, New Delhi, p. 1–241.

Lakshminarayana and Someshekar, R.K., 2001. *Ecology of Polluted Water*, (Ed.) Arvind Kumar. APH Pub. Corp., New Delhi, 1(2): 51–60.

Pande, R.K. and Mishra, Asha, 2000. Water quality study of freshwaters of Dehradun (Sahastradhara stream and Mussoorie Lake). *Aquacult.*, 1: 57–62.

Santhanam, R., Velaynatham, P. and Jegatheesan, G., 1989. *A Manual for Freshwater Ecology*. Daya Publishing House, Delhi, p. 1–92.

Singh, S.R. and Swaroop, K., 1980. Studies on primary production of phytoplankton in Surha lake (Ballia India). Int. Revue. Ges. *Hydrobiol.*, 65(5): 706–717.

Sumitra, V., 1973. Seasonal variation in primary production in three tropical ponds. *Hydrobiologia*, 48: 117–123.

Trivedi, R.K. and Goel, P.K., 1984. *Chemical and Biological Methods for Water Pollution Studies.* Environmental Publication, Karad, pp. 1–251.

Varshney, P.K., Nair, R. Vijayalakshmi and Abid, S.H.H., 1983. Primary productivity in near shore water of Thai Maharashtra coast, India. *J. Mar. Sci.*, 12: 27–30.

Yeragi, S.G. and Shaikh, Nisar, 2003. Studies on primary productivity of Tansa River. *J. Natcon*, 15(1): 125–130.

Water and Wastewater Management in 2 Vols (2011) *Pages 281–285*
Editors: **D.R. Khanna, R. Bhutiani & Gagan Matta**
Published by: **DAYA PUBLISHING HOUSE, NEW DELHI**

Chapter 43

Fenvalerate: A Synthetic Pyrethroid Toxicity to Freshwater Fishes

K.S. Tilak and K. Yacobu
Department of Zoology, Nagarjuna University, Nagarjuna Nagar – 522 510, A.P.

ABSTRACT

Synthetic Pyrethroids are the new major class of broad spectrum organic insecticides used in Agriculture. Among the Pyrethroids, Fenvalerate (RS a cyano 3 Phenoxy Benzyl RS 2-[4 Chlorophenyl] 3-Methyl Butyrate) is used extensively and is most effective Pyrethroid. The technical isomer is a mixture of four optical isomers 2 S, a S, 2R and a S. The 2S isomers are more toxic than 2R. They are toxic to aquatic organisms. When they are being transported into aquatic environment. The toxic action of the toxicant will depend on metabolism and mode of action in aquatic organism. Hence to study it static and continuous flow through tests were conducted to determine LC_{50} values to the freshwater fish. The possible mechanism of toxic action will be presented.

Keywords: *Fenvalerate, Toxicity, Freshwater fishes, Toxicological mechanism.*

Introduction

The pyrethroids are the esters of cyclopropane carboxylic acids with alkenyl methyl cyclopentenolene alcohols. The activity depends on the ester and is influenced by the absolute configuration of the asymmetrical carbon atom at C1 of the cyclopropane ring and at C4 of the cyclopentenolene ring. All very active pyrethroids have R configuration at C 1 and 'S' configuration at C4. The presence of gem dimethyl group at C2 of the cyclopropane ring is essential for a high insecticidal activity. The substituents at C3 of the cyclopropane ring either as C1s or trans position are not an absolute requirement because even without substituents the insecticidal action is present but only saturated side chains give high activity. The replacement of the isobutenyl group by an isobutyl decreases activity whereas a butadienyl group may enhance potency (Wouters and Berchen 1978). The important pyrethroids are Allethrin, Bioresmethrin, Permethrin, Decamethrin, Cypermethrin and Fenvalerate. All of them are highly active and stable insecticidal pyrethroids, hence being used extensively in agriculture, to control the insect pests, to control the vectors of endemic diseases, to protect the seeds during storage and to fight household insects.

Due to this widespread usage, they will be transported into aquatic environment, which is the ultimate sink for all pesticides, and aquatic toxicity has become an integral part of the process of environmental hazard evaluation of toxic chemicals. Generally, the potential impact of pollutants is more on the aquatic organisms, than terrestrial environment (Murty, 1986).

Among the Pyrethroids, the neurotoxic organic insecticide fenvalerate (alfa cyano-3-phenoxy-phenyl) methyl-4chloro-(lmethyethyl Benzene acetate) is used to control boll warms on cotton, tobacco plants and vegetables. It is being extensively used in the local areas and keeping the view of the transportation, accumulation, concentration, the toxic impact of the toxicant to the freshwater fish is attempted. The portenous, palatable precious table item can be linked to the human food chain and the possible deleterious effects can be given a due consideration.

Materials and Method

The freshwater fishes (sizes of the fish are given in Table 43.1) were brought from local fish farms and by fisherman. The fishes were acclimatized to the laboratory conditions in well-aerated non-chlorinated tap water at test medium conditions. During the period of acclimatization and experimentation the fish were not fed. If the number of deaths exceeded 5 per cent in any batch of fish during acclimatization, the concerned batch was discarded.

The studies on toxicity were conducted for technical as well as 20 Emuslifiable Concentrate (Ee) employing static and continuous flow through systems as recommended in the report of the committee on methods of toxicity tests with aquatic organisms (Amon, 1975).

The solutions of desired concentrations were prepared in acetone to yield a concentration of 100 mg/ml. The control fish received an equal quantity of acetone to that of the highest concentration of the toxicant tested.

For flow through system, test solutions of desired concentrations were prepared once in five hours in glass reservoirs and were let into the test containers through thin-walled polythene tubes. The flow rate was adjusted with regulators such that four litres of water passed through containers in one hour. The conditions of the test medium were temperature 28±2°C, oxygen 6-8 ppm, Hardness 80 mg/L, Alkalinity–425 mg/L and pH 8.3. All the precautions laid down in the report of committee on toxicity tests were followed (Amon, 1975).

The pilot experiments were conducted to determine the concentrations causing mortality of test fish in' the range of 10 to 90 per cent. For each concentration, 10 fish were tested and the experiment was repeated thrice. Probit analysis (Finney, 1971) as recommended by Roberts and Boyce (1972) was followed to calculate the LC_{50} values. The technical fenvalerate and 20 per cent EC were supplied by Searle (India) Limited, Mumbai.

Results and Discussion

The earlier reports on the toxicity of fenvalerate to the fishes are: Bradbury and Coats (1982), Bradbury *et al.* (1985, 86, 87) Coats and Jeffery (1979), Zitko *et al.* (1977, 79), Hya (1989) and

Anitha *et al.* (1999a and 1999b). The LC_{50} values were given in the Table 43.1. The toxic action of fenvalerate is accounted in earlier reports due to its change as metabolites or due to certain biochemical changes in the tissue.

Table 43.1: The LC_{50} Values and their 95 per cent Confidence Limits of Fenvalerate Technical Grade to Freshwater Fishes for 96 Hours in Static and Continuous Flow Through System

Name of the Fish and Toxicant	*Length of the Fish*	*LC_{50} Value for 96 Hours*	
		Static	*Continuous Flow through System mg/l*
Labeo rohita			
Fenvalerate (Technical)	2-4 cm	0.012	0.0031
20 per cent EC	2-4 cm	0.016	0.0037
Catla Catla			
Fenvalerate (Technical)	2-4 cm	0.004	0.0018
20 per cent EC	2-4 cm	0.017	0.0006
Cirrhinus mrigala			
Fenvalerate (Technical)	2-4 cm	0.002	0.0001
20 per cent EC	2-4 cm	0.016	0.0050
Aplochelus panchax			
Fenvalerate (Technical)	2-4 cm	0.027	0.014
20 per cent EC	2-4 cm	0.002	0.009
Ctenopharyngodon idellus			
Fenvalerate (Technical)	3-5 cm	1.9	1.90
20 per cent EC	3-5 cm	2.60	1.70

The values are tested for chisquare and found to be not significant at $P<0.05$. The 95 per cent confidential limits are also calculated.

Generally, there are three main chemical reactions involved in the degradation of a toxicant *i.e.* isomerization, hydrolysis and oxidation (Demoute, 1989). The fenvalerate has the common degradation products in mammals, birds and amphibians. The basic metabolic reactions are similar in all, *i.e.* cleavage of ester bond and oxidation of released acid and alcohol fragments. The initial transformations are followed by conjugation with sugars or amino acids which allows faster exertion.

Same type of degradation is reported in fish also, Bradbury *et al.* (1987) studied the toxicokinetics of fenvalerate in rainbow trout and reported the uptake of the gill. The qualification of the residues of the gill tissues; confirmed the isomer of venvalerate which is toxic *i.e.* 2s isomer the degradation product. The qualification and quantification of the residues proves that fenvalerate is highly toxic for fish and is species specific. In vertebrates except fish pyrethroids as opposed to organochlorines, which have the same ligophilicity are readily metabolized and excreted. The half-life periods never go beyond a few days. Birds have a larger capacity to eliminate their products than mammals. The high toxicity to fish is partly explained by their poor ability to metabolize them. Takimotto *et al.* (1986) reported that the lower rates of elimination and metabolism of fenvalerate in trout is the factor responsible for higher toxicity.

Their metabolism is largely oxidative in fish (Edwards *et al.*, 1987). The fenvalerate acid moiety is very different from the other groups of pesticides and the ester cleavage releases the acid fragment, which is mainly excreted or as a glucorinide conjugate. The cyanide ion is transformed into thiocyanate and also carbon dioxide. Because of this, the freshwater fish are affected due to toxic action of fenvalerate.

The toxicity varies in LC_{50} value of 96 hours in continuous flow through system as well as static. The residue analysis by TLC and GLC (Anitha Susaan *et at.* 1996 and 1999) showed the deposition in fish tissues as residues.

In the Table 43.1, toxicity data of the experiments confirmed that the commercial formulation is more toxic than the technical compound. The metabolites formation, storage of the pesticide after manufacture and as there was no strict quality control all of them are causing synergistic effect to the fish apart from the above mentioned reasons of earlier reports. Although, the pyrethroids are safest insecticides, when administered orally. But they may become highly toxic when they reach central nervous system in sufficient concentration. The sensitivity of the fishes in the present study to technical grade fenvalerate and 20 per cent EC formulation was in the order of *Cirrhinus mrigala* > *Catla catla* > *Labeo rohita* > *Aplocheilus panchex* > *Ctenopharyngodon idellus*.

Hence the formulation products must be viewed seriously before any pesticide is given a representation to be introduced into the environmental usage.

References

Anita Susan, T., Veeeraiah, K. and Tilak, K.S., 1999a. Biochemical changes in the tissues of the fish *Catla catla* exposed to fenvalerate a synthetic pyrethroid. *J. Ecobiology,* 11(2): 109–116.

Anita Susan, T., Veeeraiah, K. and Tilak, K.S., 1999b. A study on the bioaccumulation of fenvalerate a synthetic pyrethroid in the whole body tissue of *Labeo rohita, Catla catla, Cirrhinus mrigala* by GLC. *Enviromedia,* 18(1): 57–59.

Amon, 1975. *Committee on Methods of Toxicity Tests with Fish, Macroinvertebrates and Amphibians.* EPA Oregon, 61 pp.

Bradbury, S.P. and Coats, J.R., 1982. Toxicity of fenvalerate to bobwhite quail including brain and liver residues associated with mortality. *J. Toxicol. and Environ. Health,* 10: 307–319.

Bradbury, S.P., Coats, J.R. and Mckim, J.M., 1985. Differential toxicity and uptake of two fenvalerate formulation in fathead minoows (*Pimephales promelas*). *Environ. Toxicol. Chem.,* 4: 533–541.

Bradbury, S.P., Coats, J.R. and Mckim, J.M., 1986. Toxicokinetics of fenvalerate in rainbow trout *Salmo gairdneri. Environ. Toxicol. Chem.,* 5: 567–576.

Bradbury, S.P., McKim, J.M. and Coats, J.R., 1987. Physiological response of rainbow trout to acute fenvalerate intoxification. *Pestic. Biochem. Physiol.,* 27: 275–288.

Coats, J.R. and O'D Jefery, N., 1979. Toxicity of four synthetic pyrethroid insecticides to rainbow trout. *Bull Environ. Contam. Toxicol.,* 23: 250–255.

Demoute, J.P., 1989. A brief review of the environmental fate and metabolism of pyrethroids. *Pestic. Science,* 27: 375–385.

Edwards, R. and Millburn, P., 1985. The metabolism and toxicity of insecticides in fish. In: *Progress in Pesticides Biochemistry and Toxicology, Vol. 5: Insecticides,* (Eds.) D.H. Hutson and T.R. Roberts. John Wiley and Sons Ltd., New York, pp. 249–274.

Finney, D.J., 1971. *Probit Analysis.* Cambridge University Press, 333 pp.

Haya, K., 1989. Toxicity of pyrethroid insecticides to fish. *Environ. Toxicol. Chem.,* 8: 381–391.

Murty A.S., 1986. *Toxicity of Pesticides to Fish, Vols. I and II.* CRC Press Inc.

Roberts and Boyce, C.B.C. 1972. In: *Methods of Microbiology, Vol. 7A,* (Eds.) Norris and D.W. Ribbons, Academic Press, New York.

Takimotto, Y., Matsyda, O.M.T. and Miyamoto, M., 1986. Metabolism of several pyrethroid insecticides by carp *Cyprinus carpio.* In: *Abstract, The Sixth International Congress of Pesticides Chemistry,* August 10–15.

Wouters, W. and Bercken, J.V.D., 1979. Action pyrethroids. *General Pharmacology,* 9: 387–398.

Zitko, V., Carson, W.G. and Metcalfe, C.D., 1977. Toxicity of pyrethroids to juvenile Atlantic salmon. *Bull. Environ. Contem. Toxicology,* 18: 35–41.

Zitco, V., McLeese, D.W., Metcalfe, C.D. and Carlson, W.G., 1979. Toxicity of perrnethrin, decamethrin and related pyrethroids to Salmon (*Salmo Saler*) and lobster. *Bull. Environ. Contam. Toxicol.,* 21: 338–343.

Water and Wastewater Management in 2 Vols (2011) *Pages* **286–293**
Editors: **D.R. Khanna, R. Bhutiani & Gagan Matta**
Published by: **DAYA PUBLISHING HOUSE, NEW DELHI**

Chapter 44

Pesticidal Use in Swampy and Derelict Agro System in Champaran District, Bihar, India: A Detrimental Menace to Fish Farming

Bijay Bhushan Prasad[1] and Krishna Mohan Singh[2]
[1]*Aquaculture Laboratory, Department of Zoology, S.N.S. College, Motihari – 845 401, Bihar*
[2]*Department of Zoology, J.L.N.M. College, Ghorashan, East Champaran, Bihar*

ABSTRACT

In the new millennium and with the advent of high yielding varieties of paddy, the use of various pesticides, some of which may be in minute quantities are highly toxic to aquatic life. Fish culture, therefore, is not more compatible with paddy farming wherever the latest high yielding varieties of paddy are cultivated. Champaran (Bihar) is known to be an area of high quality yielding rice and are even exported to several countries. There are numerous swampy and derelict lentic water bodies in this area. In recent years, however, attempts are being made in the fish culture to utilize the vast swampy and derelict water area for immediate use for the benefit of the poor, without getting involved in to costly reclamation process. Despite of poor and unorganized capture fishery, the per cent yield was estimated to be good. The fishes like *Channa, Heteropneust, Anabas. Clarias* and even some Carps are being cultured in the paddy fields as well as in the water bodies of this area. By virtue of their hardy nature and air breathing habits in many cases, they are excellent material not only for utilizing swampy and derelict water bodies but also in paddy fields as they permit high stocking density and respond to supplementary feeding. Their production potential is by and large directly proportional to inputs and intensity of operational management.

Despite of good inputs of these commercial fishes, the potential toxic compounds in various chemicals used for agricultural and other domestic purposes get distributed by a variety of means and accumulate concentrations in the soil and water. These accumulated organic compounds lower the DO level and does not support fish life. Fish species are far from being biochemical inert. As a matter of facts several pesticides are known to induce microsomal enzyme systems in the liver of fish. Physiology, biochemistry and pathology has enabled us to assess the toxicity of the water body with fish and in these circumstances, fish can be used as an indicator of water quality.

Keywords: *Channa marulius, Swampy, Derelict, Pesticide, Mortality,* LC_{50}.

Introduction

East Champaran district has a wide range of paddy fields with several lentic water bodies, which remained unnoticed. In the area of paddy fields, unmanaged lakes and ponds which retain water for 4 to 8 months aquaculture (especially the air breathing fishes) can be carried out. These water bodies are major sources of air breathing culture for the people of this economically and socially backward area. The culture of the fishes in the area which remain flooded even after the paddy is harvested and lentic water bodies after rainy season serve as an off season occupation for farmers of this area.

In the recent years, however, with the advent of high yielding varieties of paddy, the use of various pesticides, many of which even in minute quantities are highly toxic to aquatic life, has become widely prevalent. Champaran is known to be an area of high quality yielding rice, but apart from this, the people of this area are still suffering either from malnutrition or under-nutrition.

The effect of different chemicals is different and quite complicated to understand without scientific study. The potentially toxic compounds in the various chemicals used for agricultural and other domestic purposes are distributed by a variety of means and get accumulated in the soil and water. These accumulated organic compounds have to be decomposed. Some chemicals do not show the active toxicity to fish in the small concentration as they show in higher animals. It is totally incorrect to infer that this is due to the fact that fish lack some of the more common enzyme systems present in the mammals and that the chemical concerned to not therefore pose a serious hazard to them. On the contrary, fish species are far from being biochemical inert system and as a matter of fact several pesticides are known to induce system of microsomal enzyme system in the liver of fish. Thus, fish can be used as an indicator of the quality of water which may end our search for an "acting model"

Determination of median concentration (LC_{50}) value of a pesticide can be used to regulate the discharge of pesticide into nearby water system to protect aquatic life; whereas the safe concentration are actually fractions of LC_{50} values and therefore, would be useful in regulating their discharge.

Keeping all the above mentioned facts in mind, the present work has been aimed to find out the mortality rate and LC_{50} values of most common organophosphate (Abate) and carbamate pesticide (Baygon) individually for common freshwater air breathing fish *Channa marulius* (Bloch). which lives in the swamps, and derelict water bodies having low oxygen and high CO_2 conditions, yet has a great nutrient values as well. Such studies are necessary for the maintaining of healthy stock of the fish in our inland water, which can ensure that the species present are not adversely affected in any way by water pollutants.

Material and Methods

Healthy and living specimens of *Channa marulius* (Bloch) commonly known as 'saur' of 38.10±3.25 g weight groups were collected from the swampy and derelict lentic water bodies of East Champaran district of Bihar, India.

After the collection, fishes were brought to laboratory and subjected to treatment, quarantine and acclimatization processes following the methodology given by APHA (1985). They were treated with Terramycin solution (15 mg/l) in laboratory tap water for 48 hrs. in round mouthed stainless steel drums (Donald *et al.*, 1983) of 30 litre capacity. At the end of 48 hrs., the fishes were washed and then treated with $KMnO_4$ solution (2 mg/l) to remove ectoparasites, fungi and other dermal infection. Mortality was noted during both treatment periods. Now the fishes were subjected to 3 days of quarantine and 7 days of acclimatization periods before being subjected to bioassay "tests of all selected toxicants individually and were patterned following methods described by Doudoroff *et al.* (1951) and APHA (1985). The desired concentration range for toxic solution were done as described in APHA (1985) where as LC_{50} so values were observed by interpolation method and probit analysis method (Finney, 1981).

Results and Discussion

The LC_{50} values of two selected pesticides *i.e.* Abate and Baygon at 24, 48, 72 and 96 hrs of exposure to *Channa marulius* (Bloch.) were determined individually by exposing the fish to various selected concentrations and mortality of such two replications at selected time intervals. No mortality in the fish was recorded either in control or in the fish exposed to 1.3 ppm Abate and 1.3 ppm Baygon concentrations.

Externally, the response of the pesticides on the physical activities of the fish were looked almost similar, but the fish exposed to Abate were comparatively more sensitive than that of Baygon pesticide. The fishes which were exposed to Abate ($0.0.0^1.0^4$)–tetramethy]-0-0'thiod-p-phenylene phosphorothioate) showed restlessness and hyper activity during initial stage followed by erratic swinmling, erratic opercular movements and respiratory distress.

The fishes gradually become inactive, sank to bottom of the container followed by the death without any clear cut convulsions. Fishes secrete mucous on their body surface with a slight change in the colouration of the skin. However, the fish exposed to Baygon (Tsopropoxy phenyl-Methyl carbonate) shows hyperactivity, vigorous swimming, jumping, loss of equilibrium, irritability, convulsions, coma and death. In few cases, haemorrhage, was noticed in the eyes. The mucous secretion on the body surface increases with an increase in the concentration and the formation of coagulated mucous coercing over the entire body surface just prior to death was a common feature in all the fish exposed to selected pesticides. Death was noted when the fish was not responding on prodding with a glass rod and not showing any respiratory movements.

From the result of bioassay tests, LC_{50} values at 24, 48, 72 and 96 hrs. exposure of Abate and Baygon to *Channa marulius* (Bloch) were determined individually by straight-line graphical interpolation method as well as probit analysis method.

The per cent mortality of the fish in different concentrations of the respective pesticides are put on graph and the mortality data and LC_{50} values at the selected hours of exposure, with their slop values and regression equations are presented in Tables 44.1 and 44.2 respectively.

The LC_{50} values obtained by two methods are found almost same as the difference in the values obtained for a particular pesticide at a particular hour of exposure are not found statistically significant in any case, but it is only the probit analysis method, by which we can get standard error and regression equation. Accordingly the mean LC_{50} values for 24, 48, 72 and 96 hrs. exposure are found to be 4.6385±0.1084, 4.1318±0.1060, 2.9036±0.1078 and 2.1768±0.1067 ppb for Abate and 24.0860±1.0673, 21.3900±1.057) 21.3900±1.0517, 14.7465±1.0682 and 9.5289±1.0598±ppm for Baygon respectively (Table 44.2).

Table 44.1: Mortality of *Channa marulius* at Different Concentrations of Abate and Baygon Pesticides at 24, 48, 72 and 96 hrs. of Exposure (Both per cent and Probit kill)

Concentration	Log Concentration	No. of Fish Dead (24 hrs.)	Per cent Kill	Probit Kill	No. of Fish Dead (48 hrs.)	Per cent Kill	Probit Kill	No. of Fish Dead (72hrs.)	Per cent Kill	Probit Kill	No. of Fish Dead (96 hrs.)	Per cent Kill	Probit Kill	Pesticide Used
1	2	3	4	5	6	7	8	9	10	11	12	13	14	15
1.3	0.114	0	0	–	0	0	–	0	0	–	0	0	–	A
5.0	0.699	0	0	–	0	0	–	0	0	–	0	0	–	B
1.7	0.230	0	0	–	0	0	–	0	0	–	2	20	4.1584	A
6.5	0.813	0	0	–	0	0	–	0	0	–	0	0	–	B
2.2	0.342	0	0	–	0	0	–	3	30	4.4756	6	60	5.2533	A
8.5	0.929	0	0	–	0	0	–	0	0	–	2	20	4.1584	B
2.9	0.462	0	0	–	0	0	–	5	50	5.2533	9	90	6.2813	A
11.0	1.041	0	0	–	0	0	–	1	10	3.7184	7	70	5.5244	B
3.7	0.568	1	10	3.7184	2	20	4.1584	9	90	6.2813	10	100	8.7190	A
14.3	11.85	0	0	–	0	0	–	4	40	4.7407	10	100	–	B
4.8	0.681	3	30	4.4756	7	70	5.5244	10	100	8.7190	10	100	–	A
18.6	1.269	2	20	4.1584	4	40	4.7467	8	80	5.8416	10	100	–	B
6.3	0.799	7	70	5.5244	100	10	8.7190	10	100	8.7190	10	100	–	A
24.1	1.382	6	60	5.2533	9	90	6.2816	10	100	–	10	100	–	B
8.1	0.908	10	100	8.7190	10	100	–	10	10	–	10	100	–	A 0
31.4	1.497	9	90	0.2816	10	100	8.7190	10	10	–	10	100	–	B 0
40.8	1.011	10	100	8.7190	10	100	–	10	100	–	10	100	–	A

No of fish exposed 10; A: Abate; B: Baygon.

Table 44.2: LC_{50} Value Obtained by Different Methods along with Regression Equation for Abate and Baygon Pesticides Individually at 24, 48, 72 and 96 hr. of Exposure to *Channa marulius*

Pesticide	*Hours of Exposure*	*LC_{50} Value by Graphical Method*	*LC_{50} Value (+SE)*	*-y = y-bx+bx*	*Regression Line (b)*	*Variance (v)*	*Standard Error (SE)*	*Average LC_{50} Value (Mean of 3 and 4)*
1	*2*	*3*	*4*	*5*	*6*	*7*	*8*	*9*
Abate	24	4.65	4.6270 +0.1084	Y = -0.8601 +7.9446x	7.9416	0.001234	0.0351	4.6385 +0.1084
	48	4.10	4.1629 +0.1060	Y = 2.7130 12.0972x	12.972	0.000646	0.0254	4.1318 +0.1060
	72	2.90	2.9067 +0.1078	Y = 1.7652 +7.7746x	9.1645	0.00108	0.0329	2.9036 +0.1078
	96	2.15	2.2039 +0.1067	Y = 2.0787 +9.1645x	9.1645	0.0008	0.0283	2.1768 +0.1067
Baygon	24	24.10	24.0720 + 1.0673	Y = 7.6879 +9.3465x	9.3465	0.0008	0.0283	24.0860 + 1.0673
	48	21.40	21.3800 +1.0517	Y = 12.4951 +13.5876x	13.5876	0.00048	0.022	21.3900 +1.0517
	72	14.75	14.7430 +1.0682	Y = 6.0206 +9.2384x	9.2384	0.00082	0.0287	14.7465 + 1.0682
	96	9.50	9.5499 + 1.0598	Y = -7.1635 +12.1877x	12.1877	0.00064	0.0252	9.5289 +1.0598

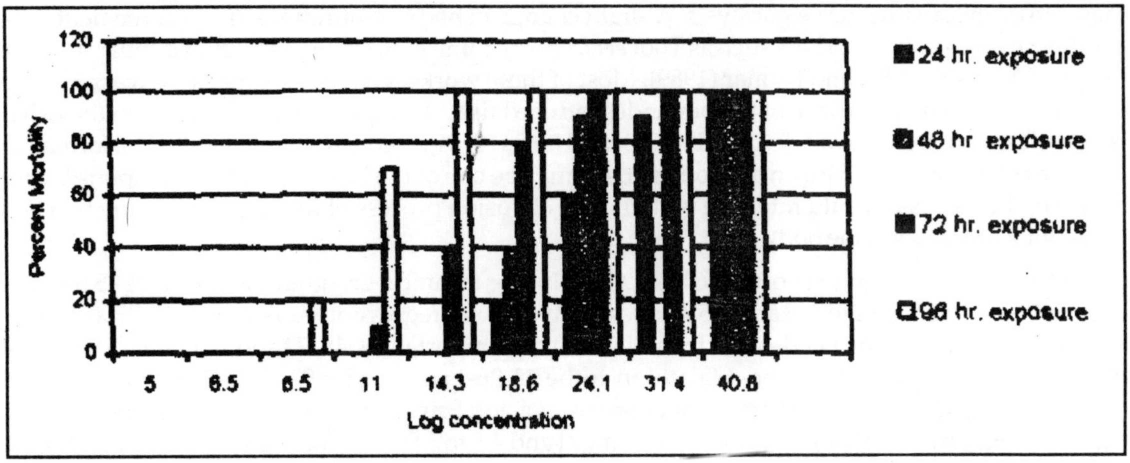

Figure 44.1: Per cent Mortality of *Channa marulius* at Different Concentration of Abate Pesticide

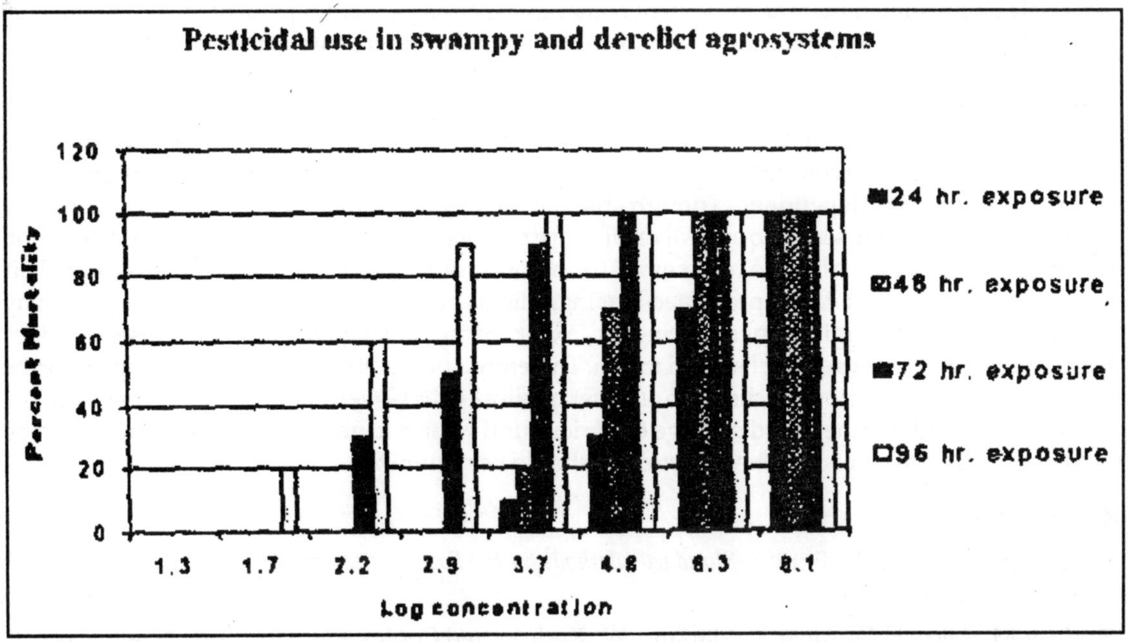

Figure 44.2: Per cent Mortality of *Channa marulius* at Different Concentration of Baygon Pesticide

In the present study, the per cent survival of the fish in both the pesticides decreased with an increase in the concentration of the pesticides, as the fish exposed to the lower concentrations showed less reaction than that of the fish exposed to higher concentration as evidenced by 100 per cent mortality in highest concentration might be due to the hyper activeness of the pesticides exposed fish as reported by Herbert and Sharben (1964).

The abnormal behaviour of the fish exposed to various concentrations of Abate and Baygon as observed above indicates that abate is acting more like a respiratory poison where the Baygon acts as neurotoxicant causing nerve paralysis. A slight change in body colourations are in agreement with the findings of several workers such as Toor *et al.* (1973). Lunn *et al.* (1976), Haider and Inbaraj (1986), Ray and Munshi (1988) and Kumar (1989). Most of these workers have also reported excess muscus secretion on the body surface of the pesticides induced fish. "Coagulation film anoxia" in the fish is also well known from the works of Ellis (1973) and Jones (1964). Hence in the present study, formation of a coagulated mucous film on the body surface may be one of the important factors responsible for death of the fish as it would interfere with the gas diffusion process of the respiratory function and other vital processes of the body.

Devi *et al.* (1981) have reported 96 hrs. LC_{50} value for technical grade endosulfan and 35 per cent EC for *Channa punctatus* to 4, 8 and 2.5 ppb respectively with regression equation, Y = 7.75 X–15.8 Y– +2.58 X-1.16 respectively. Haider and Inbaraj (1986) have reported 24, 48, 72 and 96 hr. LC_{50}- values of technical grade Endosulfan and Malathion to be 11.26-11.43, 9.68-9.82, 7.56-7.96 and 5.78–6.13 mg/l respectively. Other workers notably Srivastava and Srivastava (1988) observed 7.30 ppm for *H. fossilis*. Singh *et al.* (1984) found 1.52 to 5.0 mg/l and 4.5 to 5.0 mg/l for *C. punctatus* and *H. fossilis.*

The present findings, when compared with the findings of the earlier workers, reflect that the pesticides of commercial formulations are generally more toxic than their technical grades. Thus, it can safely be presumed that the ingredients other than actual pesticides present in the commercial formulation may contribute towards synergism by enhancing the toxicity of the actual pesticides.

Conclusions

The pesticide toxicity of Abate and Baygon on the fish *Channa marulius* (Bloch) at various hrs. of exposure and at different concentrations resulted that the mortality was high at higher concentrations and is concentration dependent. Though, fish showed some similar characteristics at higher concentration, yet Abate seems to act as a respiratory poison, whereas Baygon act's as neurotoxicant. The result of LC_{50} reflects that the commercial formulation of the pesticides are more toxic than their technical forms. Hence, it can be presumed that ingredients present in the commercial pesticides may contribute towards synergism by enhancing the toxicity of the actual pesticides. Both these pesticides Abate and Baygon are extremely harmful to fish. Therefore, some restrictions would be required to be put on their use and a periodical check up of water quality by the Government or some private agency would be obtained. Further the discharge of agricultural wastes to near by water system should be regulated so that the menace of these pesticides should be minimized.

References

APHA, AWWA and WPCF, 1985. *Standard Methods for the Examination of Water and Wastewater*, 16th edn. APHA, Washington D.C.

Dinald, R.M., Allen, H.G., Alian, L., Mariann, W., Karli, H. and Copius, P.S., 1983. Female reproductive system. In: *Reproductive and Developmental Toxicity of Metals*, (Eds.) T.W Clarkson, G.F. Norberg and P.R. Sager. Plenum Press, London, pp. 43–91.

Doudoroff, P., Anderson, B.G., Burdic, G.E., Galtsroff, P.S., Hart, W.B., Patrick, P., Strong, S.R., Surber, F.W. and Van Horn, W.M., 1951. In: *Environmental Pollution by Pesticides*, (Ed.) C.A. Edward. Plenum Press, London, NY, pp. 218.

Devi, A.P., Rao, D.M.R., Tilak, K.S. and Murty, A.S., 1981. Relative toxicity of the technical grade material isomers and formulations of endosulfan to the fish *Channa punctatus*.

Ellis, M.M., 1937. Detection and measurement of stream pollution. *U.S. Bur. Fish Bull.* (*Bull. No.* 2), 48: 365–437.

Finney, D.J., 1981. *Probit Analysis*, 3rd Edn. Cambridge University Press, N.Y. (Indian Reprint, 1981).

Haider, S. and Inbaraj, R.M., 1986. Relative toxicity of technical material and commercial formulation of malathion and endosulfan to a freshwater fish *Channa punctatus* (Bloch). *Ecotoxicol. and Environ. Safety*, 11: 347–351.

Herbelt, D.W.M. and Sharben, D.S., 1964. The toxicity to fish of mixtures of poisons. *Ann. Appl. Biol.*, 53: 33–41.

Jones, J.R.E., 1964. *Fish and River Pollution*. Butterworth, London.

Kumar, M., 1989. Effect of some pesticide on an air breathing and water breathing fish: A comparative study. *Ph.D. Thesis*, Bihar University, Muzaffarpur.

Lunn, C.R., Toews, D.P. and Pree, D.J. 1979. Effects of three pesticides on respiration coughing and heart rate of rainbow trout *Salmogairdneri richardson*. *Can. J. Zool.*, 54: 214–219.

Pickring, Q.H., Handerson, C. and Lenka, A.K., 1962. The toxicity of organophosphorus insecticides to different species of warm water fish. *Trans. Amer. Fish. Soc.*, 91: 175–184.

Ray, P.K. and Munshi, J.S.D., 1988. Oxygen consumption and ventilation rate of a freshwater major carp *Cirrhinus mrigala* (Ham) in fresh and malathion treated water. *J. Environ. Biol.*, 9(1): 5–13.

Saitha, D., Sahib, I.K.A. and Rao, K.V.R., 1981. Toxicity of technical and commercial grade malathion to fish, *Tilapia mossambica. Proc. Indian Acad. Sci.*, 90(1): 87–92.

Singh, V.P., Gupta, S. and Saxena, P.K., 1984. Evaluation of acute toxicity of carbaryl and malathion to freshwater teleosts, *Channa punctatus* (Bloch) and *Heteropneustes fossilis* (Bloch). *Toxicol. Letters*, 2: 271–276.

Srivastava, A.K. and Srivastava, A.K., 1988. Effects of aldrin and malathion on blood chloride in the Indian Catfish *H. fossilis. J. Eviron. Biol.*, 9(1 supl.): 91–95.

Toor, H.S., Mehta, K. and Chinna, S., 1973. Toxicity of insecticides (Commercial formulation) to the exotic fish common carp, *Cyprinus carpio communia* (L). *J. Resp. (Pan) Indian*, 10: 341–345.

Water and Wastewater Management in 2 Vols (2011) *Pages* **294–297**
Editors: **D.R. Khanna, R. Bhutiani & Gagan Matta**
Published by: **DAYA PUBLISHING HOUSE, NEW DELHI**

Chapter 45

Studies on the Ecology and Fish Fauna of Gopalpura Tank of Guna District, Madhya Pradesh

Renu Jain, D.K. Sharma and Dushyant Sharma
Department of Zoology, Government P.G. College, Guna, M.P.

ABSTRACT

Gopalpura Tank is located near village Gopalpura in Distt. Guna of M. P. It is one of the artificial water bodies of Guna (M. P.) used for irrigation as well as for pisciculture. No known source of pollution enters the water body. Physico-chemical studies were undertaken to enhance the limnological knowledge about the tank and to explore possibilities for better management of pisciculture.

This chapter deals with the some fishes of the Gopalpura Tank at Guna Distt. (M.P.) in relation to the abiotic factors *i.e.* physical (Temperature, Conductivity, Transparency, Turbidity, Colour) and chemical (pH, DO, Free CO_2, COD, Chloride, Total alkalinity, Hardness). Altogether the total numbers of fishes described are eleven in number from this Tank.

Keywords: *Gopalpura tank, Physico-chemical factors, Fish fauna.*

Introduction

Gopalpura Tank is located near village Gopalpura in district Guna of M.P. Guna is situated between latitude 23°90'–25°15' north and 76°85'–78°20' longitude east towards north western corner of Malwa in southern Gwalior division of M. P. The Gopalpura Tank is situated at a distance of just 2 km from Guna town, it is one of the artificial water bodies of Guna.The tank has a catchment area of about 6 sq kms. The tank is not very large in its area, but it serves very important function of irrigation for the villagers. It is also used for pisciculture of major carps and locally available fishes by fisherman and villagers. The present work is to investigate the physico-chemical parameters of the tank and their effects upon fish life.

Materials and Method

Collection of water and fish samples was done on monthly basis from August 2000 to January 2001. The monthly water samples for limnological studies were collected between 8 AM and 10 AM.

Standard methods of APHA (1985), Trivedi and Goel (1986) were employed for physical and chemical analysis of the tank water.

Fish samples were collected from fishermen catches monthly and were identified using Day (1989), Jhingran (1975), Shrivastava (1986) and Qureshi (1983).

Results and Discussion

Total 13 physico-chemical parameters were studied every month from August 2000 to January 2001 and presented in Table 45.1, Fishes are enlisted in Table 45.2.

The temperature varied from 16.5 to 25.3°C. The minimum temperature was recorded 16.50° during Jan 2001 and during August 2000 the temperature was maximum being 25.30°C.

Table 45.1: Physico-Chemical Analysis of Gopalpura Tank

Parameters	*Aug 2000*	*Sep 2000*	*Oct 2000*	*Nov 2000*	*Dec 2000*	*Jan 2001*
Temperature (°C)	25.3	24.1	22.1	20.5	18.2	16.5
pH	7.9	7.6	7.8	8.0	8.1	8.2
Conductivity (mohs/cm)	261.0	254.0	180.0	130.8	117.0	121.8
Color	Turbid	Turbid	Turbid	Light Green	Light Green	Light Green
Turbidity (NTU)	76.0	75.8	58.2	46.5	36.0	33.8
Transparency (cm)	14.8	16.0	25.0	40.9	46.2	50.7
Dissolved Oxygen (mg/l)	7.5	7.2	8.2	8.3	8.5	8.7
Free CO_2 (mg/l)	0.3	0.3	0.2	0.2	0.1	0.1
Chloride (mg/l)	12.6	6.5	6.4	5.9	5.8	5.0
Hardness (mg/l)	56.8	36.3	35.1	33.8	33.1	31.5
Total Alkalinity (mg/l)	160	151	135	238.6	238	21.5
BOD (ppm)	2.0	2.1	1.86	1.69	1.48	1.32
COD (ppm)	4.42	4.5	4.26	4.24	4.20	4.18

The value of pH fluctuated between 7.6 and 8.2.

The conductivity was recorded minimum 117 mohs/cm during December 2000 and maximum 261 mohs/cm during August 2000.

The color of water showed variation during the period of observation. The water was turbid during August, September and October 2000 and it becomes light green during November, December, 2000 and January, 2001.

Table 45.2: List of Fishes of Gopalpura Tank

Sl.No.	Present Scientific Name	Local Name	Family	Fin Formula
I.	*Labeo rohita*	Rohu	Cyprinidae	DI6(3/13), P17 V9 A7(2/5) C19 L1 40 Ltr 6½-7½/9 Barbels 1 Pair
2.	*Cirrhinus*	Nain	Cyprinidae	D16 (3/13), P18 V9 A8 (2/6)C15L1 42 Ltr 6½/6½ Barbels 1 Pair
3.	*Catla catla*	Katla	Cyprinidae	D18(3/15)P19 V9 A8(3/5)CI9 L143 Ltt 7½/16½
4.	*Puntius ticto*	Sidhari	Cyprinidae	D11(3/8)P13 V9 A8(3/5)C19L125Ltr 5½/16½
5.	*Notopterus notopterus*	Patola	Notopteridae	D8(117)P17 V6 A100 C19L1 225
6.	*Wallago attu*	Parhin	Siluridae	D5P1/14V10 A86(4/82) C17 Barbels 2 Pairs
7.	*Ompok bimaculatus*	Puma	Siluridae	D4P1/13V8 A65(2/63) C18 Barbels 2 Pairs
8.	*Mystus seenghals*	Tengar	Bagridae	D8P1/9V6 A11(3/8) C19 Barbels 4 Pairs
9.	*Heteropneusties fossilis*	Singhi	Sacobranchidae	D6P1/7V6 A62C19 Barbels 4 Pairs
10.	*Channa marnlius*	Saur	Ophiocephalidae	D46P18 V6 A32 LI65 Ltr 6/11
11.	*Mastacembelus armatus*	Bamam	Mastacembelidae	D37/78 P23 A3/75

Turbidity ranged from 33.8 mg/l to 76.00 mg/l. Transparency was reported to be increasing continuously (August to January). It ranged between 14.8 and 50.7 cm.

Dissolved Oxygen and free CO_2 in the Tank ranged between 7.2 and 8.7 ppm and 0.1 mg/l and to 0.3 mg/l respectively.

Values of Chloride ranged from 5.0 to 12.6 ppm. Hardness values ranged from 31.5 to 56.8 mg/l.

Alkalinity in the Gopalpura Tank fluctuated from 135.0 to 238.6 ppm. BOD and COD were recorded from 1.32 to 2.1 ppm and 4.18 to 4.5 ppm respectively.

Variations in the physico-chemical parameters of Gopalpura tank are quite favourable for the nature piscine life. Decreasing fish production of the Tank is certainly affected by the changing physico-chemical characteristics resulted from the addition of different chemicals. Other major causes are illegal fishing and dynamiting practices *etc.*

The fluctuation in water temperature has relationship with the air temperature. It was well with in limit for survival of fishes. Similar fluctuation are reported by Khanna *et al.*, 1999.

During the study period we found that pH variation lies mostly with in the alkaline range. According to Jhingran (1975) pH range between 7 and 9 which is considerable good for fish culture. Thus, our results indicate that the Gopalpura Tank having pH from 7.6 to 8.2 is suitable for fish culture.

During the course of study low specific conductivity values support the fact that Tank is very productive (Rao, 1993).

Turbidity and Transparency are also quite favourable for fish fauna. Transparency of water ranged between 14.8 and 50.7 cm during the course of study and had a negative correlation with turbidity. Similar results are reported by Dagaonkar and Saxena (1992).

The Dissolved Oxygen ranged from 7.2 to 8.7 ppm having moderate fluctuation. The presence of free CO_2 is due to incomplete utilization in photosynthesis and respiratory activity. Free CO_2 and

Dissolved Oxygen showed inverse relationship to one another as similar to Welch (1952), Hutchinson (1957), and Hynes (1970). The value is in the range favourable for fish life.

Chlorides were low indicating no organic pollution in Tank. Chloride concentration was recorded from 5.0 to 12.6 mg/l showed a decreasing trend during the study but in the month of August it increased up to a comparatively higher level, certainly the reason was the adding of chemicals *viz.* bleaching powder *etc.*

Chloride and Hardness showed a positive relationship. Similar type of relationship was observed by Khanna *et al.*, 1999. Hardness also showed low values, similar results were reported by Rao 1993.

Hardness and Alkalinity showed negative relationship to one another and their range are quite favourable for fish life as indicated by Mathur (1982).

BOD and COD showed a positive relationship to one another while both showed a negative relationship to DO as reported by Chopra and Patrick (1994). The value of both BOD and COD are in the range of favourable piscine life.

References

APHA, 1985. *Standard Methods for the Examination of Water and Wastewater*, 15th edn. American Public Health Association, Washington DC, p. 1268.

Chopra, A.K. and Partick Nirmal, J., 1994. Effect of domestic sewage on self purification of Ganga water at Rishikesh. I. Physico-chemical parameters. *Advances in Biosciences*, 13(11): 75–82.

Day, F., 1989. *Fauna of British India, Including Ceylon and Burma Fishes.* Taylor and Francis, London, 1: 548.

Dagaonkar, A. and Saxena, D.N., 1992. Physico-chemical and biological characterization of a temple Tank, Kaila Sagar, Gwalior, M.P. *J. Hydrobiol.*, 8(1): 1–19.

Hutchinson, G. E., 1975. *A Treatise on Limnology Vol. 1.* John Wiley and Sons, New York, London, pp. 1015.

Hynes, H. B.N., 1970. *The Ecology of Running Waters.* Liverpool University press. Liverpool, p. 1–555.

Jhingran, V.G., 1975. *Fish and Fisheries of India.* Hindustan Publishing Corporation (India), p. 1–954.

Khanna, D.R., Seth, T.R., Malik, D.S. and Singh, R.K., 1999. Fishes and their ecology of the river Suswa at Raiwala Dehradun. In: *Sustainable Ecosystem and Environment*, (Eds.) D.R. Khanna, Ashutosh Gautam and Aditya Gautam, p. 105–110.

Qureshi, T.A. and Qureshi, N.A., 1983. *Indian Fishes.* Brej Brother, Bhopal

Rao, M. Babu, 1993. Studies on the ecology and fish fauna of an oligotrophic lake Himayat Sagar, Hyderabad (A.P.) In: *Recent Advances of Freshwater Biology*, (Ed.) K.S. Rao, 2: 73–98.

Shrivastava, G.L., 1986. *Fishes of Eastern U.P.* Vishwavidyalaya Prakashan, Varanasi, p. 1–163.

Trivedi, R.K. and Goel, P.K., 1986. *Chemical and Biological Methods for Water Pollution Studies.* Environmental Publication, Karad.

Welch, P.S., 1952. *Limnology.* McGraw Hill Book Co., New York, pp. 593.

Water and Wastewater Management in 2 Vols (2011) *Pages* **298–302**
Editors: **D.R. Khanna, R. Bhutiani & Gagan Matta**
Published by: **DAYA PUBLISHING HOUSE, NEW DELHI**

Chapter 46

Fish Fauna and Fish Production of Tribal District West Nimar (Khargone), M.P.

S.K. Pathak[1] and S.K. Pathak[2]
[1]*Department of Zoology, Govt. P.G. College, Khargone – 451 001, M.P.*
[2]*Assistant Director of Fisheries, District West Nimar, Khargone*

ABSTRACT

A survey of fish fauna was conducted between July 1998 and December 1999. The fishes were collected from the local market of block head quarters of the tribal district West Nimar (Khargone). In all 40 species belonging to 23 genera and 10 families have been found in this area. They are *Cirrhinus, Catla, Chela, Labeo, Oxygaster, Puntius, Garra, Barbus, Rasbora* and *Tor* belonging to Cyprinidae family; Lepidocephalichthys, Nemacheilus belonging to Cobitidae family; Ompok and Wallago belonging to Siluridae family; Mystus and Rita belonging to Bagridae family; Notopterus, Heteropneustus, Clarias, Xenetodon, Channa and Matacembelus belonging to Notopteridae, Heteropneustidae, Claridae, Belonidae, Ophiocephalidae and Matacembelidae family respectively. It is also observed that non culturable fishes are most widely distributed and the commercial culturable fihses are commonly found in the region. In this chapter, the fish fauna of the district West Nimar is discussed.

Keywords: Fish fauna, West Nimar, Fish production.

Introduction

The complete account of the fish fauna of Madhya Pradesh is not available. Day (1889) has published fish fauna of British India including Ceylon and Burma. Since then there have been several publications on the fish fauna of various parts of India. Noteworthy publications are of Hamilton (1922), Mism (1952), Menon (1954) and Srivastava (1968). As far as Madhaya Pradesh is concerned several studies have been done in the past. The fish fauna of Mahanadi was published by Horn (1940). Swamp (1953) and Qureshi and Qureshi (1970) studied the fishes of Sagar lake. Soni (1959) gave an account of fishes of lower lake of Bhopal. Dubey and Mehra (1959) described the fishes at Jabalpur. Misra (1962) gave an account of fishes available in the M. P. The present communication is an attempt to report the fish fauna of tribal district of West Nimar.

The tribal District West Nimar is situated in the west southern part of Madhya Pradesh, between the latitudes 21° 25' and 22° 35' North and longitude 74° 25' and 76° 140' East. The total area of this district is 13285 sq. km. and it is on the 7th position in the State. The elevation varies from 500 feet to 1365 feet above MSL. The average annual rain fall is 830 mm. The lowest temperature is 12°C in winter and maximum temperature is 46 °C in summer season. The total catchment area of river Narmada is 121 km in this region. The total water area is 4199.583 ha. Map of District West Nimar is shown in Figure 46.1.

Materials and Method

The fishes were collected with the help of nets from all possible habitats and from market of different block headquarters such as Khargone, Seagone, Zimiya, Niwali, Bhagwanhura, Barwah Pansmal, Pati, Barwani, Rajpur, Thilcri, Gogawa, Kasrawad, Bhikangoan, and Maheswar between July 1998 and December 1999 after first monsoon showers. The 10 per cent formalin solution was

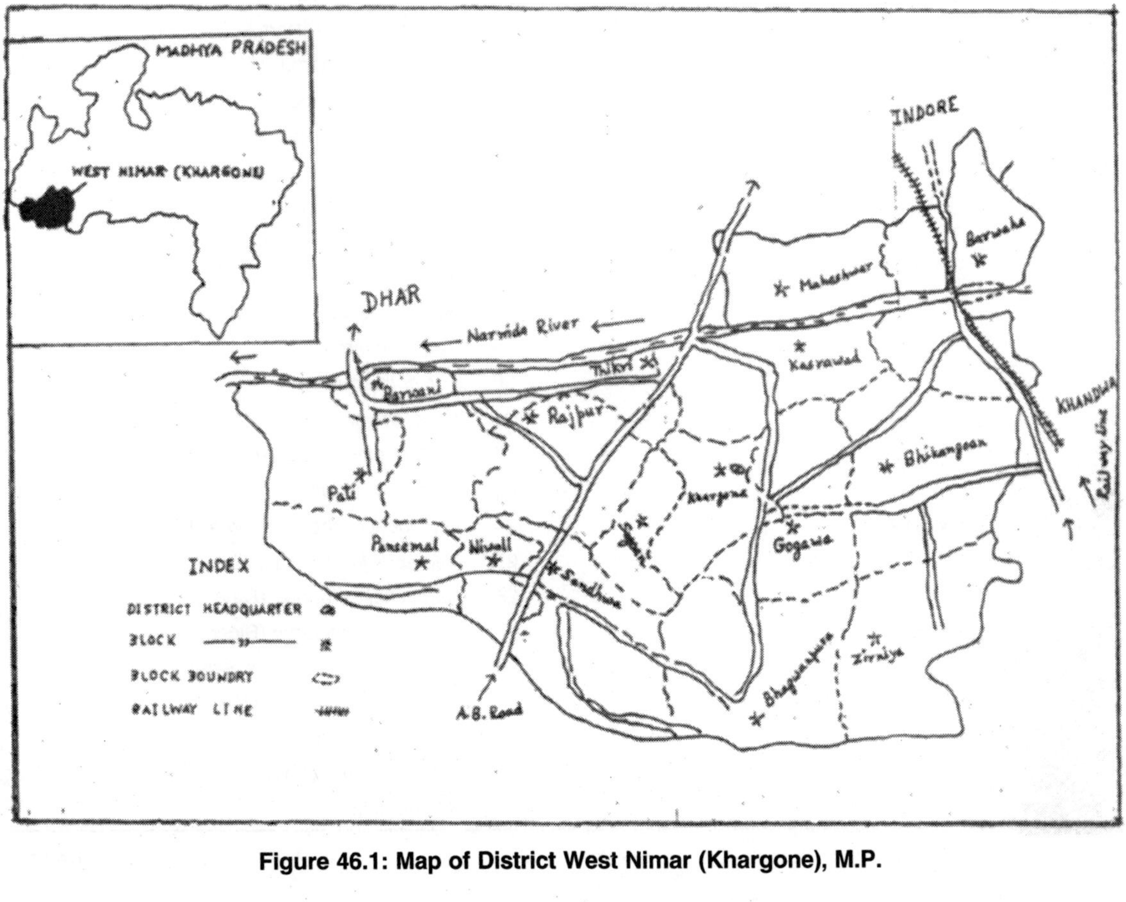

Figure 46.1: Map of District West Nimar (Khargone), M.P.

injected in the body. The samples were identified with the help of standard literature available. The specimen were deposited in the museum of Zoology Department of Post-Graduate College, Khargone.

Results and Discussion

In all, 40 species of fish fauna belongings to 23 genera and 10 families were identified from this region these are shown in Table 46.1.

Table 46.1: The Fish Fauna of the Tribal District West Nimar

Sl.No.	Family		Species
1.	Notopteridae	1.	*Notopterus notopterus* (Pallas)
		2.	*Cirrhinus mrlgala* (Ham.)
2.	Cyprinidae	3.	*Cirrhinus reba* (Ham.)
		4.	*Barilus bola* (Ham.)
		5.	*Catla catla* (Ham.)
		6.	*Chela laubuea* (Ham.)
		7.	*Chela atpar* (Ham.)
		8.	*Barilus barna* (*Ham.*)
		9.	*Labeo rohita* (Ham.)
		10.	*Labeo calbasu* (Ham.)
		11.	*Labeo bata* (Ham.)
		12.	*Labeo boga* (Bloch)
		13.	*Labeo genius* (Ham.)
		14.	*Oxygaster bacaila* (Ham.)
		15.	*Puntius ticto* (Ham.)
		16.	*Puntius sarana* (Ham.)
		17.	*Puntius sophore* (Ham.)
		18.	*Rasbora doniconius* (Ham.)
		19.	*Tor tor* (Ham.)
		20.	*Tor putitora* (Ham.)
		21.	*Gara gotyala* (Ham.)
		22.	*Barbus pinnauratus* (Ham.)
		23.	*Barbus tor* (Ham.)
3.	Cobitidae	24.	*Lepidocephlichthys guntea* (Ham.)
		25.	*Nemacheilus beavani* (Day)
		26.	*Nemacheilus aurlus* (Day)
4.	Siluridae	27.	*Ompak bimaculatus* (Bloch.)
		28.	*Wallgo attu* (Bouch and Sohn.)
5.	Bagridae	29.	*Mystus seenghala* (Sykes.)
		30.	*Mystus aor* (Ham.)
		31.	*Mystus bleekri* (Day.)
		32.	*Rita rita* (Ham.)

Contd...

Table 46.1–Contd...

Sl.No.	Family		Species
6.	Hetropneustidae or Saccobranchidae	33.	*Heteropneustus fossilis* (Bloch.)
7.	Claridae	34.	*Clarias batrachus* (Linnaeus)
8.	Belonidae	35.	*Xenentodon cancila* (Ham.)
9.	Ophiocephlidae	36.	*Channa punctatus* (Ble.)
		37.	*Channa gachua* (Ham.)
		38.	*Channa marulius* (Ham.)
10.	Matacembelidae	39.	*Matacembelus pancalus* (Ham.)
		40.	*Matacembelus armatus* (Laceoede)

Dubey and Mehra (1959) in the survey of Chambal river, reported 71 different species of fishes. Soni (1959) described 35 different type of fishes in lower lake of Bhopal. Dubey and Verma (1965) gave an account of 104 species of fishes of Madhya Pradesh. Badola (1975) reported the fish fauna of Garhwal Himalaya, Khanna and Badola (1990) reported the ichthyofauna of the river Ganga at the foot hills of Garhwal Himalaya. In the present survey, however, 40 species belonging to Notopteridae, Claridae; Belonidae, Cobitidae, Siluridae, Bagridae, Heteropneustidae, Claridae, Belonidae, Opheocephalidae and Matacembelidae families were identified from this region (Table 46.1). It was also observed that the non culturable fishes are most widely distributed and the commercial culturable fishes are common in this region.

Table 46.2: The Year-wise Fish Production in Tribal District West Nimar

Sl.No	Years	Fish Production (MT)
1.	1990–91	419
2.	1991–92	367
3.	1992–93	485
4.	1993–94	279
5.	1994–95	300
6.	1995–96	1396
7.	1996–97	1625
8.	1997–98	1660
9.	1998–99	1728

Sources: Annual report of District Fisheries Office.

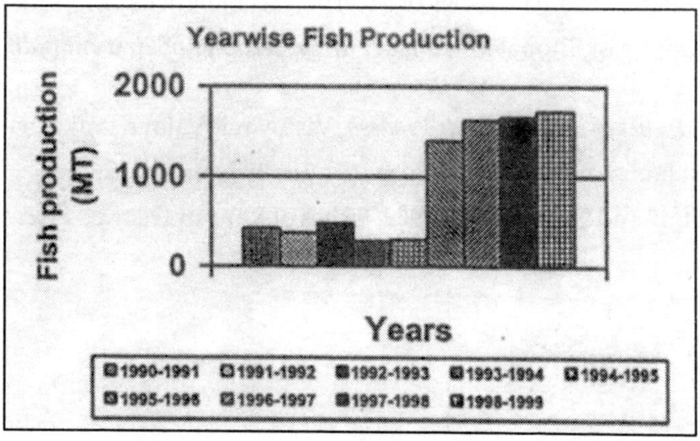

Figure 46.2: Year-wise Fish Production of District West Nimar

The year wise total fish production in the district is shown in Table 46.2 and Figure 46.2. It was lowest in 1993-94 while in the year 1998-99 it was highest. The reason behind the lowest production in the year 1993-94 was low rainfall on one hand and non availability of good seeds on the other hand. The maximum production defeating all the previous records resulted due to the implementation of Rajiv Gandhi Mission and the availability of the good seeds, which was missing in the previous years. Need not to say that Rajiv Gandhi Mission motivated the people to come forward and to be benefited by the programmes for the upliftment of them laid down by the government.

Acknowledgements

We express our thanks to all teaching members of the Zoology Department of Govt. P.G. College, Khargone, M.P. (India) for the valuabe cooperation and encouragements throughout this investigation.

References

Badola, S.P., 1975. Fish fauna of the Garhwal Hills, Part II. *J. Zoology,* 16(1): 57–70.

Day, E., 1889. *Fauna of British India,* Vols. I and II.

Dubey, G.P. and Mehra, R.K., 1959. Fish and fisheries of Chambal river. *All India Congress of Zoology.* (Pub. 1962), p. 647–665.

Dubey, G.P. and Verma, M.N., 1965. *A Preliminary Study of the Fish Fauna of MP. The Vikram,* 8(U) Reprinted, Govt. Central Press.

Hamilton, F.B., 1922. *An Account of Fishes Found in the River Ganges and its Branches,* pp.vii+405, pls. 39.

Khanna, D.R. and Badola, S.P., 1990. Ichthyofauna of the river Ganga at the foothills of Garhwal Himalayas. *J. Nat. and Physical Sciences,* 4(1–2): 153–162.

Malviya, R.B., 1961. A list of fishes from Jabalpur M.P. *India.Proc. Nat. Acad. Sci.,* Sect. B. 31(3): 349–354.

Menon, A.G.K., 1954. Further observations on the fish fauna of the Manipur State. *Rec. India Mus.,* pp. 21–26.

Misra, K.S., 1962. *An Aid to the Identification of the Common Commercial Fishes of India and Pakistan.*

Qureshi, T.A. and Qureshi, N.A., 1970. Systematic position and description of teleostean fishes of M.P. Section of Icthyology and Fisheries, Department of Zoology, Safia College, Bhopal, p. 1–38.

Sinha, B.M. and Shiromani, 1953. The fishes of Meerut. *Rec. India Mus.,* LI, p. 61–65.

Soni, D.D., 1959. Fish fauna of Bhopal lower lake. In: *Proceedings of First All India Congress of Zoology,* pp. 639–641.

Srivastava, G.J., 1953. *Fishes of Eastern Uttar Pradesh.* Vishwavidyalaya Prakashan, Varanasi, pp. 163.

Swamp, H., 1953. Fish fauna of Sagar Lake. *Sagar University Journal,* 1: 2.

Verma, N.S., Chitray, B.B. and Saxena, D.B., 1962. Fishes of Kanpur District. *Proc. Nat. Acad. Sci.,* Sec. B. 32(3): 213–232.

Water and Wastewater Management in 2 Vols (2011) *Pages 303–306*
Editors: **D.R. Khanna, R. Bhutiani & Gagan Matta**
Published by: **DAYA PUBLISHING HOUSE, NEW DELHI**

Chapter 47

Seasonal Variations in the Plankton of Suswa River at Raiwala (Dehradun)

D.R. Khanna and R.K. Singh
Department of Zoology and Environmental Science,
Gurukul Kangri University, Haridwar – 294 404

ABSTRACT

The present chapter is an attempt to present the findings of investigation on the river Suswa. It is evident from the present study that the water quality of Suswa River has deteriorated during the recent past due to increased human influx.

Keywords: *Suswa, River, Water quality, Himalaya.*

Introduction

In India, fourteen major rivers systems share about 83 per cent of the drainage basin *i.e.* Indian rivers carry about 16,45,000 million cubic metres of water annually Rao (1975). The Himalayan rivers receive 20–30 per cent of their water from melted snow and ice during summer and monsoon seasons. The Suswa river is a spring fed tributary of the River Ganga. The water quality of Suswa river has deteriorated considerably due to increased human interferences in several ways *viz.* bathing, dumping of domestic wastes, sewage discharge, *etc.* These anthropogenic activities have not only affected the river water quality and its biota adversely, but have also increased the unusability of water for social and economical purposes. Hence, the present study was conducted to investigate the riverine ecological factors of the river Suswa at Raiwala in district Dehradun at the foot hills of Shiwalik Himalayas.

Materials and Method

The water samples were collected fortnightly from different sampling stations in morning hours (07.00 to 10.00AM). The analysis of different physico-chemical and biological parameters was done by following the methods by Welch (1948), Trivedi and Goel (1984) and APHA (1995).

Results and Discussion

The ecological conditions of a river or stream have a direct bearing upon the different producer to consumer level in aquatic life and physico–chemical parameters impart a major role in determination of the quality of any water body. The average values of physical and chemical parameters are tabulated in Tables 47.1 and 47.2. The seasonal quantitative analysis of the planktonic number and percentage of different groups among phytoplankton has shown in Tables 47.3 and 47.4.

Table 47.1: Seasonal Variations in Physical Parameters of Suswa River

Parameters	*Seasons*			
	Summer	*Monsoon*	*Winter*	*Average*
Temperature (°C)	26.25±1.68	23.52±0.24	21.80±1.45	23.86±0.045
Velocity (m/sec)	1.16±0.169	0.95±0.021	0.65±0.19	0.92±0.036
Total Solids, TS (ppm)	219.30±29.74	170.04±5.09	142.39±24.64	177.24±4.18
Total Dissolved Solids, TDS (ppm)	211.53±27.43	166.59±4.34	140.06±23.01	172.73±4.14
Total Suspended Solids, TSS (ppm)	7.77±2.31	3.45±0.75	2.33±1.54	4.51±0.053

Table 47.2: Seasonal Variations in Chemical Parameters of Suswa River

Parameters	*Seasons*			
	Summer	*Monsoon*	*Winter*	*Average*
pH	7.47±0.23	7.53±0.18	8.36±0.40	7.79±0.008
Free CO_2 (ppm)	3.59±0.22	3.36±0.63	2.86±0.29	3.27±0.013
Dissolved Oxygen (ppm)	8.31±0.31	8.84±0.06	9.10±0.25	8.75±0.022
Mineral Acidity (ppm)	61.12±3.24	55.89±0.46	52.61±2.77	56.54±0.56
Total Acidity (ppm)	3.76±0.06	3.71±0.028	3.53±0.098	3.67 :!: 0.025
Alkalinity (ppm)	288.90±12.08	302.17±2.69	326.86±14.76	305.98±0.91
Chloride (ppm)	21.99±2.75	16.84±0.89	15.48±1.85	18.10±0.13
Hardness (ppm)	100.83±2.65	97.45±0.26	92.95±2.92	97.08±0.39
Calcium (ppm)	85.14±1.76	82.40±0.18	80.39±1.59	82.65±0.36
Magnesium (ppm)	15.69±0.89	15.05±0.43	12.56:!: 1.32	14.43±0.036
BOD (ppm)	2.05±0.21	1.77±0.14	1.42±0.23	1.75±0.037
COD (ppm)	4.49±0.11	4.36±0.014	4.18±0.11	4.34±0.013

In the Suswa river at Raiwala, a difference in the fluctuation of water temperature was maximum (26.25° C±1.68) observed in summer and minimum in winter. The maximum velocity (1.16m/sec±0.169) was recorded in summer. The velocity starts continuous decreasing from summer to winter. The velocity and the residue shared positive relationship. Badola and Singh (1981) reported similar trends in the river Alaknanda. High pH (8.36±0.40) in winter may be due to higher algal population in the river. The pH and dissolved oxygen showed a positive relationship. The maximum value of free CO_2 was recorded (3.59 ppm±0.22) in summers. The dissolved oxygen was found maximum (9.10 ppm±0.25) in winter due to the photosynthetic and respiratory activities of the biota as also observed by Hynes (1970), Dobriyal and Singh (1981) and Khanna (1993). The maximum value of alkalinity (326.86 ppm±14.76) was observed in winter due to decomposition of the organic matter. Similar type of findings was observed by Venkatesvarlu and Jayanti (1968) in the river Sabarmati.

Table 47.3: Seasonal Quantitative Analysis of the Plankton of the River Suswa

Seasons	*Parameters*		
	Number of Plankton per Litre of Water		*Total Plankton per Litre of Water*
	Phytoplankton	*Zooplankton*	
Summer	629.92±150.39	94.25±15.56	724.17±165.95
Monsoon	804.00±27.30	111.83±3.13	915.83±..30.43
Winter	1093.92±177.70	142.67±18.68	1236.59±196.38
Average	842.61±191.38	116.25±20.01	958.86±211.40

Table 47.4: Number and Percentage of Different Groups Among the Phytoplankton of the River Suswa during Different Seasons

Seasons	*Parameters*				*Percentage of Different Groups*		
	Total Plankton/l	*Total Diatom/l*	*Total Green Algae/l*	*Total Blue Green Algae/l*	*Diatoms*	*Green Algae*	*Blue Green Algae*
Summer	629.92± 150.39	477.52± 135.74	123.78± 12.11	28.62±2.55	75.96	19.49	4.55
Monsoon	804.00± 27.30	638.61± 21.84	136.05± 3.43	29.34±2.04	79.12	17.14	3.74
Winter	1093.92± 177.70	892.35± 157.59	162.88± 15.54	38.69±4.58	81.83	14.61	3.56
Average	842.61± 191.38	669.49± 170.76	140.90± 16.32	23.22±4.30	78.97	17.08	3.95

The maximum value of hardness (100.83 ppm±2.65) was recorded in summer and showed a positive relationship with chloride as also observed by Chopra and Patric (1994) in the river Ganga at Rishikesh. The amount of calcium in the river Suswa was found to be maximum (85.14 ppm ± 1.76) during winter. The river may be regarded as rich in calcium contents (80.39 ppm ± 1.59 to 85.15 ppm ± 1.76). Similar type of relationship was observed by Atkin and Harris (1924) in freshwater bodies. BOD and COD showed a positive relationship with one another. Similar type of relationship was reported by Chopra and Patric (1994) in the river Ganga at Rishikesh. BOD and COD have showed a negative relationship with dissolved oxygen as observed by Verma *et al.* (1984) in eastern Kalinadi.

The plankton are the heterogenous assemblage of minute organisms present in the natural waters. The total planktonic concentration was recorded maximum 91236.59 units/l ± 196.38) in winter, due to the blanketing, effect caused by velocity. Allen (1920) has shown that the water current above the moderate speed is directly inhibitory to plankton development. Maximum percentage of different groups were noted as Bacillariophyceae (Diatoms 81.83 per cent) in winter, Chlorophyceae (Green algae) 19.49 per cent in summer and Cyanophyceae (Blue green algae) 4.55 per cent in summer. Among the total plankton, Bacillariophyceae dominated over Chlorophyceae, Cyanophyceae and Zooplankton. Same trend of domination was reported by Khanna (1993) and Joshi *et al.*, (1996) in the river Ganga at Haridwar.

Acknowledgements

The authors are grateful to Professor B. D. Joshi and Professor A.K. Chopra of Dept. of Zoology and Environmental Science, Gurukul Kangri University, Haridwar for their guidance and valuable suggestions throughout this investigation.

References

Allen, W.E., 1920. A quantitative and statistical study of the plankton of the San Joaquin river and its tributaries near Stockton California, 1913. *Univ. Calif. Publ. Zool.*, 22(1): 1– 24.

APHA, AWWA, WPCF. 1995. *Standard Methods for the Examination of Water and Wastewater*, 19th edn. American Public Health Association, 1015 Fifteen Street, New Washington.

Atkin, W.R.G. and Harris, G.T., 1924. Seasonal changes in the water and heloplankton of fresh ponds. In: *Sc. Proc. Roly. Dub. Soc.*, 23(N.S.): 1–21.

Badola, S.P. and Singh, H.R., 1981. Fish and fisheries of river Alaknanda. *Proc. Nat Acad. Sci.*, 15(B): 133–142.

Chopra, A.K. and Patric, Nirmal J., 1994. Effect of domestic sewage on self purification of Ganga water at Rishikesh. I. Physico-chmical parameters. *Ad. Bios.* 13(11): 75–82.

Dobriyal, A.K. and Singh, H.R., 1981. Diurnal variation in some aspect of limnology of the river Mandakini from Gharwal Himalaya. India. *U.P. J. Zool.*, 3: 30–40.

Joshi, B.D., Bisht, R.C.S. and Joshi, Namita, 1996. Planktonic population in relation to certain physico-chemical factors of Ganga canal at Jwalapur (Haridwar). *Him. J. Env. Zool.*, 10: 75–77.

Khanna, D.R., 1993. *Ecology and Pollution of Ganga River.* Ashish Publishing House, New Delhi.

Rao, K.L., 1975. *India's Water Wealth.* Orient Longman, New Delhi, pp. 255.

Trivedi, R.K. and Goel, P.K., 1984. *Chemical and Biological Methods for Water Pollution Studies.* Enviromental Publications, Karad, p. 1–251.

Venkatesvarlu, T. and Jyanti, T.V., 1968. Hydrobiological studies of the river Sabarmati to evaluate water quality. *Hydrobiologica*, 31: 332–448. .

Verma, B.R., Sharma, P., Tyagi, A., Rani, S., Gupta, A.R. and Dalela, R.C., 1984. Pollution and saprobic status of eastern Kalinadi. *Limnologica (Berlin)*, 15(1): 69–133.

Welch, P.C., 1948. *Limnological Methods.* McGraw Hill Book Co., New York.

Water and Wastewater Management in 2 Vols (2011) *Pages* **307–311**
Editors: **D.R. Khanna, R. Bhutiani & Gagan Matta**
Published by: **DAYA PUBLISHING HOUSE, NEW DELHI**

Chapter 48

Water Quality of Rampur Reservoir of Guna District, Madhya Pradesh

Renu Jain and Dushyant Sharma
Department of Zoology, Government P.G. College, Guna, M.P

ABSTRACT

Rampur reservoir is one of the artificial water bodies of Guna (M.P.). The villagers use it mainly for irrigation as well as for pisciculture. Physico-chemical studies were undertaken to enhance the limnological knowledge about the tank and to explore possibilities for better management of pisciculture. In the present study, various physico-chemical factors such as water temperature, transparency, turbidity, conductivity, pH, alkalinity, dissolved oxygen and chloride were investigated. The recorded parameters showed interrelationship and also indicated the suitability of reservoir for the purpose of drinking and pisciculture. Proper awareness of pisciculture to the villagers is required for the proper utilization and exploitation of Rampur reservoir.

Keywords: *Rampur reservoir, Water quality, Pisciculture.*

Introduction

The Rampur reservoir is one of the artificial water bodies of Guna, situated at a distance of just 34 miles from Guna town. The work on the tank was started in 1908 and was completed in 1917. Rampur is the nearest village. Direction of the reservoir is north west of Guna, it lies between 24° 47' N and 77°10' E. The reservoir has a catchment area of approximately 102 sq miles and it is a part of Chambal Jamuna Basin. The reservoir serves as a source of irrigation for the villagers. Fishermen and villagers also use it for drinking purposes and pisciculture of major carps and locally available fishes. No scientific study has been conducted on the reservoir so far. Therefore, the present study was undertaken to gain a basic knowledge of Rampur reservoir. This would not only enable the preliminary limnological knowledge but also help in exploring the possibilities for better management, development and augmentation of pisciculture.

Materials and Method

Monthly sampling was carried out at two different sampling stations, namely A and B for one year (from April 1999 to March 2000). The samples were collected at 10–11 A. M. The physico-chemical parameters were analysed as per the procedure given in APHA (1981) and Trivedy and Goel (1986).

Results

Various physico-chemical parameters are shown in Tables 48.1 and 48.2. The summary of the results is as under.

Table 48.1: Physico-chemical Characteristics of Rampur Reservoir at Spot A

Months	*Colour Appearance*	*Temperature (°C)*	*pH*	*Turbidity (NTU)*	*Conductivity (mhos/cm)*	*Dissolved Oxygen (mg/l)*	*Transparency (cm)*	*Chloride (mg/l)*	*Total Alkalinity (mg/l)*
April 1999	Light brown	27.5	8.5	38.5	140.2	11.5	54.9	8.4	261.5
May 1999	Light brown	29.0	8.7	40.0	141.7	13.0	53.2	9.0	289.0
June 1999	Light brown	27.0	8.8	44.9	199.8	12.0	49.1	7.5	258.0
July 1999	Turbid	26.5	8.9	50.0	201.5	10.5	41.6	7.0	254.5
August 1999	Turbid	25.9	7.7	62.5	208.2	7.5	32.1	6.8	232.8
September 1999	Turbid	24.5	7.6	68.0	240.9	7.0	24.0	5.9	215.9
October 1999	Turbid	24.1	7.1	54.1	192.2	7.9	38.0	5.5	165.0
November 1999	Light green	24.0	8.1	41.5	132.0	8.0	45.0	5.7	104.9
December 1999	Light green	22.0	8.3	30.9	124.0	8.3	51.5	4.0	104.0
January 2000	Light green	18.0	8.0	33.5	134.5	8.5	55.9	6.5	167.0
February 2000	Light green	20.0	8.2	31.0	135.4	8.5	60.5	7.8	172.0
March 2000	Light brown	25.0	8.4	37.2	138.9	8.7	58.0	8.2	220.0

Colour

The water is almost colourless. The water was turbid during monsoon. The colour as reflected from the reservoir was light green during winter and light brown during summer.

Temperature

The temperature varied from 18.0°C to 29.5°C. The minimum water temperature was recorded at station A *i.e.* 18.0°C during January 2000 and maximum at station B *i.e.*, 29.5°C during May 1999. During December 1999 the temperature was recorded 22.0°C at both the stations.

Transparency

The transparency was recorded maximum during February–March at both the stations (60.5 cm Station A and 63.0 cm Station B). The minimum values were recorded during September at both the stations (24.0 cm station A and 25.0 cm station B).

Table 48.2: Physico-chemical Characteristics of Rampur Reservoir at Spot B

Months	*Colour Appearance*	*Temperature (°C)*	*pH*	*Turbidity (NTU)*	*Conductivity (mhos/cm)*	*Dissolved Oxygen (mg/l)*	*Transparency (cm)*	*Chloride (mg/l)*	*Total Alkalinity (mg/l)*
April 1999	Light brown	28.0	8.7	40.0	141.2	11.9	59.1	8.9	265.0
May 1999	Light brown	29.5	8.9	44.9	144.8	12.0	51.5	9.2	320.0
June 1999	Light brown	26.5	8.1	52.1	200.0	10.5	46.8	7.7	289.0
July 1999	Turbid	26.0	8.2	59.2	204.6	7.6	40.0	7.4	261.5
August 1999	Turbid	25.4	7.8	62.5	210.2	7.0	35.1	6.2	254.5
September 1999	Turbid	24.0	7.1	66.0	235.8	7.8	25.0	6.1	230.9
October 1999	Turbid	23.6	8.0	52.9	180.7	8.2	36.2	6.3	212.9
November 1999	Light green	23.2	8.1	44.6	128.0	8.2	43.0	6.8	105.0
December 1999	Light green	22.0	8.3	30.9	120.0	8.3	51.5	5.7	108.0
January 2000	Light green	18.1	8.1	34.1	123.8	8.6	56.0	7.0	170.0
February 2000	Light green	20.5	8.3	32.9	134.5	8.6	61.2	7.6	215.0
March 2000	Light green	25.5	8.5	30.1	136.9	9.2	63.0	8.6	235.0

Turbidity

Turbidity is a measure of extent to which light is either scattered or absorbed by the suspended material in water. The turbidity ranged from 30.1 in March to 66.0 NTU in September at spot A and 30.9 in December to 68.0 NTU in September at spot B.

pH

Rampur reservoir maintained an alkaline pH throughout the investigation period and pH was recorded between 7.1 and 8.9. The pH was minimum *i.e.* 7.1 in the month of September 1999 and October 1999 at spot A and B respectively. The maximum value *i.e.* 8.9 was recorded in the month of May 1999 and July 1999 at spot A and B respectively.

Conductivity

The conductivity was recorded minimum during December 1999 *i.e.* 120 mhos/cm at spot B and maximum during September 1999 *i.e.* 240.9 mhos/cm at spot A.

Alkalinity

The alkalinity showed higher value during summer and lower value during monsoon and winter.

Dissolved oxygen

Maximum dissolved Oxygen level (13.0 mg/l at station A and 12.0 mg/l at station B) at the Rampur reservoir was recorded during May 1999. The minimum dissolved oxygen. level of 7.0 mg/l was recorded at spot A and B during the month of September and August respectively.

Chloride

The chloride values varied from 4.0 mg/l (spot A during December) to 9.2 mg/l (spot B during May). From June onwards the values showed a decreasing trend up to December.

Discussion

In the present study water temperature of Rampur reservoir varied from 18.0°C to 29.5°C throughout the year. The fluctuations in the water temperature have relationships with the air temperature being maximum during summer and minimum during winter. Rao (1955) and Saba and Pandit (1986) also reported same results.

Present observations showed minimum transparency during monsoon seasons and it increases subsequently. If we compare it with turbidity we find that it is inversely proportional to the turbidity. The low depth of visibility during monsoon is attributed to the highest turbidity of water caused by suspended silt and organic debris, similar trend was observed by Balkhi (1987) and Khanna *et al.* (1999). Bhatt and Negi (1985) also reported higher values of turbidity during monsoon for a river ecosystem.

Similarly, the highest value of conductivity during monsoon (September 240 mhos/cm at station A and 235.8 mhos/cm at station B) may also be due to high turbidity, as indicated by Acre and Boyd (1980(that the conductivity of water increases with concentration of solids in the water.

The pH of Rampur reservoir water was found alkaline during the study period; Verma and Shukla (1967) also reported that freshwater bodies remain alkaline in nature.

Total alkalinity showed higher values during summer and a deep fall during monsoon. Goldman and Wetzel (1963) and Sharma (1992) have reported similar results.

Dissolved Oxygen is very essential for metabolism of all aquatic organisms for aerobic respiratory biochemistry (Wetzel, 1975). The present investigation showed minimum value of dissolved oxygen during September and maximum during May. The higher values of oxygen during summer are associated mainly with the higher phytoplanktonic population (Bhatt and Negi, 1985).

Present results showed lower concentration of chloride (4.0-9.2 mg/l) throughout the year indicating that the water is unpolluted from artificial sources. The low chloride content in natural water 4-10ppm only indicates the purity of water that is free from pollution (Shreenivasan, 1965).

References

APHA, 1981. *Standard Methods for the Examination of Water and Wastewater*, 11th edn. APHA, Washington D.C.

Acre, R.B. and Boyd, C.E., 1980. Water chemistry of Alabaena ponds Aubm Univ. (Ala) Agi. Ecot. *Sta. Bull.* p. 322–335.

Balkhi, M.H., 1987. Hydrobiology of Anchor lake, Kashmir. *Physiol. Ecol.*, 12(3): 131–139.

Bhatt, S.D. and Negi, Usha, 1985. Physico-chemical and phytoplankton population in a subtropical pond. *Comp. Physio. Ecol.*, 10(2): 85–88.

Goldman, C.R. and Wetzel, R.B., 1963. A study of primary productivity of clear lake, Lake country California. *Ecology*, 44: 285–294.

Khanna, D.R., Malik, D.S., Seth, T.R. and Rupendra, 1999. Correlation between abiotic factors and planktonic population in river Ganga at Rishikesh (U.P.).

Rao, C.B., 1955. On the distribution of algae in a group of six small ponds. II. Algae periodicity. *J. Ecol.*, 43: 291–308.

Saba, L.C. and Pandit, B., 1986. Comparative limnology of Bhagalpur ponds. *Comp. Physiol. Ecol.*, 14(1): 213–216.

Sharma, M., 1992. Hydrobiological studies of Halali reservoir with reference to zooplankton and fishery prospects. *Ph.D. Thesis*, Barkatullah University, Bhopal.

Sreenivasan, A., 1965. Limnology of tropical impoundments. III. Limnology and productionof Amravati reservoir Madras State, India. *Hydrobiologia*, 26: 501–516.

Trivedi, R.K. and Goel, P.K., 1984. *Chemical and Biological Methods for Water Pollution Studies.* Environmental Pub., Karad, p. 7–251.

Verma, S.R. and Shukla, G.R., 1967. Limnological studies of Haridwar in relation to fish and fisheries. *Indian Env. Health*, 9: 317–326.

Wetzel, R.G., 1975. *Limnology*. W.B. Saunders Co., Philadelphia, U.S.A.

Water and Wastewater Management in 2 Vols (2011) *Pages 312–317*
Editors: D.R. Khanna, R. Bhutiani & Gagan Matta
Published by: DAYA PUBLISHING HOUSE, NEW DELHI

Chapter 49

Modeling to Assess Effect of Light on Phytoplanktonic Growth Dynamics: A Review

D.R. Khanna, R. Bhutiani and K.S. Chandra
Department of Zoology and Environmental Sciences,
Gurukul Kangri University, Haridwar

ABSTRACT

Light plays an important role in the life of organisms. Light penetration is controlled by the amount and kind of materials that are dissolved and suspended, ultimately results in turbidity and transparency of water. Mathematical models are very much related and useful for light dependency of phytoplankton.

Keywords: Phytoplankton, Light, Transparency, Turbidity, Modeling.

Introduction

Light plays an important role in the life of organisms including the whole diversity of plants ranging from microscopic phytoplankton to giant trees, because it is the sole source of energy which effects the growth dynamics and structure of most aquatic and terrestrial communities. The unidirectional nature of light gives rise to a vertical gradient of light intensity as a function of depth. Since phototropic organisms absorb light to make a living, this vertical gradient is not static but dynamic (Reynolds, 1984).

Turbidity and transparency are the major factors in water column, which affects the growth of phytoplankton, because light penetration is controlled by the amount and kind of materials that are dissolved and suspended in the water. Two processes diminish light under water column: absorption and scattering (Kirk, 1994). Absorption and scattering interact in a complex and nonlinear manner to govern the attenuation of light under water.The equation governing the propagation of light underwater, called the radiative transport equations, have no exact solution: but several computer programs have been written to solve the equations by various numerical methods.

Light Limited Growth Models

It seems that Kok (1952) and Sverdrup (1953) used mathematical models for the first time about five decades ago. Thereafter a number of models were suggested to study the problem and for an excellent discussion on various models. One may refers to Talling (1957), Platt *et al.* (1990), Kirk (1994), Huisman, and Wessing (1994), Huisman (1999) *etc.* Light-limited growth models use mathematical and numerical techniques to simulate the physical and biological processes that affect the growth dynamics of phytoplankton. A comprehensive light limiting growth modeling approach require the following component:

1. Physico- chemical and biological data.
2. Basic equation governing light behavior.

Physico-chemical and Biological Data

Physico-chemical and biological parameters plays an important role in population dynamics of phytoplankton. Various parameter that effects the population dynamics of phytoplankton may be broadly classified as light intensity, photoperiod, nutrient availability, chlorophyll content, transparency, turbidity, temperature, velocity, water column depth. Light intensity, duration, and water column depth are the basic parameter and others are the derived ones.

Basic Equations Governing Light Behaviour

Light downwelling in water column is expressed by two basic exponential equation:

1. The Beer-Lambert's law
2. Bouguer's law.

The attenuation of light due to absorption and scattering along the path at a distance S is given by the equation:

$$L(s) = L(0)e^{-\int_0^s K(s)ds} = L(0)e^{-KS}$$

where,

K(s) is a function of local parameters of the medium, and Ks =is the opacity or optical thickness.

The light intensity, I, decreases with depth S according to Lambert- Beer's law:

$$I(s) = I_{in}e^{-(kws + k_{bg}s)}$$

where,

I_{in} is the incident light intensity, k is the specific light attenuation coefficient of the phytoplankton, and K_{bg} is the total background turbidity due to non- phytoplankton components. The light intensity at the bottom of the water column, I_{out}, is given by $I_{out} = I(Z)$.

Depth integrals are used to calculate depth- integrated production by Monod equation:

$$P(I)=\frac{P_{\max} I}{(P_{\max}/\alpha)+I}$$

where,

P_{mex} is the maximum rate of specific production, and is the slope of the P(I) function at I =0.The advantage of the Monod equation is that it has a simple analytical solution of its depth integral.

Combining the Beer-Lambert's law of absorption and Monod Equation, the following dynamic system (Huisman and Weissing 1994, Weissing and Huisman 1994).

$$\frac{dw}{dx}=\frac{1}{z}\frac{Kw}{\left(Kw+K_{bg}\right)}\int_{I_{out}}^{I_{in}}\frac{p(I)}{kI}dI-Dw$$

where,

P(I) is a function of the local light intensity I. This model predicts that there is a critical value of I_{out}, where we have called the critical light intensity, at which the phytoplankton population should remain stationary. The phytoplankton population should increase as long as I_{out} is above it's critical light intensity, whereas the population should decrease as soon as I_{out} is below its critical light intensity. These population dynamics leads to a steady state. A phytoplankton population should grow unite, at steady state, it has reduced the light intensity at the bottom of the water column to its critical light intensity.

The light resource index for phytoplankton growth can be given as (Hamiliton and Schladow, 1997):

$$R_I=\frac{I}{I_{sat}}\exp\left[1-\frac{I}{I_{sat}}\right]$$

where,

I = light irradiance in a stationary water zone (STA-WZ) with its saturation value for reproduction as I_{sat}.

Steele's Equation (1962):

$$F_{Lp}=\frac{I(Z)}{K_{Lp}}e^{1-\frac{I(z)}{K_{Lp}}}$$

where,

KLp = the PAR at which phytoplankton growth is optimal [ly/d]. This function can combined with the Beer-Lambert's law and integrated over water depth to yield

$$\phi_{Lp}=\frac{2.718282}{K_e H}\left[e^{-\frac{I(0)}{K_{Lp}}e^{-K_e H}}-e^{-\frac{I(0)}{K_{Lp}}}\right]$$

Smith's Function (1980):

$$F_{Lp} = \frac{I(z)}{\sqrt{K^2{}_{Lp} + I(z)^2}}$$

where,

KLp = the Smith parameter for phytoplankton [ly/d]; that is, the PAR at which growth is 70.7 per cent of the maximum. This function can be combined with the Beer-Lambert's law and integrated over water depth to yield. FLp = phytoplankton growth attenuation due to light and KLp = the phytoplankton light parameter.

$$\phi_{Lp} = \frac{1}{k_e H} In \left[\frac{I(0)/K_{Lp} + \sqrt{1+\left(I(0)/K_{Lp}\right)^2}}{\left(I(0)/K_{Lp}\right)e^{-k_e H} + \sqrt{1+\left(\left(I(0)/K_{Lp}\right)e^{-k_e H}\right)^2}} \right]$$

Light Intensity and Phytoplanktonic Modeling

Various types of model are developed in past few decades. Models which are developed for light intensity and its effect on phytoplankton growth are based on some basic assumptions such as Lambert-Beer's law, Steele's equation and Smith's equation for light and its behavior in absorbing and scattering column and Monod equation for the depth integration.

Symbol	*Interpretation*	*Unit*
S	Depth of the water column	m
z	Total depth of water column	m
w	Biomass of water column	g/m_2
W	Total biomass of the water column	g/m_2
I	Light irradiance	$J.m^{-2}.s^{-1}$
I_{in}	is the incident light intensity	$J.m^{-2}.s^{-1}$
k	is the specific light attenuation coefficient of the phytoplankton	m^{-1}
K_{bg}	is the total background turbidity due to non-phytoplankton components	$m^{2.}g^{-1}$
I_{out}	is the light intensity at the bottom of the water column	$J.m^{-2}.s^{-1}$
KLp	The Smith parameter for phytoplankton	ly/d
FLp	Phytoplankton growth attenuation due to light	m^{-1}
K(s)	A function of local parameters of the medium	–
Lp	The phytoplankton light parameter	–
I_{sat}	Saturation value for reproduction	–
P(I)	is a function of the local light intensity	–
P_{mex}	The maximum rate of specific production	1/s

The specific light transmission coefficient, k was given by Bindloss (1976). Reynolds (1984) gave the growth rate models for phytoplankton in presence of light. George and Edwards (1976) gave the model for surface distribution of phytoplankton chlorophyll. Huisman and Weissing (1994. 1995), and Weissing and Huisman (1994) developed an analytically tractable model to analyze the effects of a dynamic light gradient on phytoplankton competition and community structure. Kok (1952), Sverdrup(1953) assume to be the founder of light intensity and its effect on the phytoplanktonic growth dynamics. Cullen (1990) developed a model on the growth and phototosynthesis in phytoplankton. Talling (1957), Platt *et al.* (1990), Kirk (1994) also developed some dynamic model for light and phytoplanktonic population growth. Numerical expressions for approximation for depth integral is also given by Platt *et al.* (1991). Several investigators Talling (1965), Jewson (1977), Reynolds (1984), have used the concept of a euphotic depth to calculate phytoplankton carrying capacities under light-limited conditions and also summarized the equation to give the model for phytoplankton and its growth limited by light. Jassby and Platt (1976) formulate a mathematical model for the relationship between light and phytoplankton. Kemp and Mitsch (1979) gave a general model for phytoplankton. Evers (1991) developed a model for light-limited continuous cultures for plankton.

References

Bindloss, M., 1976. The light-climate of Loch Leven, a shallow Scottish lake, in relation to primary production of phytoplankton. *Freshwater Biology*, 6.

Cullen, J.J., 1990. On models of growth and photosynthesis in phytoplankton. *Deep Sea Research*, 37: 667–683.

Evers, E.G., 1991. A model for light-limited continuous cultures: Growth, shading and maintenance. *Biotechnology and Bioengineering*, 38: 254–259.

George, D.G. and Edwards, R.W., 1976. The effect of wind on the distribution of chlorophyll–a and crustacean plankton in a shallow eutrophic reservoir. *J. Appl. Ecol.*, 13: 667–690.

Hamilton, D.P. and Schaldow, S.G., 1997. Prediction of water quality in lake and reservoirs. Part II: Model calibration, sensitivity analysis and application. *Ecol. Mod.*, 96: 111–123.

Kemp, W.M. and Mitsch, W.J., 1979. Turbulence and phytopankton diversity: A general model of the paradox of plankton. *Ecological Modelling*, 7: 201–222.

Kirk, J.T.O., 1994. *Light and Photosynthesis in Aquatic Ecosystems*. Camb. Univ. Press, England.

Kok, B., 1952. On the efficiency of *Chorella* growth. *Acta Botanica Neerlandica*, 1: 445–467.

Platt,T., Sathyendranath, S. and Ravindran, P., 1990. Primary production by phytoplankton: Analytic solution for daily rates per unit area of water surface. *Proc. of the Royal Society of London*, B241: 101–111.

Platt, T., Bird, D.F. and Sathyendranath, S., 1991. Critical depth and marine primary production. *Proceedings of the Royal Society of London*, B246: 205–217.

Reynolds, C.S., 1984. *The Ecology of Freshwater Phytoplankton*. Cambridge Univ. Press, UK.

Smith, R.A., 1980. The theoretical basis for estimating phytoplankton production and specific growth rate from chlorophyll, light and temperature data. *Ecological Modeling*, 10: 243–264.

Steele, J.H., 1962. Environmental control of photosynthesis in the sea. *Lim. Oce.*, 7: 137–150.

Sverdrup, H.U., 1953. On conditions for the vernal blooming of phytoplankton. *Journal du Conseil Permanent International pour l'Exploration de la Mer*, 18: 287–295.

Talling, J.F., 1957. The phytoplankton population as a compound photosynthetic system. *New Phytologist*, 56: 133–149.

Talling, J.F., 1965. The photosynthetic activity of phytoplankton in East African lakes . *Internationale Revue Der gesamten Hydrobiologia*, 50: 1–32.

Huisman, J. and Weissing, F.J., 1994. Light-limited growth and competition for light in well-mixed aquatic environments: An elementary model. *Ecology*, 75: 507–520.

Huisman, J. and Weissing, F.J., 1995. Competition for nutrients and light in a mixed water column: A theoretical analysis. *American Naturalist*, 146: 536–564.

Huisman, J., 1999. Population dynamics of light-limited phytoplankton: Miocrocosm experiments. *Ecology*, 80(1): 202–210.

Jassby, A.D. and Platt, T., 1976. Mathematical formulation of the relationship between photosynthesis and light for phytoplankton. *Limnology and Oceanography*, 21: 540–547.

Jewson, D.H., 1977. Light penetration in relation to phytoplankton content of the euphotic zone of Lough Neagh, North Ireland. *Oikos*, 28: 74–83.

Weissing, F.J. and Huisman, J., 1994. Growth and competition in a light gradient. *Journal of Theoretical Biology*, 168: 323–336.

Water and Wastewater Management in 2 Vols (2011) *Pages* **318–323**
Editors: **D.R. Khanna, R. Bhutiani & Gagan Matta**
Published by: **DAYA PUBLISHING HOUSE, NEW DELHI**

Chapter 50

Sources and Quality of Drinking Water in Relation to its Impact on Health with Reference to Tribal Groups of Chotanagpur Plateau

B.N. Panday, P.K.L. Das, A.K. Jha and A.K. Ojha
Eco-Genetical Research Lab, P.G. Department of Zoology, Purnia College, Purnia, Bihar

ABSTRACT

Sources and quality of drinking water were studied in the tribal belts of Chotanagpur plateau of Bihar. The main sources of drinking water are: supply water, wells handpumps, streams, ponds, rivers and pitwater. Physico-chemical parameters are within the permissible limits. But, the bacterial contamination was recorded in all the water sources. Due to the use of such contaminated water, the population suffers from diseases like diarrhoea, jaundice, amoebic dysentry, typhoid *etc.* It is advisable that such water must be properly treated before use.

Keywords: Chotanagpur plateau, Tribal, Drinking water, Sources, Quality and Water borne diseases.

Introduction

Water is the pre-requisite of life. A regular supply of neat and clean water is essential for different metabolic activities. The general scarcity of safe potable water has become a global problem. Supply of potable water has been affected by contamination of water sources in most developing countries. Rivers, lakes and groundwater have been polluted by industrial effluents, run-off and urban waste. Groundwater is the major source of drinking water in many parts of India. Due to direct or indirect interferences of human beings, the water has been found to contain various pollutants that cause harmful effects on drinking water and adverse impact on human health and aquatic life as well (Telliard and Rubin, 1987). The increase in human population and fast development has led to the scarcity of drinking water. Much of the ill health in the underdeveloped countries is largely due to lack of safe drinking water. In India, only 12 per cent of people get good clean water (Kudesia, 1986). Chotanagpur platue belt of Bihar is highly industrialised area. The area is mainly dominated by tribes. The objective of present study is to analyse the sources and quality of drinking water in relation to its impact on health.

Materials and Method

Water resources were seen with the personal visit of the sites as well as interview with villagers. Water samples from different areas were collected and analyzed for different parameters according to APHA (1975) and Trivedy and Goel (1984). Ten water samples were collected from each study area. Sterilized tubes were used for collection of water samples for bacterial count.

Sources of Drinking Water

During our field survey we found following sources of drinking water in this zone.

Supply Water

This mode of drinking water is confined to industrial and bureaucratic areas like Jamshedpur, Rakha, Jadugoda, Ghatshila, Dhalbhumgarh, Musabani, Chakylia, Bharagora, *etc.* Water is of high quality. In Jamshedpur, supply water comes from two sources, one from Dimna lake and another from Swarnarekha river.

Wells

The most common source of drinking water is well. This may be either Kuccha or Pukka. Pukka wells are made by government or by other agencies. Generally, Kuccha wells dry in summer. Kuccha are unhygenic in conditions. Kai, Lichens, Mosses, generally grows on the mouth of wells and during the water use, used water diffuses into it. Tribals are since not aware of hygiene, this causes severe gastrointestinal problem. In Chakuria itself there are 78124 wells out of which only 464 are used for drinking purposes.

Ponds

This area since gets 150-200 cm rainfall annualy. This becomes a good source of drinking water in remote villages. During the visit, we observed number of ponds filled with water. Some of the ponds are perennial while other dries in summer. Water of these ponds is used to prepare food at home by nearby residents. Number of ponds are mostly in Potka block.

Hand Pump

In most of the area Government has provided hand pumps as a source of drinking water. But during our survey work we found that most of the hand pumps were not working in lack of proper maintenance. The tribal people are so poor that they are even unable to provide valve and washer for the functioning of the hand pump. As the water level in this region is very low, most of the pumps become dry during summer season.

Streams

East Singhbhum is fortunate to have many small streams. These streams are the major (better to say only one) source of drinking water in remote tribal villages. Water of Bakai and Jog streams caters the need of the villagers of Kulamara, Rowam and Kumirmuri region.

Pit Water

The situation becomes worst during summer season when most of the small rivers and hand pump as well as wells dry up. Villagers have to travel miles to fetch drinking water. During our visit, we noticed that in many areas villagers dig river beds to a depth of 5' to 6' for water. Many villagers reported that water of such pit in the bed of river becomes so dirty that it is not possible to drink it. They filtered water with the help of cloth and then use it. We noticed that such stagnant water becomes a very suitable breeding ground of mosquito.

Artesian Well

These are very rare sources. We have seen (rare sources) only one artesian well in Kulamara village. Water from this well flows and after tresspassing paddy fields joins with Bakai stream. We have noticed that people were using water for daily purpose.

Results and Discussion

The pH of water samples varies between 7.10 and 7.69 which is well within the safe limits. Turbidity is also within permissible limits. Total hardness was below 300 mg/l is considered potable. In the present study hardness is within the permissible limits. Total alkalinity and TDS are also within the permission. The sulphate content of natural water is an important consideration in determining their suitability for public water supplies. Higher concentration of sodium sulphate in water can cause misfunctioning of alimentary canal. The chloride values varies between 3-60 mg/l which is within the permissible limits (40-160 mg/l). There has been a considerable research work related to nitrate contamination of groundwater due to point, non-point and natural sources. Increasing nitrate contamination in groundwater is associated with a disease known as Methamoglobinema (infant cynosis), if it is present in concentration above 44 mg/l (WHO,1984). The nitrate value in the present investigation is within the permissible limits (28-35 mg/l).

High iron concentration has been associated with gastrointestinal disturbances (Babcock, 1951 and Sharma, 1994). Beside this, high iron concentration is also associated with hepatic and pancreatic troubles as well as abortion and mental disorders. High concentration of iron in drinking water has been reported by many workers in various parts of India (Aowal, 198l; Kakati *et al.*, 1990; Das *et al.*, 1992; Pandey *et al.*, 1992, 1994 and 1998). Endemic flurosis continues to be a challenging problem. It is one of the common oral health problem of humanity in developing countries. Water with high content of fluorides is the main cause of fluorosis. According to Rao (1998) about 225 million Indian scattered in different status are affected with fluorosis. In the present study the fluoride value ranged between 0.6-0.65 mg/L which is within the permissible limits.

In the present study the MPN values are found to vary from 5-62 MPN/100 ml. The highest MPN value was found in Kulamara. It is an established fact that the very high mortality rate is found among the population using pond/reservoirs/streams/rivers and earthen wells as the main source of drinking water (Ghosh, 1985; ICMR, 1985). Further, water contamination is an integral factor, which recognizably increased with urbanization and industrialization by direct and indirect means. This poses a threat to water sources which have evolved into a formidable factory in the spread of human and animal diseases. The causitive agents for typhoid, fever, bacillary dysentery, amoebic dysentry, paratyphoid fever and cholera spread by water (Charles and Alice, 1957). The prevalence of water

Table 50.1: Average Physico-chemical and Bacteriological Quality of Tribal Villages of Chotanagpur Plateau

Sl.No.	Spots	Parameters											
		Colour	Turbidity	pH	Total Hardness	Total Alkalinity	Fluoride	Chloride	Nitrate	Sulphate	Total Dissolved Solids	Iron	MPN/ 100 ml
	E. Singhbhum												
1.	Deosol	Clear	7.0	7.69	145	140	0.80	148	34	35	275.0	Trace	32
2.	Bomdih	Clear	22	7.38	150	158	0.60	048	28	40	375.0	0.20	25
3.	Kulamara	Clear	20	7.58	170	172	0.70	160	38	60	550.0	0.20	62
4.	Borakata	Clear	8.0	7.50	155	160	0.60	145	35	50	525.0	Trace	45
	Gumla												
5.	Chainpur	Clear	6.0	7.10	100	140	0.80	040	–	05	180.6	0.25	5.0
6.	Kurumgarh	Clear	5.0	7.35	80	80	0.70	060	–	05	185.5	0.12	10
7.	Tigawal	Clear	8.0	7.45	70	70	0.65	040	–	05	163.4	1.00	12
8.	Tetardipa	Clear	7.0	7.25	120	120	0.70	120	–	03	240.6	0.25	18
9.	Jate	Clear	5.0	7.20	60	60	0.60	050	–	20	0.120	0.12	09

All the values are in mg/l except pH, colour and turbidity.

borne diseases in these tribal population is not surprising because major sources of drinking water in these tribal villages are river, lakes, ponds, pits and wells.

The coliform values in the present study are found to exceed the standard permissible limits and, therefore, indicates a high degree of pollution at these points. The water of these villages are potentially hazardous to the health of the public. During the course of present study we have noticed that the tribal population residing in these areas are victims of a large number of diseases such as Diarrhoea, Jaundice, Amoebic dysentery, Cholera *etc.* The percentage of these diseases in the local tribal population have been shown in the Table 50.2.

Table 50.2: Distribution of Water Borne Diseases Among Tribal Populations of Chotanagpur Plateau

Sl.No.	*Tribal Population*	*Population (no.)*	*Diseases*			
			Diarrhoea	*Jaundice*	*Amoebic Dysentry*	*Typhoid*
	E. Singhbhum (Jamshedpur)					
1.	Santal	844	22.86 per cent	16.94 per cent	37.44 per cent	22.74 per cent
2.	Oraon	134	22.38 per cent	14.92 per cent	26.11 per cent	36.56 per cent
3.	Bhumij	257	23.34 per cent	19.45 per cent	37.35 per cent	19.84 per cent
4.	HO	302	19.20 per cent	17.21 per cent	43.04 per cent	20.52 per cent
5.	Kharia	117	25.64 per cent	17.09 per cent	34.18 per cent	23.07 per cent
6.	Munda	87	28.73 per cent	17.04 per cent	32.18 per cent	21.83 per cent
	Gumla and Ranchi					
7.	Oraon	371	11.90 per cent	7.01 per cent	15.90 per cent	3.50 per cent
8.	Munda	238	19.33 per cent	7.14 per cent	21.43 per cent	5.46 per cent
9.	Korwa	103	21.36 per cent	9.71 per cent	28.16 per cent	7.77 per cent
10.	Kharia	47	27.66 per cent	8.51 per cent	34.04 per cent	6.38 per cent

Coliform MPN/100 ml of sample, in most of the areas is very high which indicates the poor quality of water in turn indicating possible fears of sewage mixing with water. On the whole, it can be concluded from the present study that tribals are prone to several diseases due to consumption of such polluted water. If the adequate use of alum and other disinfectants is made and villagers are made fully aware about hygenic use of potable water, then only water borne diseases can be controlled. The Government must also take care of ecologically undeveloped tribals of Chotanagpur plateau to improve the quality of drinking water.

Acknowledgements

The financial help for this work was provided by Ministry of Environment and Forests, Government of India, under the project entitled "Pattern of Human Settlement in and around the North Gangetic Belt of Bihar with particular reference to certain Schedule Tribes under MAB Programme" Ref. N.14/5/94–MAB/RE, under the guidance of Dr. B.N. Pandey, Principal Investigator, P.G.Department of Zoology, Purnia College, Purnia, Bihar (India) Pin-854301 is gratefully acknowledged.

References

Aowal, A.F.S.A., 1981. Design for iron elemination for hand tubewells. *J.L.W.W.A.*, 13(1): 65–80.

APHA, 1975. *Standard Methods of Examination of Water and Wastewater*, 14th edn. APHA, Washington, USA.

Babcock, R.H., 1951. Iron and manganese in water supplies and method of removal. *Water Sewerage Work*, 98(10): 442.

Charles, F.C. and Alice, L.S., 1957. *Principles of Microbiology*. The C.V. Mobsy Company, St. Louis, pp. 579–583.

Ghosh, S., 1985. Dimension of morbidity and mortality among children. In: Paper Presented at Workshop on *Genetic Epidemiological Approach to Health Care,* National Institute of Health and Family Welfare, New Delhi.

ICMR, 1985. *Diarrhoeal Diseases in Infants and Children.* Indian Council of Medical Research, New Delhi.

Kakati, G.N. and Bhattacharya, 1990. Trace metals in surface water of Greater Gauhati. *Ind. J. Env. Health,* 32: 197.

Kudesia, V.P., 1986. *Water Pollution.* Pragati Prakashan, Meerut.

Pandey, B.P., Das, P.K.L. and Jha, A.K., 1992. Physico-chemical analysis of drinking water of Purnea district, Bihar. *Acta. Ecol.,* 14(2): 108–114.

Pandey, B.N., Das, P.K.L., Jha, A.K. and Tripathi, R.N., 1994. An assessment of quality of drinking water of Katihar, North Bihar. *Acta. Ecol.,* 16(2): 144–149.

Pandey, B.N., Mishra, S.K., Yadav, S. and Sharma, P.D., 1998. An assessment of drinking water quality of Purnea district (North Bihar) in relation to its impact on health. *J. Env. and Poll.,* 5(4): 259–263.

Rao, R., 1998. Rasam to the rescue. *The Tribune,* pp. 6.

Sharmah, A., 1994. A study on the presence of iron in rural water sources and its removal. *B. Tech. Project Report,* NERIST, Arunachal Pradesh.

Teliard, W.A. and Rubin, M.D., 1987. Control pollutants in water. *J. Chromatographic Sci.,* 25: 322–327.

Trivedy, R.K. and Goel, P.K., 1984. *Chemical and Biological Methods of Water Pollution Studies.* Environmental Publications, Karad.

WHO, 1984. *International Standards for Drinking Water.* WHO, Geneva.

Water and Wastewater Management in 2 Vols (2011) *Pages 324–332*
Editors: D.R. Khanna, R. Bhutiani & Gagan Matta
Published by: DAYA PUBLISHING HOUSE, NEW DELHI

Chapter 51

Study of Water Quality of River Suswa Near Raiwala, Uttarakhand

D.R. Khanna[1], S.K. Pathak[2], R. Bhutiani[1] and Kumar S. Chandra[1]
[1]Department of Zoology and Environmental Science, Gurukul Kangri University, Haridwar
[2]Department of Zoology, Govt. P.G. College, Mhow, M.P.

ABSTRACT

The present study was performed to estimate the quality of Suswa river at Raiwala. Test for both, physico-chemical and heavy metals were carried out at different sites of the Suswa river near Raiwala.

Keywords: *Suswa, Pollution, Physico-chemical, Status, Water quality.*

Introduction

Suswa is spring fed river confluencing into the river Ganga. The water of the river Suswa is used by the people mainly to meet their daily water requirements. Raiwala is situated at the right bank of this river. The population of Raiwala is approximately 60,000 to 70,000. Recently, few hotels and recreational centers have been set up at Raiwala, just near the river bank. A few meat shops are also situated in Raiwala. The city garbage and the leftout wastes from these meat shops, hotels, restaurants and shopping complexes find their way into the river along with the domestic sewage without any prior treatment. Therefore, the quality of Suswa water gets affected, which needs a continuous monitoring to assess the impact of all the above stated activities. This assessment is a necessary step for sustaining the ecology of the Suswa river. Various scientists have investigated the quality of water of different rivers in this region. But, no such study has been conducted far to assess the water quality of Suswa river. The present chapter is an attempt to present the findings of the recent investigation carried out between 2003 and 2004. This may serve as a data base for future investigations. There different sites were selected on the basis of assessibility, uniformity of concentration and well mixed zone. The selected sites are as below

1. Site A: 5 kms upstream from Raiwala town.
2. Site B: Near Raiwala town
3. Site C: 5 kms downstream from Raiwala town.

Materials and Method

The physico-chemical parameters and heavy metals (Cu, Pb, Cr, and Zn) were estimated by the following standard methods of APHA (1989) and Trivedy and Goel (1984). The samples from all the three sites were collected on monthly basis, at the same date and time. The test for heavy metals were performed by the atomic absorption spectroscopy method using AA S.

Results and Discussion

The heavy metals and physico-chemical parameters are represented in Figure 51.1 and Figures 51.2A–F. In the Suswa river at Raiwala, a difference in the fluctuation of water temperature was

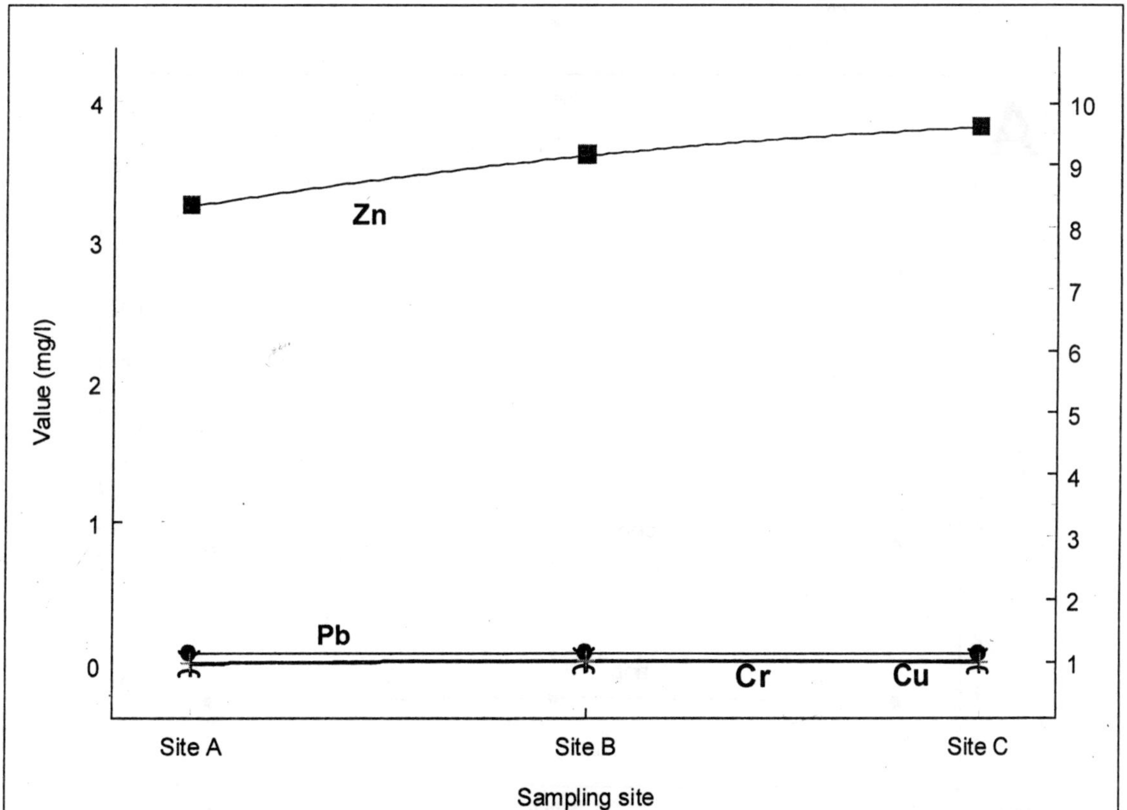

Figure 51.1: The Graph of Different Heavy Metals of River Suswa at Three Sampling Sites in the Year 2003–2004

Figure 51.2: The Graph of Different Physico-chemical Parameters of River Suswa at Three Sampling Sites in the Year 2003–2004

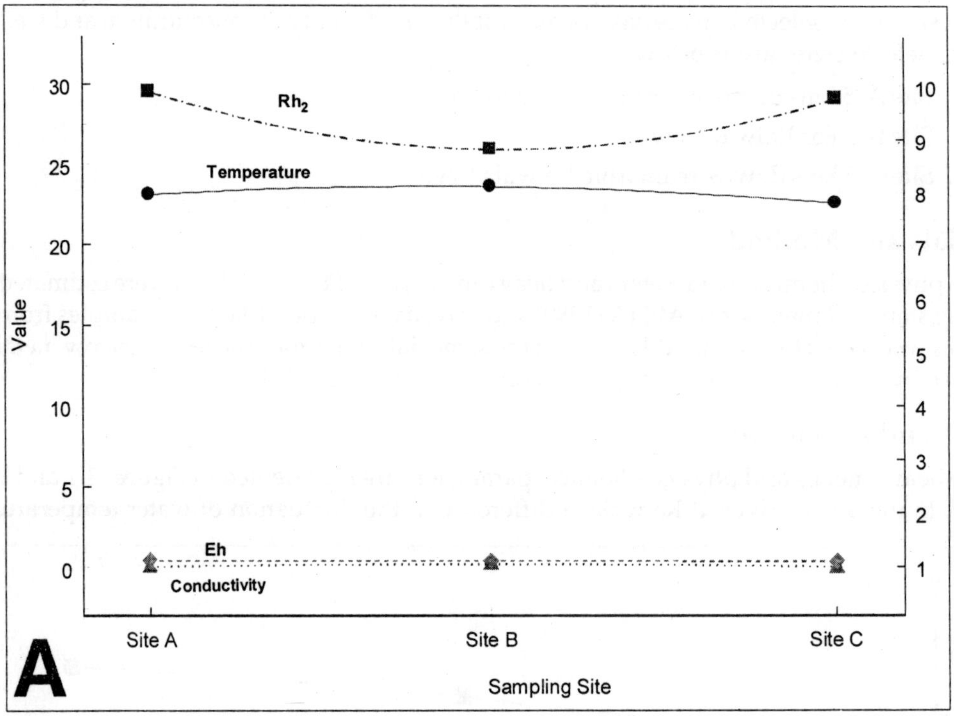

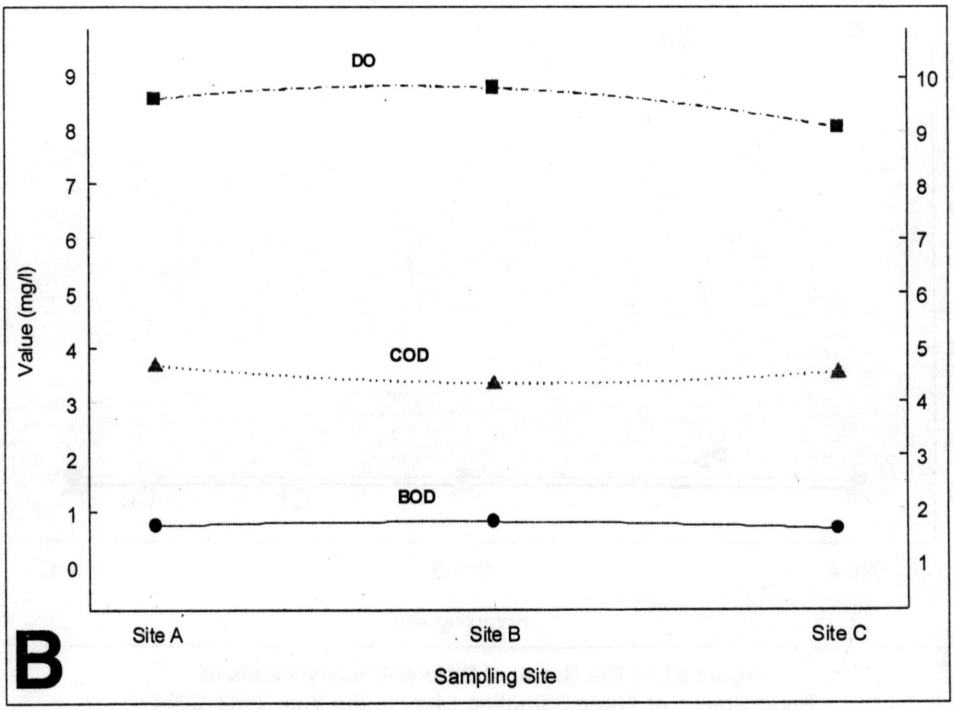

Contd...

Figure 51.2–Contd...

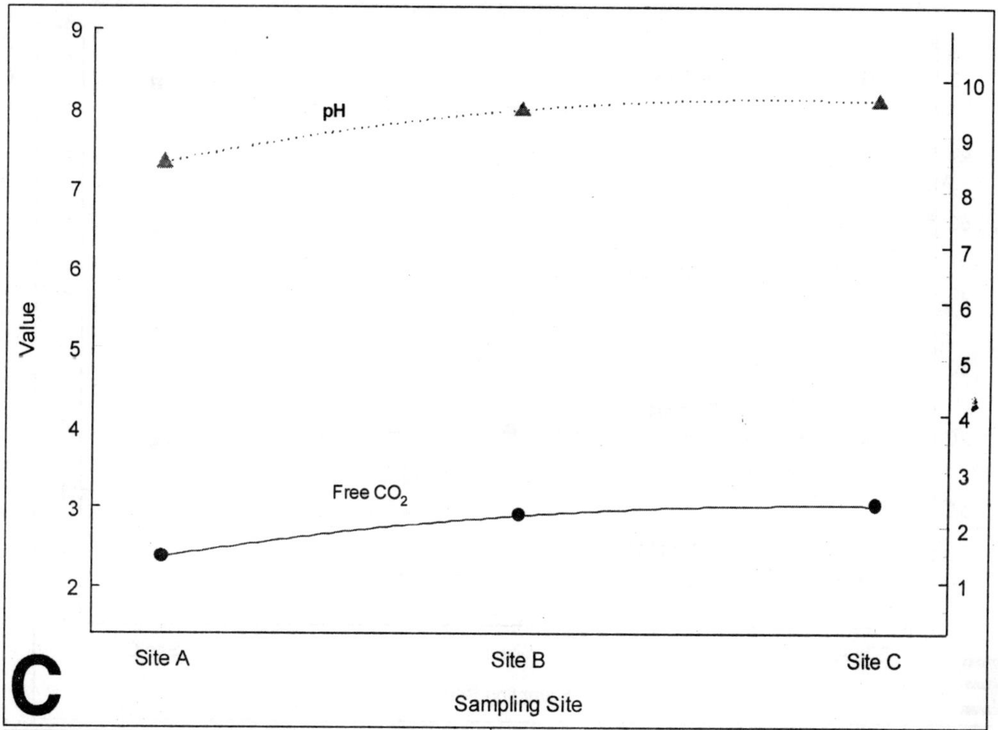

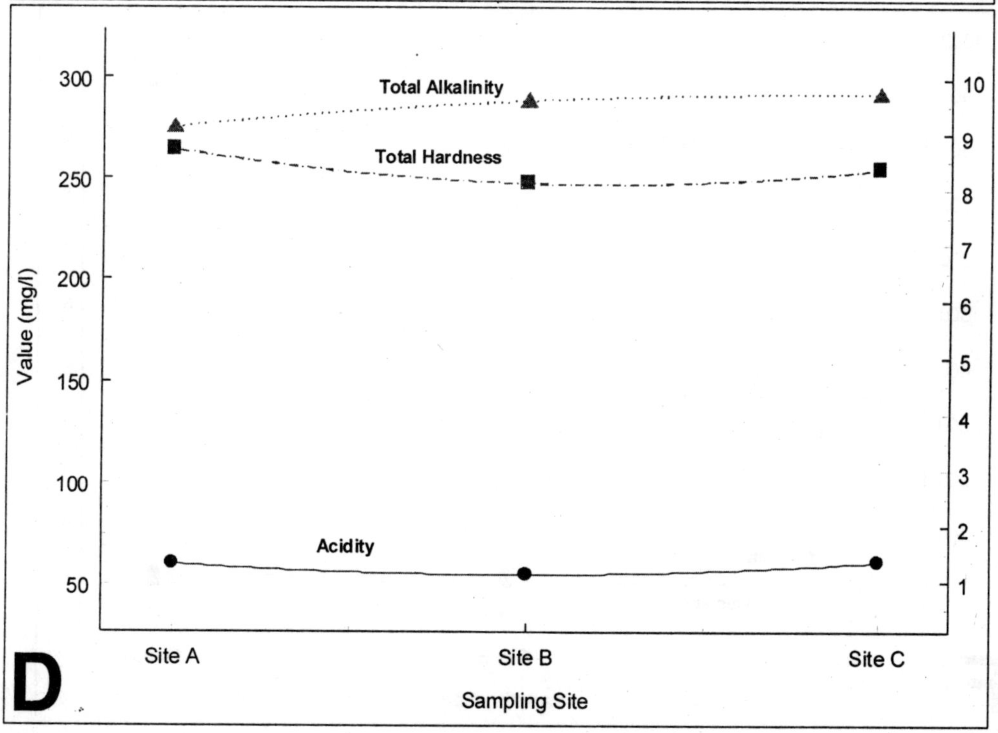

Contd...

Figure 51.2–Contd...

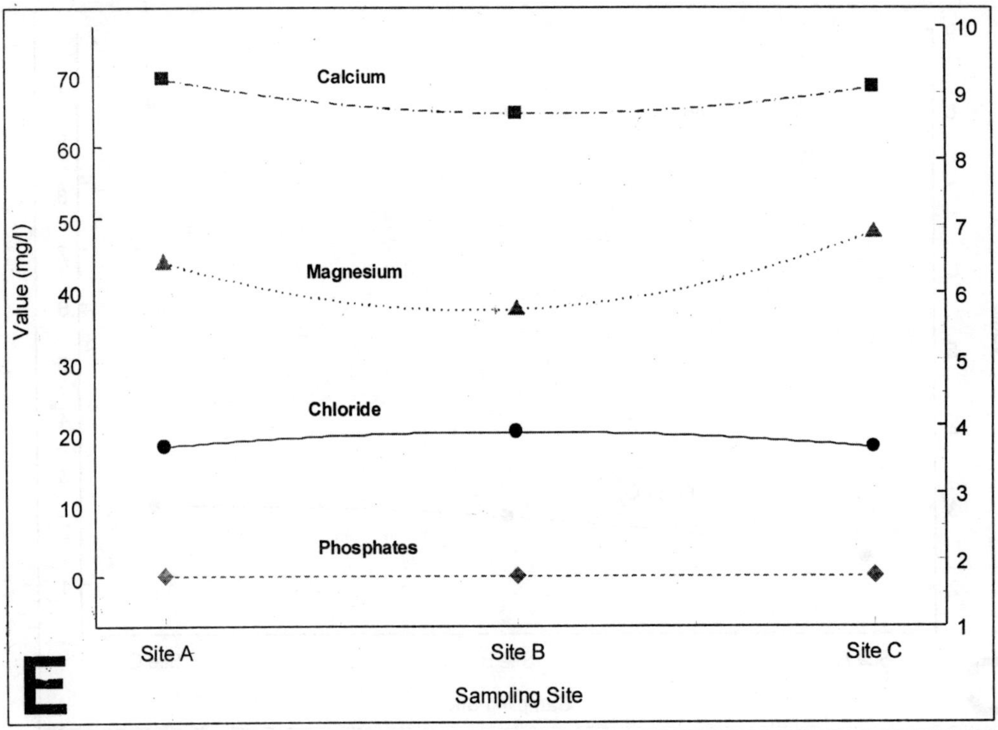

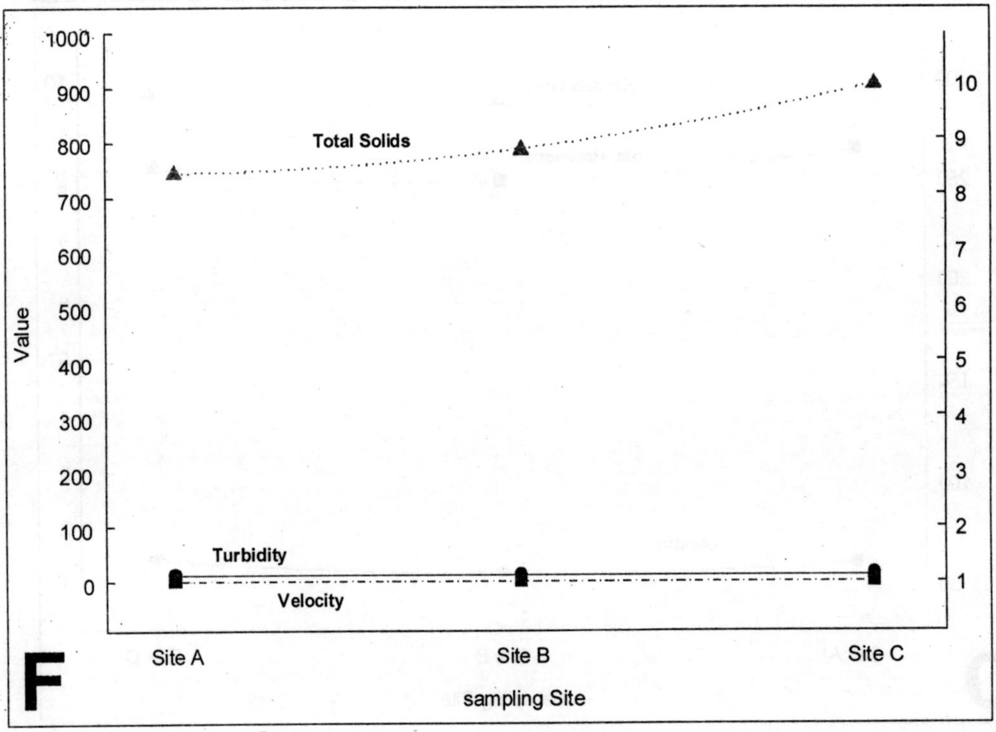

maximum (23.65°C ± 3.44) observed at sampling site-B and minimum (22.65°C ± 4.12) at site-C. The water temperature showed an upward trend from winter season to summer season followed by a downward trend from monsoon season onwards. A more or less similar trend has been observed in the river Yamuna by Chakrabarty *et al.* (1959). The fluctuation of temperature was well within the limit for survival of fishes. Similar types of fluctuations were reported by Dwivedi *et al.* (1995) in two ponds at Patna.The temperature showed negative relationship with the dissolved oxygen almost throughout the study as also reported by Das and Srivastava (1956a), Dobriyal (1985), Khanna *et al.* (1993) and Chugh (2000). Maximum average dissolved oxygen was recorded (8.74 mg/l ± 0.24) at sampling site-B and minimum (8.01 mg/l ± 0.46) at site-C. The dissolved oxygen reduced gradually from summer to monsoon season due to turbidity, which retarded the photosynthetic activity of aquatic flora.This trend was also reported by Singh *et al.* (1982) in the river Brahamputra of Guwahati. Again the dissolved oxygen increases from monsoon to winter season it was due to the activity of plankton in water, which are found more active when temperature was not too high. Similar trend was also reported by Badola and Singh (1981) in the river Alaknanda, Khanna (1993), and Chugh (2000) in river Ganga. It has been recommended that a minimum of 4 mg/l of dissolved oxygen should be maintained in water for healthy growth of fish and other microbial population. Highest values of dissolved oxygen were recorded in river Cauvery by Somashekar *et al.* (1984), and in river Kosi by Bhatt *et al.* (1984).The maximum average velocity was recorded (0.39 m/s ±0.05) at sampling site- B and minimum (0.31 m/s ± 0.02) at sampling site–A. The velocity started increasing after winter season and beyond monsoon season, the velocity started decreasing in winter.

In the present study it has been observed that the velocity and the total solids showed positive relationship. Most of Indian rivers showed a similar tendency with respect to fluctuations of total solids (Kudesia and Verma, 1985). The maximum average total solids were noted (907.12 mg/l ± 255.1) at sampling site-C and minimum (745.10 mg/l ± 3.14) at site- A. The total solids were recorded minimum in winter due to gradual sedimentation of the filterable residue, at the bottom and also due to the minimum velocity of the river, which favored effective sedimentation. The total solids were maximum in monsoon season due to increased velocity, which favored soil-erosion. Similar condition was also reported by David (1956) in Bhadra river. The BOD (Biochemical oxygen demand) was observed maximum (0.81 mg/l ± 0.04) at sampling site-B and minimum (0.68 mg/l ± 0.29) in winter season. A negative relationship has been observed between BOD and DO contents. A similar pattern has been reported by Khanna (1993), Chugh (2000). BOD determination is a most useful technique to assess the level of organic pollution in river system. The maximum average value of COD (Chemical oxygen demand) was observed (3.68 mg/l ± 0.87) at sampling site-A and minimum (3.35 mg/l ± 0.65) at site- B. The carbon dioxide content of the water depends upon the temperature of water, depth of water, rate of respiration and decomposition of organic matter in water. The carbon dioxide content of the water increases with increase in temperature. Free Carbon dioxide average value was observed maximum (3.02 mg/l ± 2.88) at sampling site-C, and minimum (2.37 mg/l ± 0.74) at site-A. Chakrabarty *et al.* (1959) recorded the maximum free CO_2 in Jamuna during monsoon at Allahabad. The dissolved oxygen and free carbon dioxide are usually inversely related to one another because of the photosynthetic and respiratory activities of the biota. Conductivity determines the total amount of ionisable salts in water. It is due to ionization of dissolved inorganic solids. The fluctuations in conductivity were caused mainly by variations in the ionic precipitations and the diluting effect of rains (Welch, 1948). In the present study maximum conductivity was observed (0.22 siemens/cm ± 0.01) at sampling site-B and minimum (0.04 siemens/cm± 0.008) at site-C. The maximum turbidity was found (12.47 JTU ± 18.23)at sampling site-C. At sampling site-B the average value of turbidity was found minimum(10.99 JTU ± 19.56). The turbidity and total solids were closely interrelated with one another and cause common effect upon the river and aquatic life as also stated by Verma *et al.* (1984). Bhatt *et al.* (1984) attributed that during monsoon months, the river water contained large amount of silt, fine sand particles, organic matter and clay. According to Basu *et al.* (1973), the measurement of pH has great importance because chemical and biochemical reactions in an aquatic body take place at a particular pH. The maximum average value of pH was recorded (8.11 ± 0.08) at sampling site-C and the minimum was observed (7.52 ± 0.42) at site-A. Meshram (1996) reported lower pH values during summer and higher pH during rainy season.The oxidation-reduction potential (Eh) and oxidation-reduction index (rH_2) showed its higher values during summer and monsoon, while lower during winters. Similar observations were also reported by Gautam (1990).The maximum value of alkalinity was observed (291.12 mg/l ± 32.43) at sampling site C and minimum (275.33 mg/l ± 39.78) at sampling site-A. Similar types of findings were observed by Venkateshwarlu and Jayanti (1968) in the river Sabarmati. pH and alkalinity showed a positive correlation with one another as suggested by Freiser and Fernando (1966) in the ionic equilibria in analytical chemistry.

According to Hays and Anthony (1958), the decomposition of organic matter leads to the high alkalinity of the waters. The maximum and minimum values of hardness were recorded (264.25 mg/l ± 21.73) and (247.32 mg/l ± 19.02) at sampling sites-A and B respectively. Hardness showed a negative relationship with chloride. Chopra and Patrick (1994) observed positive relationship between chloride and hardness in river Ganga at Rishikesh. Hardness showed a positive relationship with alkalinity while Chopra and Patrick (1994) observed negative relationship in river Ganga at Rishikesh. Maximum amount of calcium in Suswa was found as (69.49 mg/l ± 13.73) at site-A and minimum (64.66 mg/l ± 7.73) at site-B. pH and calcium showed positive relationship to one another. The value of magnesium was observed maximum (48.13 mg/l ± 4.17) at site C and minimum (37.33 mg/l ± 3.82) at site B. Singhai (1986) reported a positive correlation between magnesium and total hardness as also observed in present study. Maximum value of chlorides was observed (20.14 mg/l ± 6.58) at site B and minimum (17.99 mg/l ± 1.35) at site-B. Chloride is one of the important chemical indicator of pollution. Sengar *et al.* (1985) and Raina *et al.* (1984) also showed significant levels of chloride content.The observed low value of phosphorus are not to be taken as indication of low productivity.In the present study all the heavy metals as Lead, Copper, Chromium and Zinc were taken for observation and revealed that all heavy metals are having partial or some how positive relation with velocity and negative relation with plankton life. Concentrations of all the heavy metals in different seasons were quite favorable for the biota and not in favorable to the water quality as suggested by Mathur (1982). Heavy metals get contaminated into aquatic systems as a result of various natural activities (weathering of soils and rocks from volcanic eruptions) and from a variety of human activities involving the mining, processing or use of metals or substances.

References

Adebisi, A.A., 1981. The physico-chemical hydrology of a tropical seasonal river–Upper Ogun river, Nigeria. *Hydrobiol.*, 79(2): 157–165.

APHA, AWWA, WPCF, 1998. *Standard Methods for the Examination of Water and Wastewater*, 20th edn. Washington D.C., New York.

Atkin, W.R.G. and Harris, G.T., 1924. Seasonal changes in the water and Heleoplankton of freshwater ponds. *Proc. Roly. Dub. Soc.*, 8(N.S): 1–21.

Badola, S.P. and Singh, H.R., 1981. Hydrobiology of river Alaknanda of Garhwal Himalayas. *Indian J. Ecol.*, 8(2): 269–276.

Basu, A.K., Rao, C.G.S., Dhaneshwar, G., Mukherjee, M.K.A., Kutty, D., Ray and Trivedi, R.C., 1973. Study on Hooghly estuaries pollution in the vicinity of tissue mill. *Proc. Symp. on Environ. Poll.*, p. 24–46.

Bhatt, S.D., Bisht, Y. and Negi, U., 1984. Ecology of Limnoflora in river Koshi of the Kumaun Himalaya (U.P.). *Proc. Indian Natu. S.C.*, 50(4): 395–405.

Chakraborty, R.D., Ray, R.P. and Singh, S.B., 1959. A quantitive survey of plankton and physiological conditions of the river Jamuna at Allahabad in 1954–55. *Indian. J. Fish*, 61: 186–203.

Chopra, A.K. and Patrik, Nirmal J., 1994. Effect of domestic sewage on self putrification of Ganga water at Rishikesh. I. Physico-chemical parameters. *Ad. Bios.*, 13(1): 75–82

Chugh, Tarun, 2000. Seasonal variation in the microbial ecology of River Ganga at Haridwar. *Ph.D. Thesis*, Gurukul Kangri University, Haridwar.

Das, S.M., 1961. Hydrogen ion concentration, plankton and fish in freshwater eutrophic lakes of India. *Nature*, 191(4787): 511–512.

Das, S.D. and Srivastava, V.K., 1956a. Some new observations on freshwater plankton. I. Plankton of fish tanks of Lucknow. *Sci. and Cult.*, 21(8): 466–467.

David, A., 1956. Studies on the pollution of the Bhadra river at Badrawati fisheries effluents. *Proc. Nat. Inst. Set.*, India, 93(3): 132–160.

Dobriyal, A.K., 1985. Ecology of limnofauna in the small streams and their importance to the village life in Garhwal Himalaya. *Uttar Pradesh J. Zool.*, 5(2): 139–144.

Dwivedi, H.N., Pandey, B.N., Manoj, Kumar and Alam, Naushad, 1995. Limnological studies in relation to fish production of two ponds of Patna. *Him.J. Env. Zool.*, 9: 7–10.

Ellis, M.N., 1937. Detection and measurement of stream pollution. *Bull. U.S. Bur. Fish*, 48: 356–437.

Freiser, H. and Fernando, Q., 1996. *Ionic Equilibria in Analytical Chemistry*. Wiley and Sons, New York.

Gautam, Ashutosh, 1990. *Ecology and Pollution of Mountain Waters*. Ashish Publishing House, New Delhi, p. 1–209.

Hays, E.R. and Anthony, E.H., 1958. *Limnol. Oceanogr.*, 3(3): 297–307.

Heper, B., 1962. Primary production in the fish pond and its application to fertilization experiments. *Limnol. Oceanogr.*, 7(2): 131–136.

Hutchinson, G.C., 1957. *A Treatise on Limnology, Geography, Physical and Chemical.* John Wiley and Sons, Inc., New York, 1: 1–1015.

Khanna, D.R., 1993. *Ecology and Pollution of Ganga River*, Ashish Publication House, Delhi, p. 1–241.

Khanna, D.R., Badola, S.P. and Dobriyal, A.K., 1993. Plankton ecology of river Ganga at Chandighat Haridwar. In: *Advance in Limnology*, (Ed.) H.R. Singh. Narendra Publishing House, New Delhi, p. 171–174

Kudesia, V.P. and Verma, S.P., 1985. A study of industrial pollution on Kali river. *J. Env. Sci.*, 1(2): 41–49.

Mathur, R.P., 1982. *Water and Wastewater Testing*. Neemchand and Bros. Publishers, Roorkee, p. 1–54.

Meshram, C.B., 1996. Limnological studies of Wadali lake, Amravati, Maharashtra. *Ph.D. Thesis*, Amravati University, Amravati.

Raina, V., Saha, A.R. and Ahmed, S.R., 1984. Pollution Studies on river Jhelum. I. An assessment at water quality. *Indian J. Env. Hlth.*, 26: 187–210.

Sengar, R.M.S., Sharma, K.D. and Pathak, P.D., 1985. Studies on distribution of algal flora in polluted and non-polluted regions in Yamuna river at Agra (U.P.). *J. Indian Bot. Soc.*, 64: 365–376.

Singh, H.P Choudhry, M. and Kalekaur, V., 1982. Seasonal and diurnal changes in physico-chemical features of the river Brahamputra at Guwahati. *Ind. J. Zool.*, 2(82): 77–84.

Singhai, S., 1986. Hydrobiological and ecological studies of newly made Tawa reservoir at Ranipur. *Ph.D. Thesis*, H.S. Gaur University, Sagar.

Somashekhar, R.K., 1984. Phytoplanktons as indicators of water quality: A study of river Cauvery. *Int. J. Environ. Studies*, 24: 115–123.

Subbamma, D.V. and Ramasarma, D.V., 1992. Studies on the water quality characteristics of a temple pond near Machilipatnam, Andhra Pradesh. *J. Aqua. Biol.*, (1–2): 22–27.

Trivedi, R.K. and Goel, P.K., 1984. *Chemical and Biological Methods for Water Pollution Studies*. Environmental Publication, Karad, p. 1–251.

Venkateshwaru, T. and Jayanti, T.V., 1968. Hydrobiological studies of the river Sabarmati to evaluate water quality. *Hydrobiologia*, 33(3–4): 442–448.

Verma, S.R., Sharma, P., Tyagi, A., Rani, S., Gupta, A.K. and Dalela, R.C., 1984. Pollution and saprobic status of eastern Kalinadi. *Limnologica*, 15(1): 69–133.

Welch, P.S., 1948. *Limnological Methods*. The Blakiston Co. Philadelphia, p. 1–381.

Water and Wastewater Management in 2 Vols (2011) *Pages 333–338*
Editors: D.R. Khanna, R. Bhutiani & Gagan Matta
Published by: DAYA PUBLISHING HOUSE, NEW DELHI

Chapter 52

Diurnal Variation in Some Aspects of Limnology of Gopalpura Tank of District Guna, Madhya Pradesh

Dushyant Sharma and Renu Jain
Department of Zoology, Govt. P.G. College, Guna, M.P.

ABSTRACT

Gopalpura tank is one of the artificial water bodies of Guna (M. P.). It is used mainly for irrigation and pisciculture by the nearby villagers. Diurnal variations in physico-chemical factors such as water temperature, pH, dissolved oxygen, free CO_2, alkalinity and plankton (phytoplankton as well as zooplankton) of the tank were studied. Fluctuations were found in water temperature, pH, dissolved oxygen, free CO_2 and alkalinity. Free CO_2 and alkalinity exhibited a negative relationship with the temperature. Abundance of plankton also showed a distinct diurnal variation.

Keywords: Gopalpura tank, Diurnal variation, Limnology, Plankton.

Introduction

Gopalpura tank is one of the artificial water bodies of Guna, situated at a distance of just 2 km from Guna town. The work on the tank was started in December 1971 and completed in March 1977. Surrounded by village from all the sides, the reservoir has a catchment area of approximately 6 sq. ft. Though, the tank is not very large but it serves as a source of irrigation for the villagers. The water is also utilized for pisciculture by the fishermen and local villagers.

Diurnal variation study of the freshwater bodies of India has been made by few workers. A few to mention are Ganapati (1955), George (1961), Michael (1966), Verma (1967), Surnitra (1971),

Mishra *et al.* (1975 and 1976), Bohra *et al.* (1979), Sharma and Gupta (1981) and Dobriyal and Singh (1981). Since no such study was carried out so far on Gopalpura tank, it was considered desirable to study the diurnal variation in physico-chemical factors as well plankton of the tank.

Materials and Method

Investigation on the diurnal variations of Gopalpura tank was made on Feb. 10 and 11, 2001. Samples were collected from different sites-Station A, Gopalpura village side and station B, Cantt side, at an interval of 3 hours, starting at 9 AM on Feb. 10, 2001. Standard methods by APHA, 1985 and Trivedi and Goel (1986) were followed for the analysis of physico-chemical factors and plankton studies. The water for hydrobiological analysis was collected from the shore and little below the surface. The water was filtered and preserved in 4 per cent formaline. The qualitative and quantitative analysis of plankton was made in Sedgwick Rafter Counting chamber.

Results and Discussion

The results are shown in Figures 52.1 and 52.2 and Tables 52.1 and 52.2. The details are given in foregoing paragraphs.

Physico-Chemical Factors

Water Temperature

The water temperature varies from 12.5°C (at station 'A') to 20.5°C (at both stations) being maximum at 15.00 hours on Feb. 10, 2001 and minimum at 03.00 hours on Feb. 11, 2001. The maximum temperature

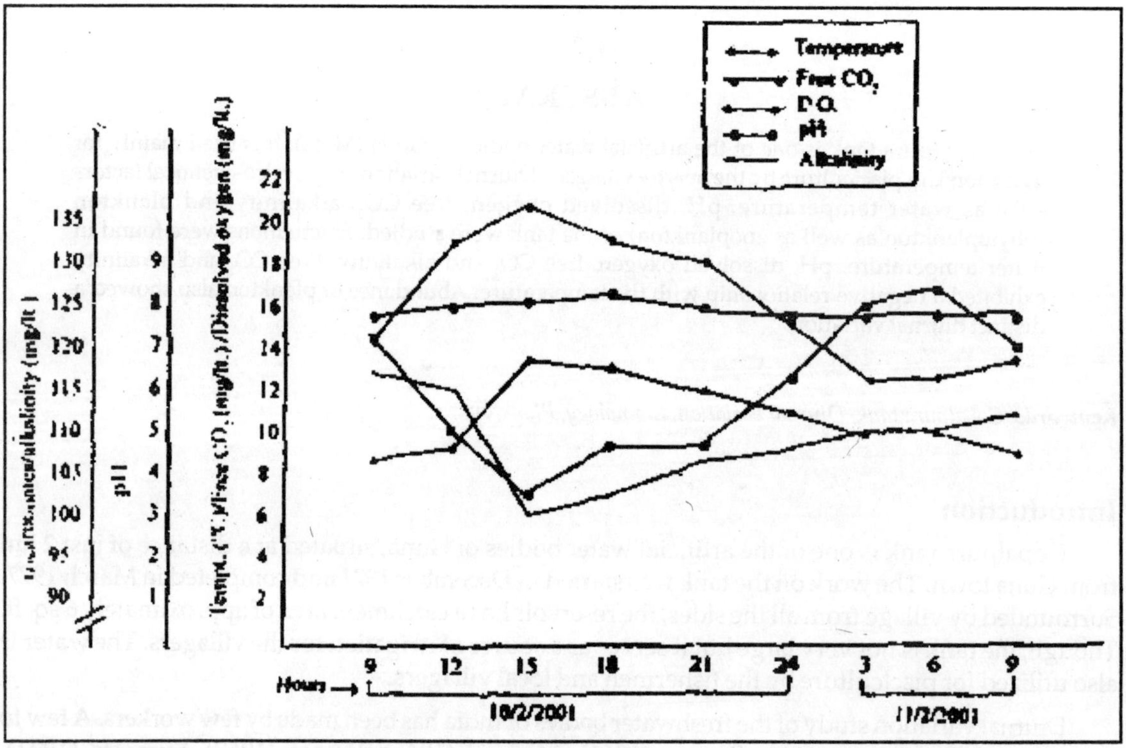

Figure 52.1: Diurnal Variation in Physico-chemical Factors of Gopalpura Tank–Spot A

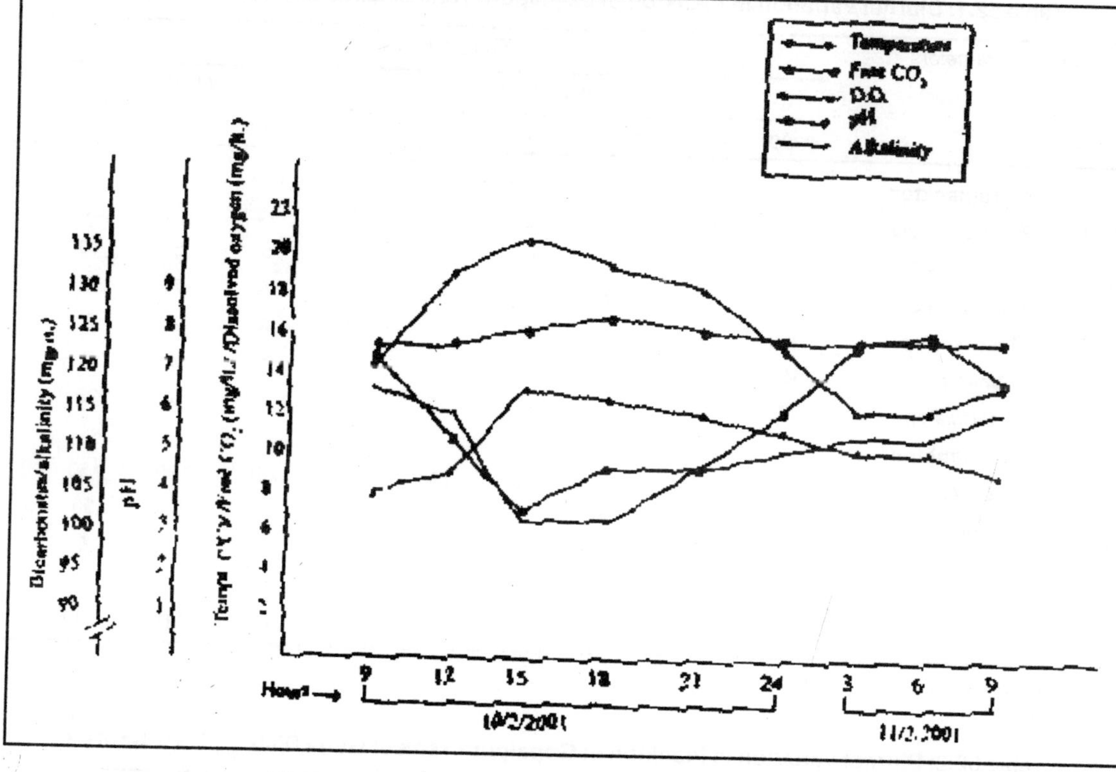

Figure 52.2: Diurnal Variation in Physico-chemical Factors of Gopalpura Tank–Spot B

obtained during noon corresponds to the atmospheric temperature, which was also highest during day hours and falls during night. Similar results were obtained by Sumitra (1971), Bohra *et al.* (1979), Sharma and Gupta (1981) and Dobriyal and Singh (1981). The maximum temperature observed during noon can be explained by the fact that the rate of cooling and heating of water temperature depends on atmospheric temperature.

pH

In case of pH, we obtained a slight increase (0.2) in pH during day hours at both the stations. The difference between pH values of day and night hours may be due to photosynthetic activity (Consumption of CO_2), whereas during night it was slightly less due to respiratory activity. George (1961) observed similar diurnal variation in pH in his study on Delhi pond. Verma (1967) also reported similar observations.

Dissolved Oxygen

Dissolved oxygen varied from 8.5 mg/l to 13.5 mg/l (station A) and 9.00 mg/l to 13.5 mg/l (station B). The minimum values were obtained at 9.00 hours while maximum values were obtained at 15.00 hours on 10.2.2001. The dissolved oxygen was found to be high during day hours. There are two main sources of dissolved oxygen in water (*i*) Diffusion from air and (*ii*) Photosynthetic activity within water.

Table 52.1: Diurnal Variation in Plankton of Gopalpura Tank of Guna District, M.P.–Spot–A

Sl.No.	*Parameters*	*Time In Hours*								
		10-02-2001						*11-02-2001*		
		0900	*1200*	*1500*	*1800*	*2100*	*2400*	*0300*	*0600*	*0900*
A.	**Phytoplankton**									
1.	*Euglena oxyuris*	30	40	42	25	10	4	10	10	20
2.	*Oscillatoria* sp.	15	14	14	10	10	4	5	6	10
3.	*Closterium* sp.	6	16	20	14	4	4	5	7	8
4.	*Anabaena* sp.	4	4	10	6	1	2	3	2	3
5.	*Synedra* sp.	15	17	20	14	9	4	4	3	7
6.	*Navicula* sp.	45	15	20	20	10	5	5	10	25
B.	**Zooplankton**									
1.	Daphnia	200	20	20	90	120	300	210	180	150
2.	Cyclops	30	9	17	10	28	20	35	39	30
3.	Brachianus	5	7	6	10	15	35	10	10	5
4.	Nauplius larvae	10	4	2	16	20	30	32	2	10
5.	Eggs	90	15	15	80	155	240	175	115	100

All the values are n/l.

Table 52.2: Diurnal Variation in Plankton of Gopalpura Tank of Guna District, M.P.–Spot–B

Sl.No.	*Parameters*	*Time In Hours*								
		10-02-2001						*11-02-2001*		
		0900	*1200*	*1500*	*1800*	*2100*	*2400*	*0300*	*0600*	*0900*
A.	**Phytoplankton**									
1.	*Euglena oxyuris*	32	45	47	22	12	4	10	10	25
2.	*Oscillatoria* sp.	17	15	14	10	10	5	5	6	10
3.	*Closterium* sp.	6	15	20	15	4	5	5	7	8
4.	*Anabaena* sp.	4	4	12	7	2	2	4	2	2
5.	*Synedra* sp.	16	17	22	14	9	4	4	4	10
6.	*Navicula* sp.	47	15	22	22	10	6	6	10	32
B	**Zooplankton**									
1.	Daphnia	210	25	25	90	120	280	210	175	150
2.	Cyclops	35	10	177	10	30	20	40	42	42
3.	Brachinus	5	7	7	10	18	35	10	10	8
4.	Nauplius larvae	10	5	6	16	25	30	32	12	10
5.	Eggs	95	20	20	80	150	220	175	125	100

All the values are n/l.

Since photosynthetic activity of Phytoplanktons is highest during day (noon) hours, peak values of dissolved oxygen were obtained during that time (15.00 hours to 18.00 hours). During night hours the photosynthetic activity stops, it results in the consumption of accumulated oxygen and liberation of CO_2. The dissolved oxygen shows positive relationship with temperature. Though, Misra *et al.* (1976) and Dobriyal and Singh (1981) reported negative relationship of dissolved oxygen with temperature. Positive relationship was found by Itzava (1957) and Sharma and Gupta (1981).

Free CO_2

Free CO_2 varied from 7.00 to 16.00 mg/l. The minimum value was obtained at 15.00 hours, while maximum value was obtained at 3.00 hours at both stations. Free CO_2 increases during night and during day it decreases. The high values during night may be due to respiration. As the photosynthetic activity stops, the CO_2 increases during night. Thus we found an inverse relationship between free CO_2 and dissolved oxygen.

Alkalinity

The alkalinity attributed was mainly due to bicarbonates. The alkalinity was found to be maximum during early morning hours and then decreasing during noon and then again increasing, with a decrease in the temperature, during night and thus it showed negative relationship with temperature. High values of alkalinity represent nutrient rich quality of the tank.

Plankton

The plankton manifest a remarkable vertical movement. The phytoplankton included *Euglena, Oxyuris, Oscillatoria* sp., *Closterium* sp., *Anabaena* sp. *Synedra* sp. and *Navicula* sp. The Zooplanktons were represented by rotifers, cladoceran and copepods. Rotifer, cladoceran and copepod fauna of Indian freshwater have been studied by a number of workers, however, the studies on diurnal variations are scanty. Jolly (1952), George (1961), Michael (1966), Sharma and Gupta (1981) have done work on this aspect.

The phytoplankton showed high number during early morning hours and increasing trend during the day and reaching maximum during noon. They then migrated downwards. Similar findings were obtained by Khar *et al.* (1970) while working on fish pond at Shahjahanpur (D.P.). Thus, it appears that their diurnal movement is affected by phototropism. Though it cannot be said that this is the single factor affecting their movement but light, temperature, food, gravity may also be the probable cause for distribution of planktons as indicated by Pennak (1944) and Jolly (1952). The zooplankton represent the clearest variation. The maximum numbers were observed during night (21 hours to 24 hours). The minimum number in most of the zooplankton (as indicated in the Tables 52.1 and 52.2) were found during day time. Similar results were obtained by Michael (1966). It appears that they avoid high temperature and bright light during day time. Vaas and Sachlan (1953) and George (1961) also observed maximum rotifer population at night.

References

APHA, AWWA, WPCF, 1985. *Standard Methods for the Examination of Water and Wastewater,* 15th edn. American Public Health Association, Washington D.C.

Bohra, O.P., Misra, S.D. and Bhargav, S.C., 1979. Diurnal variation studies on Nakhi Lake, Mount Abu, India. *Bioresearch,* 3: 33–43.

Dobriyal, A.K. and Singh, H.R., 1981. Diurnal variation in some aspects of limnology of the river Mandakini from the Garhwal Himalayas. *Uttar Pradesh J. Zool.,* 1: 16–18.

Ganapati, S.V., 1955. Diurnal variation in dissolved gases, hydrogen ion concentration and some of the important dissolved substances of biological significance in three temporary Rock pool in stream bed at Mettur–Dam. *Hydrobiologia*, 7: 285–303.

George, M.G., 1961. Diurnal variation in two shallow pond in Delhi, India. *Ibid*, 18: 265–273.

Itzaava, Y., 1957. The diurnal variation of the oxygen content of fish culture pond water. *Bull. Jap. Soc. Sci. Fish.*, 22: 685–693.

Jolly, V.H., 1952. Diurnal surface concentration of zooplankton in Lake Tempo, New Zealand. *Hydrobiologia*, 25: 466–472.

Khan, A.A., Siddiqui, A.Q. and Nazir, M., 1970. Diurnal variation in a shallow tropical freshwater fish pond in Shahjhanpur, U.P. (India). *Ibid*, 35: 297–304.

Michael, R.G., 1966. Diurnal variation in physico-chemical factors and zooplankton in the surface layer of three freshwater fish ponds. *Indian J. Fish*, 13: 48–82.

Misra, S.D., Bhargava, S.C. and Bohra, O.P., 1975. Diurnal variation in physico-chemical factors at Padamsagar reservoir during premonsoon period of the year 1974. *Geobios*, 2: 32–33.

Misra, S.D., Bhargava, S.C. and Bohra, O.P., 1976. Diurnal variation in certain hydrobiological factors and phytoplankton pigments at Padamsagar reservoir, Jodhpur (Raj.). *Trans. Ind. Soc. Desert. Tech.*, 1: 18–19.

Pennak, R.V., 1944. Diurnal movement of zooplankton organism in some Colorado mountain lakes. *Ecology*, 25: 387– 403.

Sharma, C.S. and Gupta, B.B., 1981. Diurnal variation in hydrobiology of a shallow derelict pond at Aligarh (Uttar Pradesh). *Uttar Pradesh J. Zool.* 1: 24–28.

Sumitra, V., 1971. Studies on the diurnal variation in physico-chemical and biological factors in Teppakulum tank. *Proc. Ind. Acad. Sci.*, 74 B: 63–73.

Trivedy, R.K. and Goel, P.K., 1986. *Chemical and Biological Methods for Water Pollution Studies.* Environmental Publishers, Karad.

Verma, M., 1967. Diurnal variation in a fish pond in Seoni, India. *Hydrobiologia*, 30: 129–137.

Vass, K.F. and Sachlan, M., 1953. Limnological studies on diurnal fluctuation in shallow ponds in Indonesia. *Verh. Int. Vera. Limnol.*, 12: 309–319.

Water and Wastewater Management in 2 Vols (2011) *Pages 339–342*
Editors: **D.R. Khanna, R. Bhutiani & Gagan Matta**
Published by: **DAYA PUBLISHING HOUSE, NEW DELHI**

Chapter 53

Fluctuation in the Population Density of Macro-invertebrates of River Ganga at Pashulok Barrage, Rishikesh, Uttarakhand

D.R. Khanna, Tarun Chugh and Praveen Sarkar
Department of Zoology and Environmental Sciences,
Gurukul Kangri University, Haridwar – 249 404

ABSTRACT

The present paper describes fluctuation in the population density of macro-invertebrates of river Ganga at Pashulok Barrage, Rishikesh. This investigation was carried out between September 1998 and Feb 1999. Total Eight groups of macro-invertebrates were observed. Out of which, Diptera (33.52 per cent), Ephimeroptera (30.62 per cent), Gastropoda (13.65 per cent) and Trichoptera (10.59 per cent) dominated over the other groups. The population of macro-invertebrates found maximum in the month of February, whereas the minimum values were recorded in September.

Keywords: *Macro-invertebrates, River Ganga.*

Introduction

The population density of any ecosystem, not remains static, while changes with time. The macro-invertebrates are one of the important parts of the aquatic ecosystem. These macro-invertebrates provide food to the aquatic ecosystem. These macro-invertebrates provide food to the aquatic fauna and also acts as an indicator of natural water quality. The macro-invertebrates change their population density with the change in physico-chemical characteristics of water. Therefore, macro-invertebrates are the important constituents of aquatic food web. The study on the macro-invertebrates of the Ganga River at Rishikesh has not received proper attention. However, some workers studied the macro-invertebrates of Ganga River and its tributaries at different places (Khanna, 1993 and Joshi *et al.*, 1996). The present chapter describes fluctuation in the population density of macro-invertebrates in the river Ganga at Pashulok Barrage, Rishikesh.

Materials and Method

The study was conducted between September 1998 and February 1999 on the river Ganga at Rishikesh. The sampling station was selected at the right bank of the Ganga River which is 500 meter down stream known as Pashulok Barrage.

The macro-invertebrates were collected each month between 7 A.M. to 11 A.M. All the macro-invertebrates were collected using $1m^2$ quadrate having a fine net. Analysis was done following the standard methods of APHA (1976) and Needham and Needham (1972).

Results and Discussion

The data is showing monthly fluctuation of different macro-invertebrate groups in their population density is tabulated in Table 53.1. Total eight groups of macro-invertebrates were counted. The macro-invertebrates population mainly constituted of the Diptera, Ephimeroptera, Gastropoda and Trichoptera, whereas Plecoptera, Odonata, Hemiptera and Coleoptera were found less in density. The population of macro-invertebrates was dominated by the Diptera (total average 67.50 units/m^2 ± 56.49) and Ephimeroptera (67.67 units/m^2 ± 3 7.36) followed by Gastropoda (27.5 units/m^2 ± 21.44), Trichoptera (21.33 units/m^2 18.57), Plecoptera (7.11 units/m^2 ± 5.35). Coleoptera (5.33 units/m^2 ± 4.50). Hemiptera (3.60 units/m^2 ± 1.67) and Odonata (3.40 units/m^2 ± 2.51). The macro-invertebrates showed their highest average value in the month of February (45.38 units/m^2 ± 51.76), and lowest average value in September (3.17 units/m^2 ± 4.92). The total percentage of different macro-invertebrate groups was recorded as Ephimeroptera (30.62 per cent), Trichoptera (10.59 per cent), Diptera (33.52 per cent), Gasropoda (13.65 per cent), Hemiptera (1.769 per cent), Odonata (1.68 per cent), Plecoptera (5.46 per cent) and Coleoptera (2.65 per cent) in Table 53.1.

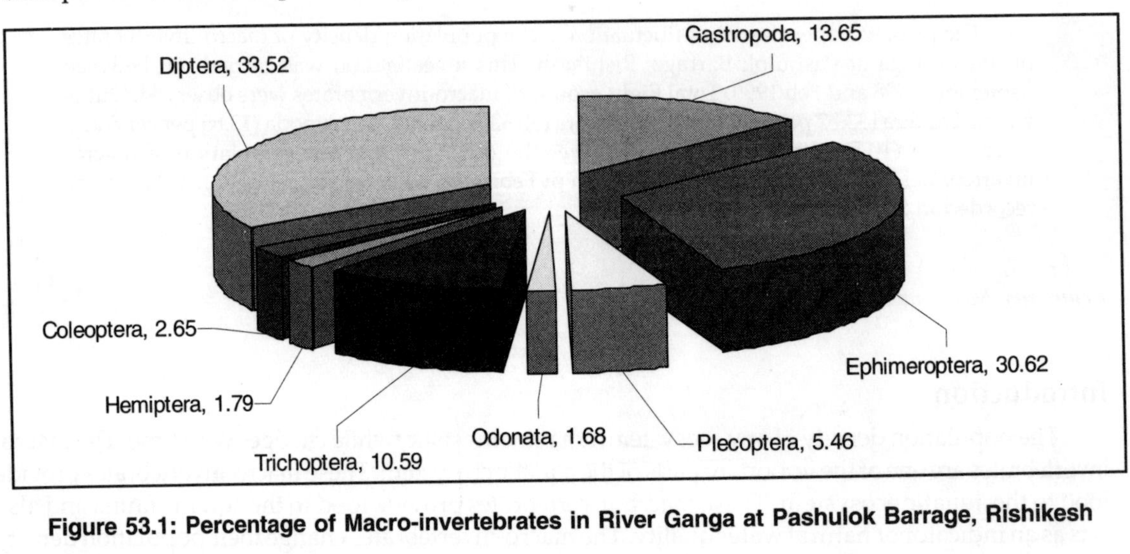

Figure 53.1: Percentage of Macro-invertebrates in River Ganga at Pashulok Barrage, Rishikesh

Table 53.1: Macro-invertebrate Population in the River Ganga Between September 1998 and February 1999 at Pashulok Barrage, Rishikesh

Month	*Parameters*								
	Ephimeroptera	*Plecoptera*	*Odonata*	*Trichoptera*	*Hemiptera*	*Coleoptera*	*Diptera*	*Gastropoda*	*Average*
September, 1998	12	–	1	1	–	1	1	3	3.17±4.92
October, 1998	26	–	2	4	2	1	14	11	8.57±9.97
November, 1998	55	5	2	12	4	2	55	19	19.25±24.63
December, 1998	90	8	–	30	4	9	70	40	35.86±36.44
January, 1999	80	16	7	48	2	11	138	30	41.50±50.25
February, 1999	107	15	5	33	6	8	127	62	45.38±51.76
Average	61.67±33.58	11.0±5.69	3.4±2.71	21.33±19.75	3.60±1.15	5.33±4.82	67.5±54.1	27.5±14.74	27.90±19.96

The minimum population density of macro-invertebrates was found in the month of September and the maximum population density was observed in the month of February. The macro-invertebrates showed fluctuation in their population density because as it is directly influenced by the biotic and abiotic factors, Pratt *et al.* (1981), Singh (1997) and Vig (1998) also showed the same phenomenan. In Rishikesh the rainy season was between late June and September. During these months there was a large amount of water with high velocity, which wash away the eggs and larval of macro-invertebrates. Richardson 1921 and Khanna 1993 has reported this activity in their work and shown that scarcity of food as another factor causing retardation in their population density during rainy season.

Due to high velocity of water, the plankton and other food material washed away. The temperature and total solids were found high in rainy season which retards the growth of biotic community of water. In this study macro-invertebrate population showed a continuous increase from September onwards, which was due to availability of food to them. There was a good planktonic population due to active photosynthesis during winter months. Thus, the macro-invertebrates are positively related to plankton population. Therefore, it may be concluded from the above study that the population density of macro-invertebrates fluctuates due to the influence of several biotic and abiotic components.

Acknowledgements

The authors are grateful to Professor B.D. Joshi and Professor A.K. Chopra, Department of Zoology and Environmental Science, Gurukul Kangri University, Haridwar, for their encouragement.

References

APHA, 1976. *Standard Method of Examination of Water and Wastewater*, APHA, New York.

Joshi, B.D., Bisht, R.C.S., Joshi, Namita and Singh, Rajendra, 1996. A study of planktonic and benthic components of three selected tributaries of river Ganga between Devprayaag and Rishikesh. *Him. J. Env. Zool.*, 10: 23–26.

Khanna, D.R., 1993. *Ecology and Pollution of Ganga River*. Ashish Publishing House, Delhi, pp. 1–241.

Needham, J.E. and Needham, P.R., 1972. *A Guide to the Study of Freshwater Biology*. Holden Day Inc., San Fransisco, California.

Pratt, M.J., Coler, A.R. and Godfrey, J.P., 1981. Ecological effects of urban storm water runoff on benthic macro-invertebrates inhabiting the Green river, Massachusetts. *Hydrobiologia*, 83: 29–42.

Richardson, R.F., 1921. Changes in bottom and share fauna of the middle Illinois river and its connecting lakes since 1913–1915, as a result of the increased southward of sewage pollution. *Illinois Nat. Hist. Surv. Bull.*, 14: 33–75.

Singh, Arun K., 1997. Abundance of macro-zoobenthic organisms in relation to physico-chemical characteristics of river Ganga at Patna (Bihar). *India J. Environ. Biol.*, 18(2): 103–110.

Vig, Monika, 1998. Limnology of "Hindon" with reference to impact of sewage and industrial effluents of benthic organisms. *Ph.D. Thesis*, Chaudhari Charan Singh University, Meerut, pp. 1–167.